Plant Transposons and Genome Dynamics in Evolution

Plant Transposons and Genome Dynamics in Evolution

Editor
NINA V. FEDOROFF
King Abdullah University of Science and Technology
Thuwal, Saudi Arabia
and Pennsylvania State University
Pennsylvania, USA

WILEY-BLACKWELL
A John Wiley & Sons, Inc., Publication

Wiley-Blackwell is an imprint of John Wiley & Sons, formed by the merger of Wiley's global Scientific, Technical and Medical business with Blackwell Publishing.

Editorial offices: 2121 State Avenue, Ames, Iowa 50014-8300, USA
The Atrium, Southern Gate, Chichester, West Sussex, PO19 8SQ, UK
9600 Garsington Road, Oxford, OX4 2DQ, UK

For details of our global editorial offices, for customer services and for information about how to apply for permission to reuse the copyright material in this book please see our website at www.wiley.com/wiley-blackwell.

Cataloging-in-Publication Data is available through the Library of Congress.

A catalogue record for this book is available from the British Library.

Wiley also publishes its books in a variety of electronic formats. Some content that appears in print may not be available in electronic books.

Cover design by Modern Alchemy LLC

Set in 10.5/12 pt Times by Aptara® Inc., New Delhi, India
Printed and bound in Malaysia by Vivar Printing Sdn Bhd

1 2013

Contents

Color plate is located between pages 142 and 143.

Contributors

Jeffrey L. Bennetzen	Department of Genetics, University of Georgia, Athens, Georgia, USA
David Botstein	Lewis-Sigler Institute for Integrative Genomics, Princeton University, Princeton, New Jersey, USA
Vicki L. Chandler	Gordon and Betty Moore Foundation, Palo Alto, California, USA
Hugo K. Dooner	Waksman Institute, Rutgers University, New Jersey, USA
Nina V. Fedoroff	King Abdullah University of Science and Technology, Thuwal, Saudi Arabia and Pennsylvania State University, Pennsylvania, USA
Mary A. Gehring	Whitehead Institute and Massachusetts Institute of Technology, Cambridge, Massachusetts, USA
Avraham A. Levy	Department of Plant Sciences, Weizmann Institute of Science, Rehovot, Israel
Damon R. Lisch	Department of Plant and Microbial Biology, University of California, Berkeley, Berkeley, California, USA
Robert A. Martienssen	Howard Hughes Medical Institutes, New York, New York, USA and Gordon and Betty Moore Foundation, Palo Alto, California, USA and Cold Spring Harbor Laboratory, Cold Spring Harbor, New York, USA
Thomas Peterson	Department of Genetics, Development and Cell Biology and Department of Agronomy, Iowa State University, Ames, Iowa, USA
Alan H. Schulman	MTT Agrifood Research Finland and Institute of Biotechnology, University of Helsinki, Helsinki, Finland
Clifford F. Weil	Department of Agronomy, Purdue University, West Lafayette, Indiana, USA
Thomas Wicker	Institute of Plant Biology, University of Zurich, Zurich, Switzerland
Jianbo Zhang	Department of Genetics, Development and Cell Biology, Iowa State University, Ames, Iowa, USA

Foreword

David Botstein

Science is both a rational and a social endeavor. The basic paradigm for scientific progress, comprising a progression from observation to theory and experimentation, has been in place for several hundred years. Mostly science moves in steps small enough so that the introduction of a new idea, theory, or point of view is followed fairly quickly by a consensus among scientists who become convinced by the evidence. Scientists are taught to value new ideas and to evaluate the evidence, even when new ideas are revolutionary and fundamental. Often they have no difficulty doing so: Einstein's astonishingly original ideas were understood and largely accepted by physical scientists within a decade or so of their publication.

Occasionally, however, there are geniuses who make observations, propose theories, and carry out convincing experiments that are somehow so far in advance of contemporary scientific understanding that the general acceptance of their ideas, even by the scientific community, lags for many decades. The work of three well-known giants in the history of biology displayed this kind of intellectual "prematurity": Charles Darwin, Gregor Mendel, and Barbara McClintock.

This book is a compendium of what is known and accepted today about transposons and genomic dynamics in plants, a field whose basic ideas manifestly derive from the work and insights of Barbara McClintock to a degree similar to the influence of Mendel on genetics and Darwin on evolution. As with Darwin and Mendel, acceptance of McClintock's ideas has taken many decades. But unlike Darwin and Mendel, the depth of her insight is still to be fully appreciated.

What stood in the way? Some of McClintock's discoveries, such as the relationship between chromosomes and linkage groups, achieved immediate acceptance. Transposition, by contrast, took decades and repeated rediscovery in organisms other than maize. Still others, such as the concept that genomes sense and respond to external stimuli, are just beginning to find experimental support and intellectual acceptance.

The barrier was not obscurity or even gender. McClintock achieved a high status early in her career on the basis of her achievements. Unlike Mendel, she and her ideas were well known and widely accepted in her community. Despite her eminence, other scientists appear simply to have failed to understand some of her ideas or the evidence on which they were based. Unlike the case of Darwin, there were no religious or ideological barriers to the acceptance of McClintock's ideas. Hers seems to have been purely a matter of having been ahead of her time.

But I also believe that two commonly held convictions had to change to make way for full acceptance of McClintock's ideas about genome dynamics in evolution, which are the subject of this book. The first was the generalization that only proteins and their regulation are really important in understanding biology and evolution. The second was the conviction that the mechanisms of evolution could not themselves evolve.

The focus on proteins was entirely understandable: it was at the heart of the molecular biology revolution. Molecular biologists explicitly taught the "central dogma" that information flows from DNA to RNA to protein and only thence to phenotype and fitness. No surprise, then, that the biological community sought to understand everything in terms of proteins and their regulation. This led to an unwanted and, and surely unintended consequence: dismissal of noncoding DNA as "junk." Of course, this "junk" included all the transposons and transposon remnants that, as readers of this book will see, are what make the genome dynamic and are the drivers of genome evolution.

The misunderstanding about evolution of evolutionary mechanisms arose from the long-running debate about Darwinism, not only among scientists, but also in society more generally. Countering religious rejection of Darwinism in favor of divine intention and teleological arguments, scientists rather vehemently rejected the legitimacy of teleology in scientific reasoning. This rejection led, perhaps unwittingly, to the dismissal of "evolvability" as a property that could be selected in evolution. For most of the twentieth century, the scientific community treated the concept of evolvability as requiring something like intention or, at least, precognition.

The idea that one genome is more fit than another because it is more mutable was an idea that, at best, was hard to imagine. The dismissal of transposons as "junk" DNA and as "parasites" whose destructiveness genomes must rigidly control also interfered with perception of their contribution to evolvability. Thus in order for McClintock's ideas about genome dynamism to be accepted, it was first necessary for the scientific community to assimilate the existence and sheer genomic abundance of transposons and then to appreciate their agency in the mechanisms of chromosome mechanics and functional evolution.

This book is a thorough examination of the current state of knowledge about the numbers and nature of transposons and retrotransposons and how they have shaped plant genomes. Progress of this more incremental variety has come through the invention and application of rapid techniques for genome and transcript analysis. The results have led to a renewed appreciation that Barbara McClintock understood much more than the basic ideas of chromosome mechanics and transposition, both of which were already widely accepted by the time of her death in 1992.

Indeed, McClintock discovered and recognized the significance of what we now call "epigenetics" – the heritable, reversible regulation of gene activity. The study of epigenetics and epigenomics has only recently become one of the hottest research fields of our time. The larger community is just now beginning to assimilate fully the notion that phenotypes reflect not only genotypes but also the epigenetic consequences of both development and response to the environment. Finally, and perhaps most importantly, McClintock understood, as the rest of us are only beginning to figure out, that there are well-orchestrated genomic stress responses that can rapidly restructure genomes – the quintessence of evolvability.

Introduction

Nina V. Fedoroff

McClintock's discovery of transposition in the middle of the twentieth century was roughly contemporaneous with Watson and Crick's landmark elucidation of DNA structure. But although Watson and Crick were recognized with a Nobel Prize within a decade, several more decades elapsed before the importance of McClintock's work on transposons was recognized with the award of an unshared Nobel Prize. The mystery of why it took so long for transposable genetic elements to be recognized as something more than a genetic oddity is dissipating as we increasingly appreciate the role of epigenetic mechanisms in silencing transposons and maintaining chromosome stability. Given the current recognition of their importance, it is curious that the study of DNA methylation and other epigenetic mechanisms has only recently advanced from the status of mildly disreputable to the cutting edge of investigation.

The term "transposable element" (TE) is generally used here to refer to both transposons that move through a DNA intermediate and retrotransposons that move through an RNA intermediate. The present volume seeks to capture and distill the veritable mountain of information that has now accumulated on the many flavors of plant TEs, their genetics, genomics, and epigenetics. It also provides an opportunity to indulge in a bit of hindsight, with its extraordinary acuity, and to reassess the larger picture of transposons in gene structure and regulation, as well as in genome and organismal evolution.

Although she is best known for her discovery of transposable genetic elements, recounted in Chapter 1, McClintock's contributions went well beyond transposition. Her seminal work on epigenetic regulation, described in Chapter 4, remains largely unrecognized, as do her insights into genome restructuring. Indeed, it became fashionable to discredit McClintock's view that transposons are gene regulators. And yet, although she did not get everything exactly right, her early insights seem remarkably prescient from a contemporary vantage point, compelling a rethinking of both regulation and the relationships among the genome's indigenous gene populations.

Early in her work on transposons, McClintock came to the conclusion that they were unmoored gene regulatory systems that had become associated with different genes by virtue of their ability to move. This view was reinforced by her growing appreciation that a single active transposon could promote excision of transposition-defective elements belonging to the same family from insertion sites in several genes simultaneously. This hierarchical relationship, in turn, reinforced her conviction that transposons were integral parts of the developmental regulatory machinery and she therefore named them "controlling elements."

She viewed transposons as bits of heterochromatin by analogy to the connection between heterochromatin and certain types of variegated gene expression in *Drosophila*. In a 1950 paper published in the *Proceedings of the National Academy of Sciences*, she wrote:

> Changes in quantity, quality or structural organization of heterochromatic elements may well alter the kind and/or degree of particular exchanges that occur, and in this way control the chromosome organization and the kind and the relative effectiveness of genic action.

This has turned out to quite close to the contemporary recognition that large blocks of silenced and recombinationally inert retrotransposons separate small "islands" of genes in many plant genomes, as discussed in detail in Chapters 2 and 10.

McClintock's intense study of the *Suppressor-mutator* (*Spm*) transposon, described in Chapter 4, produced the first detailed genetic characterization of an epigenetic regulatory system, further expanding the range of transposon regulatory attributes that could influence expression of a gene into which a transposon had inserted.

Contemporary concepts of gene regulation are rooted in the pioneering work of Jacob and Monod on bacterial genes. Although McClintock's assessment of the parallels between the prokaryotic and eukaryotic regulatory systems appears quite accurate in retrospect, her views gained little traction at the time, perhaps because the gene regulatory story was confounded by transposition, a phenomenon that had not yet been discovered in bacteria. But writing in the *American Naturalist* in 1961, McClintock said:

> It should be emphasized that, although transposition of controlling elements in maize made it possible to recognize their presence in the chromosome complement and to study the mode of operation of the component elements of a system, transposition does not necessarily characterize the behavior of a controlling element. An element previously exhibiting transposition may become fixed in a location. If it is the gene-associated element that becomes fixed, the action of the gene will then be permanently under the control of the system to which the element belongs.

We now know that precisely such regulatory captures underlie the various phenotypes of *Spm* insertion mutations (Chapter 4). More that that, we know that such captures are a regular feature of gene evolution in plants, so much so that it has been proposed that all epigenetic regulation of plant genes derives from transposons (Chapters 6–8).

By the time McClintock wrote the 1961 *American Naturalist* article comparing bacterial and maize gene regulatory systems, her understanding of the *Spm* transposon's genetic regulatory mechanism had advanced well beyond anything that had then been described in either prokaryotes or eukaryotes. In retrospect, it is clear that the complexity of McClintock's description arose primarily from the fact that the *Spm* element is regulated by both epigenetic modification and the transposon's regulatory system. As a result, insertions of different transposition-defective elements at different positions in the gene and promoter regions of the pigment biosynthetic genes that served her as reporter genes gave alleles with a remarkable variety of phenotypes (Chapter 4).

What is extraordinary is her insight that the *trans*-acting *Spm* transposon itself could undergo changes in expression that were heritable, but reversible – what we now call epigenetic. In the same *American Naturalist* article, she wrote:

> One of the most interesting and theoretically important types of expression of *Spm* consists in the sequentially occurring reversals in phase of its activity – from active to inactive and back to active.... Following such a reversal of phase, the duration of the particular phase may be long, continuing unaltered through many cell or even plant generations, or it may be short, reversal occurring again in a number of cells

> only a relatively few cell generations removed from that which initiated the preceding phase. Control of duration of a particular phase appears to be associated with the event that produces the particular reversal of phase. By selective methods, it has been possible to isolate *Spm* displaying either a long duration of an active phase or a long duration of an inactive phase.

More than that, she reported that an active *Spm* transposon could activate an epigenetically inactive one, suggesting that the transposon encodes its own epigenetic activator, subsequently identified as the transposon-encoded TnpA protein and shown to function precisely as predicted from the genetic interactions (Chapter 4).

We now know that epigenetic silencing is accomplished by a complex of mechanisms that includes histone modification, RNA interference, and RNA-directed DNA methylation. DNA methylation is reversed by both passive and active mechanisms, and reactivation of silent transposons occurs under a variety of conditions, as described in Chapter 5. Although recent experiments suggest that there is gene- and transposon-specificity in epigenetic silencing and reactivation, the *Spm* transposon's epigenetic regulatory mechanism remains among the very few that have been extensively investigated to date, either genetically or at the molecular level.

The invention and perfection of DNA sequencing techniques in the late 1970s set the stage for the subsequent rapid expansion of knowledge about the structure, gene content, and organization of genomes. During the debates that took place at the time about whether it was worth sequencing complete genomes in view of the suspicion that much of the DNA was repetitive and did not code for either proteins or the then-known structural RNAs, transposons were lumped with other kinds of repetitive sequences and given Ohno's and Dawkins labels of "junk" and "selfish" DNA based on the view that they existed solely to propagate themselves and made no contribution to genome structure or function.

Angiosperm genomes, like the genomes of other higher eukaryotes, vary widely in their haploid DNA content, even within a single species, a phenomenon long known as the C-value paradox. We now know that, indeed, this wide C-value disparity is attributable to the differential accumulation of transposons and retrotransposons, predominantly the latter (Chapters 2 and 10). At the same time, it has become increasingly clear that the organization and evolution of higher plant genomes are driven largely by transposons and retrotransposons.

Astonishingly, the vast majority of the DNA in higher plants comprises transposons and retrotransposons: 85% of the maize genome, for example, consists of TEs, predominantly retrotransposons. The typical large angiosperm genome exhibits small "islands" of genes in a "sea" of repetitive DNA, primarily consisting of retrotransposons (Chapter 10). Although there is significant constancy of total gene numbers and retention of gene complements, the colinearity of homologous genes declines with evolutionary distance and intergenic regions change rapidly (Chapter 10). Comparisons even among inbred strains of maize reveal substantial differences in gene organization and even larger differences in both the length of intergenic regions and their content of transposons and retrotransposons (Chapter 10). Whole genome comparisons across species suggest that both the movement of genes and the intergenic churn are caused by transposons and retrotransposons. Whether examining the results of transposition events involving a single transposon (Chapter 3) or viewing the contribution of transposons to the evolution of chromosomes (Chapter 10), the centrality of transposons to contemporary genome organization is inescapable.

Transposons make many subtle contributions to gene and genome evolution, as well. Transposons create genes, modify them, and program and reprogram their expression (Chapters 7–10). The traffic in genes and regulatory sequences is bidirectional: transposons pick up genes that code for proteins other than transposases and transposase genes are pressed into services other than transposition.

Transposons are central to the epigenetic phenomenon of "imprinting" that imbues genes with different expression patterns depending on whether they were transmitted through male or female gametes (Chapter 7), differences that arise during the major epigenetic reprogramming that takes place during gametogenesis (Chapter 5).

Although ideas about junk DNA have evolved substantially over the past two decades, the transposon monikers have stuck. Transposons are still referred to as "selfish" DNA, "invaders" and "parasites," with the subtext that they are largely dangerous and destructive, hence in need of taming. The idea that epigenetic mechanisms evolved precisely to "control" the destructive potential of such "parasites," advanced a decade and a half ago, has also persisted.

The value of these notions diminishes as we learn more about epigenetic mechanisms and gain insight into how transposons shape genomes. The real puzzle is that transposons accumulate in large numbers in eukaryotes, but not in prokaryotes. But this is also true of other categories of sequences, both coding and noncoding. That is, what distinguishes eukaryotic genome organization from that of prokaryotes is the ability to retain and manage large amounts of duplicated DNA. How did eukaryotes tip the balance between duplication and deletion that keeps genome size small in organisms in which homology-dependent recombination mechanisms predominate?

The answer to these questions lies precisely in the epigenetic mechanisms that eukaryotes have elaborated to a much greater extent than prokaryotes. Plants have a more complex and redundant array of epigenetic mechanisms even than animals (Chapter 5); importantly, however, transposons are not its only targets. A common denominator triggering silencing is the repetitive character of the sequence, not its identity as a transposon. Repeats are readily eliminated by unequal crossing over by homologous recombination and it is precisely homologous crossing over that is subject to increasingly stringent control in eukaryotic evolution.

The capacity to keep duplicated sequences is an essential step in the evolution of multicellular organisms, underpinning the ability to target expression of different subsets of genes to different cells and tissues. Equally key is the ability to program genes for differential expression by a variety of mechanisms, among which are the relatively stable mechanisms involving DNA and histone modification, as well as the more labile small-RNA-mediated mechanisms.

Because homology-dependent illegitimate recombination events between transposons in different locations have the potential to disrupt genomes, the very ability to suppress illegitimate recombination must inevitably favor the accumulation of transposons, the results of whose antics might otherwise be relegated to the evolutionary scrap heap. Thus it was perhaps the elaboration of epigenetic mechanisms originating in prokaryotes to moderate homologous recombination that made it possible for genomes to grow by duplication and for transposons to proliferate. This is precisely the inverse of the "parasite control" hypothesis, which posits that epigenetic mechanisms arose to control transposons.

And yet, even though epigenetic silencing mechanisms effectively minimize transposable element activity, they do not eliminate it, and the fingerprints of transposon activity are evident at every level of genome organization. That brings me to the final piece of the puzzle McClintock left us. McClintock described the suite of nuclear events, including transposon activation and various chromosome aberrations and rearrangements, that unfold in the wake of what she called genomic "shocks," such as irradiation or mutagenesis, or, as in her experiments, the introduction of two broken chromosomes by a genetic cross. It has, by now, been amply documented that plant transposons are activated in response to a variety of genomic perturbations and both biotic and abiotic stresses, including pathogen infection, the passage of plant cells through tissue culture, interspecific hybridization and allopolyploidization (Chapters 2, 6, and 9). This appears to be true, as well, in other eukaryotes, from yeast to flies to humans, and the common denominator is, of course, DNA damage. Chapter 9 proposes that dysregulation of the epigenetic machinery underlies responses to such genomic shocks. Experimental evidence has just begun to emerge from the molecular study of hybrid dysgenesis in *Drosophila*. A deeper understanding of how the epigenetic regulatory systems are themselves modulated to facilitate damage control and restore genome integrity remains for future investigations to unravel.

In sum, then, the present volume provides a rich picture of the role that TEs have played in sculpting the genomic landscape of plants at multiple levels of organization and on time scales from the generational to the evolutionary. Given their abundance in most higher-eukaryotic genomes, the ancient origins of the DNA resecting enzymes that they encode, and the clear evidence of their impact on gene and genome structure and regulation, there seems little value in continuing to view them as "parasites." As well, given the ubiquity and variety of epigenetic regulatory mechanisms, it seems increasingly implausible that they were invented to control TEs. On the contrary, it seems more probable that TEs proliferated and came to drive eukaryotic genome evolution because of and not despite epigenetic regulation. The present volume documents the many ways that transposons have contributed to the evolution of plant genes and genomes, arguably explaining their extraordinary plasticity – indeed, their very evolvability.

Finally, there are many questions to be addressed once we accept TEs as legitimate – indeed, dominant – inhabitants of the eukaryotic genome. The extraordinary size of many plant genomes suggests that the accumulation of vast numbers of TEs and other kinds of repetitive DNAs is tolerable. Whether the large and rapidly evolving blocks of retrotransposons actually confer a selective advantage is not known. How might the transposon landscape of chromosomes influence the stability and the participation of chromosomes in the mechanics of mitosis and meiosis? The observations that retrotransposon blocks are recombinationally inert and that disruptions in epigenetic regulation disrupt meiosis may well be hints that will lead us to a deeper understanding of the architecture and dynamics of contemporary genomes.

Plant Transposons and Genome Dynamics in Evolution

1 The Discovery of Transposition[1]

Nina V. Fedoroff

Introduction

The discovery of transposition can be dated quite precisely. Writing about the first case of an unstable mutation caused by insertion of the *Dissociation* (*Ds*) locus, which she had earlier identified and named for its ability to cause chromosome breakage and dissociation, McClintock observes: "At the time, I did not know that *Ds* could change its location. Realization of this did not enter my consciousness until late this spring, following the harvest of the greenhouse crop." Inked corrections in McClintock's hand on a typed manuscript, never published, from January of 1949 identify the spring as that of 1948 and the greenhouse crop as that of winter 1947–1948.

Studies on Variegation

Not surprisingly, the discovery of transposition is embedded in a larger context, both in McClintock's work and in earlier studies on what were initially called "mutable" or "unstable" genes and "ever-sporting" plant varieties that exhibit variegation for flower and leaf color. While these studies did not lead directly to McClintock's discovery of transposable elements in the sense that she was working with such materials, she was undoubtedly aware of the earlier work, particularly that of Emerson and Rhoades. Both of these maize geneticists had carried out systematic genetic studies on mutable genes in maize and contributed substantial insights into their nature and behavior. Thus, it is with the work of these authors that the discussion begins, although it is important to note as preamble even earlier mention in the literature of the peculiar behavior of mutable genes.

De Vries, for example, developed a general concept of "ever-sporting" varieties from studies in *Antirrhinum* (de Vries, 1905). He concluded that the inheritance of variegation and the occasional fully colored mutations or "sports" arising from them generally do not show what we now call Mendelian inheritance, although he did report instances of the inheritance of somatic mutations to full color. Correns, working with *Mirabilis jalapa*, and East and Hays, studying variegation in *Zea mays*, similarly noted that somatic mutations from a variegated to a fully colored phenotype

[1] Modified from Fedoroff, N.V. (1998) The discovery of transposable elements. In *Discoveries in Plant Biology*, vol. 1 (eds Shain-Dow Kung and Shang-Fa Yang), World Scientific Publishing Co. Pte. Ltd., Singapore, pp. 89–104.

Plant Transposons and Genome Dynamics in Evolution, First Edition. Edited by Nina V. Fedoroff.

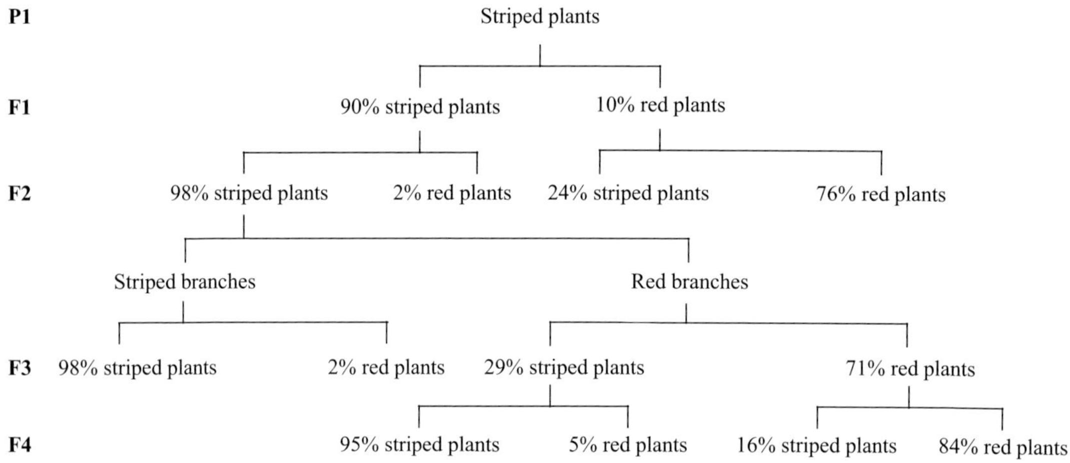

Figure 1.1 Diagram adapted from de Vries showing the inheritance of variegation patterns in *Antirrhinum*.

showed Mendelian inheritance. To set the stage, it is worthwhile reproducing a diagram from de Vries experiments on *Antirrhinum* (Figure 1.1) (de Vries, 1905).

De Vries concludes (p. 161):

> From these figures it is manifest that the red and striped types differ from one another not only in their visible attributes, but also in the degree of their heredity. The striped individuals repeat their peculiarity in 90–98 percent of their progeny, 2–10 percent sporting into the uniform red color. On the other hand, the red individuals are constant in 71–84 percent of their offspring, while 16–29 percent go over to the striped type. Or in one word: both types are inherited to a high degree, but the striped type is more strictly inherited than the red one.

In the same vein, Emerson commences his first important paper on the genetics of variegation, with the following striking statement, which he thereafter elegantly refutes (Emerson, 1914): "Variegation is distinguished from other color patterns by its incorrigible irregularity."

Emerson describes his experimental system:

> It is characteristic of the ears of certain varieties of maize known, at least in the Middle West, as "calico" corn. In these varieties, the pericarp of most of the grains has few to many narrow stripes of dark red, the remaining area being colorless or showing a sort of washed-out red. Often broad red stripes appear on some grains, a single stripe covering from perhaps one tenth to nine tenths of the grain. Not uncommonly there are entirely colorless grains (so far as pericarp is concerned) and also solid red grains scattered over the ear. Much more rarely there is found a "freak" ear with a large patch of self-red or nearly self-red grains. Or sometimes an ear is composed largely of red or almost red grains with a small patch of striped or nearly colorless grains. In such cases it is not uncommon for the margin of the red area to cut across a grain so that one side—always the side toward the red patch—is red and the other side colorless or striped. Ears that are colorless throughout, except for a single striped grain, are not unknown and there are even known ears that are red except for a single striped grain. Very rarely a plant has one self-red ear and one variegated ear on the same stalk.

Emerson commenced his study using a few "freak" ears obtained from local and national corn expositions and had no information about their parentage. He asked a different question than de Vries had asked, inquiring whether there was a relationship between the amount of red-pigmented tissue in a given kernel and the number of red ears produced upon self-pollination in subsequent generations. The clear answer emerged that the more red there was in the kernels planted, the larger the fraction of red ears in the progeny. He further found by analyzing the progeny of plants producing red ears that red kernels produced plants that were commonly heterozygous for the red and variegating traits. Emerson concludes: "The development of red in the pericarp is evidently associated with and perhaps due to a modification of some Mendelian factor for pericarp color in the somatic cells."

Thus, Emerson had captured variegation within the Mendelian paradigm, adding the important insight that a somatic change could occur in a Mendelian factor, becoming a heritable change that obeyed simple Mendelian principles. But he readily admitted that it was "... utterly impossible at the present time to conceive of the cause or even the nature of this change...." He nonetheless went on to conjecture that *V*, as he designated the factor for variegation, might be "... a sort of temporary inhibitor, an inhibitor that sooner or later loses its power to inhibit color development, a power that once lost is ordinarily never regained." Even more firmly, he suggests that "... it may be that there is present in variegated maize merely a dominant factor for self-color, *S*, that is temporarily inactive, but that sooner or later becomes permanently active." Emerson had articulated the concept that variegation was due to the association of some kind of a factor with what is now called a gene and that its loss was what allowed the gene to be re-expressed.

In a subsequent study, Emerson noted that, although the loss of the inhibitory factor showed very high heritability, occasional variegated kernels appeared on otherwise fully pigmented ears (Emerson, 1917). This suggested that, at some low frequency, the inhibitory factor might once again become associated with the gene, then called a "unit factor." Emerson viewed variegation as a reversible change in an otherwise conventional unit factor, which distinguished itself from other kinds of mutations by its high frequency. Emerson later made some puzzling observations that were to remain unexplained until McClintock's studies decades later. First, he made the counterintuitive observation that reversion to wild-type was less frequent when the variegating pericarp color gene was homozygous and therefore present in two copies than when it was heterozygous with a stable non-pigmenting allele. Second, he noted that chromosomes carrying a stable recessive non-pigmenting allele of the pericarp color locus recovered by segregation from a variegating heterozygote show some ability to suppress variegation when again used to create a variegating heterozygote. As explanations of the latter, he entertained the radical hypothesis that information is transferred between alleles either as "... a direct contamination of one allelomorph by another ... " or by transfer of "... distinct gene elements ... " from one allele to another. But he readily admits that he, the writer, "... is wholly unable to devise a consistent working hypothesis to account for his results on any such assumption ... " and suggests the alternative hypothesis of distinct modifiers of variegation, which had already been reported in *Drosophila virilis*.

Mutable Genes

Substantial work was done on mutable genes in both *Drosophila* and plants, including *Zea mays*, during the ensuing decade, and a debate arose over Emerson's view that variegation was a gene mutation process distinguishable from other types only by its high frequency. Goldschmidt had proposed that mutations are a consequence of position effects and both he and Correns held that mutable genes are sick or diseased genes and that any conclusions derived from their study were not applicable to other types of mutations (Correns, 1910; Goldschmidt, 1938). By contrast, Rhoades and Demerec shared Emerson's view that there was no clear-cut difference between stable and unstable genes (Demerec, 1935; Rhoades, 1941).

Rhoades reported a seminal observation in 1936, the isolation of a dominant maize gene from an ear of Black Mexican sweet corn that caused instability of a previously stable recessive allele of the *A1* locus (Rhoades, 1936). The original segregation ratio suggested that the "dotted" character, exhibiting somatic sectors of the deep pigmentation characteristic of the *A1* allele, required both the recessive a_1 allele and a second locus. Rhoades went on to provide definitive evidence for such a second locus, designating it the *Dotted* (*Dt*) locus (Rhoades, 1938). He further showed that the a_1 allele recovered from the ear that exhibited the variegated phenotype differed in no way from the standard stable recessive a_1 tester allele originally isolated by Emerson and used for two decades without showing evidence of variegation. Rhoades established that the *Dt* locus was not linked to the *A* locus and that both the standard and newly isolated a_1 alleles exhibited variegation or mutability only in its presence. In addition to *Dt*, whose presence was essential for mutability, Rhoades reported that there were additional genes that modified the timing or frequency of somatic mutation.

Thus, Rhoades made the argument that the difference between stable and unstable mutations may, at least in some cases, be more apparent than real and result from the genetic environment. However, his subsequent studies further convinced him that the *Dt* gene was much more specific in its effects than such general treatments as irradiation or high temperature. The presence of *Dt* neither caused a general increase in mutation frequency nor affected variegation of the pericarp color allele Emerson had studied, by then designated P^{vv}.

McClintock's Studies on Chromosome Breakage

McClintock's discovery of transposition had its origins in her studies on the behavior of broken chromosomes (McClintock, 1938, 1939, 1941a, 1941b, 1942). Her objective was to determine the behavior of chromosomes with a broken end during mitotic divisions, and the results of her experiments led to the insight that chromosomes lacking telomeres do not separate during replication, producing dicentric chromosomes. The dicentric chromosomes break, regenerating chromosomes with broken ends, creating what McClintock referred to as the "breakage-fusion-bridge" cycle. Based on her knowledge of the behavior of broken chromosomes, she invented a method for producing small terminal and subterminal deletions, using plants that developed from zygotes that had received a broken chromosome 9 from each parent.

Having identified and studied several recurring mutations, McClintock undertook a search for new mutations using F2 progeny derived from F1 plants that had received a recently broken chromosome 9 from one parent and the selfed progeny of plants that had received a newly broken chromosome 9 from each parent. She reports:

> An interesting type of chromosomal behavior has appeared in three of the broken-chromosome cultures mentioned above. In each culture, one of the broken chromosomes 9 is continually being lost from cells during development. This loss is not due to bridge formation or to ring chromosome behavior, but appears to be caused by the inability of the two halves of this chromosome to migrate to opposite poles in some of the somatic anaphase figures.

She further notes:

> In addition, some of the mutants appearing in these cultures are individually provocative. Several show variegation characterized by a change from mutant to normal-appearing tissues. For any one plant, a distinctive or basic rate of change is apparent, but this basic rate differs from plant to plant. Sectors

> with changed rates of variegation appear in all plants, especially in the later-appearing tissues. It is significant that twin sectors accompany many if not most of the alteration in rate; this is expressed by the appearance of a sector of tissue having a greatly increased rate of variegation immediately adjacent to a sector of tissue having a much reduced rate of variegation.

These, it turns out in retrospect, were the first emerging indications that transposable elements had been activated in her cultures, in all probability by processes associated with the breakage and repair of chromosomes. In the following year, McClintock followed up on both types of mutations. She begins:

> In the cultures that were grown to observe the mutations produced as a consequence of the breakage cycles of chromosome 9, an unusual and unexpected series of new mutants appeared, characterized by types of instability known in the genetic literature as mutable genes, variegation, or mosaicism.

She notes that, although reports of the appearance of new mutable genes is relatively rare in the maize literature, she had already isolated 14 new cases of such instability and observed more. Continuing with the investigation of the new type of chromosome breakage, McClintock says:

> The most unexpected expression of variegation that has appeared in these investigations is associated with the occurrence, in many somatic cells, of breakage in chromosome 9 that takes place at a particular locus in the short arm of this chromosome. This breakage results in the formation of an acentric fragment and the subsequent elimination of this fragment at a somatic anaphase.

She carried out a particularly instructive cross to plants that carried several recessive markers spanning most of the short arm of chromosome 9 and found that all of the dominant markers on the breaking chromosome were being lost simultaneously, which bore the implication that breakage occurred repeatedly within a restricted region or at a single site. Supporting cytological studies revealed both a small pycnotic chromatin body in the cytoplasm, as well as a chromosome 9 constitution in which one homolog lacked a large segment of the short arm.

McClitock concludes the report with the tantalizing observation that both variegation in the newly isolated unstable mutants and chromosome breakage showed striking differences in frequency and timing between plants and that there seemed to be a parallel between the two phenomena, mutability and chromosome breakage, in the control of frequency and timing. She ends with the following prophetic guess: "Possibly the resemblance is more than coincidental, in that the underlying phenomena are basically similar."

By the following year, McClintock was confident enough that the chromosome breakage in the new strain was happening at a single site to name it the *Dissociation* or *Ds* locus ". . . because the most readily recognizable consequence of its action is this dissociation." She had mapped the locus by both cytological and genetic methods to a position that demarcates the proximal third of the short arm of chromosome 9. Recognizing that its action was dominant, she designated the nonbreaking version of the locus *ds*. More importantly, McClintock had recognized that a second dominant locus was required for chromosome dissociation at the *Ds* locus and she named it *Activator* or *Ac*, for its ability to activate chromosome breakage at the *Ds* locus.

Turning to the investigation of the mutable loci that had surfaced in her cultures, McClintock soon realized that although some did not require a second locus for instability, others were unstable only in the presence of *Ac*. This began to provide support for her earlier guess that the chromosome breakage and mutability phenomena were related in some way. She observed that the new mutable alleles of the *C* and *Waxy* (*Wx*) loci that required the presence of *Ac* for instability had all arisen in plants that contained *Ac*. But she acknowledged that "[i]t cannot be stated that *Ac* induces the initial state of instability of a locus, although a causal connection may be suspected."

Recognition that *Ds* Transposes

It was at about this time that McClintock became aware of the ability of *Ds* to move. In her Carnegie Yearbook report for 1947–1948, this simple statement appears: "It is now known that the *Ds* locus may change its position in the chromosome after such coincidental [chromosome] breaks have occurred."

This statement is followed by a brief summary of the observations leading to this startling conclusion. In an unpublished document dated April 1949, she describes the experiments in considerable detail. Under the title "The origin of transposed *Ds*, Case I" she begins:

> The first recognized case of transposition of *Ds* arose in the cross of a plant (4108C-1) having the constitution *wd I Sh Bz Wx Ds* in one normal chromosome 9 and *Wd C sh bz wx ds* in a normal homologous chromosome 9. This plant was heterozygous for *Ac* (*Ac ac*). The types of kernels resulting from the cross of this plant to a female plant carrying *C sh bz wx ds ac* are given in table 1.

What provided the first clue that something had changed was the observation of two exceptional kernels that did not, as expected and observed for all of the other kernels containing *Ac*, lose all of the dominant markers distal to *Ds* simultaneously. The original and new patterns are illustrated in Figure 1.2. Figure 1.2A shows the colorless phenotype of kernels carrying the dominant inhibitory *I* allele of the *C* locus, which is required for aleurone pigmentation, and Figure 1.2B illustrates the pigmentation patterns resulting from loss of the entire short arm of chromosome 9 distal to the original position of *Ds*. The concomitant loss of the dominant *I* and *Bz* alleles reveals areas of the bronze pigmentation characteristic of the *C bz* combination. The colored rims between these sectors are the result of complementation between the *C* and *Bz* alleles in adjacent sectors, but there are no fully pigmented *C Bz* sectors. Figure 1.2C shows the new pigmentation pattern, which includes deeply the pigmented *C Bz* sectors expected for breaks between the *I* and *Bz* loci giving rise to colored sectors. Curiously, there were additional twin sectors, circled in Figure 1.2C, that, in turn, had lost succeeding loci proximal to *I* sequentially, including the *Bz* locus. These observations led to an intense investigation, summarized in a detailed description nearly 50 pages in length,

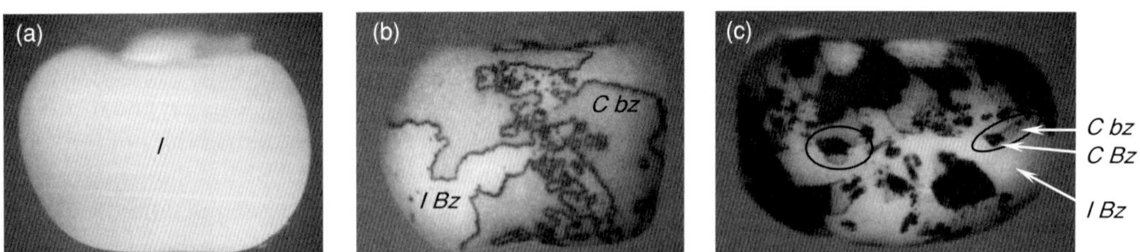

Figure 1.2 Pigment phenotypes of kernels that led to McClintock's discovery of *Ds* transposition. *I*: dominant inhibitory allele of the *C* locus; *C*: full anthocyanin pigmentation when together with the wildtype *Bz* allele of the *Bronze* locus; *bz*: recessive allele of *Bronze* locus. The *Shrunken* and *Waxy* loci are not included in these illustrations for simplicity. The circles highlight twin sectors resulting from formation of dicentric chromosomes and the subsequent unequal breakage and segregation of the dominant alleles. For color detail, see color plate section.

accompanied by more than 40 tables and diagrams, of the genetic and cytological experiments that led McClintock to the conclusion that the *Ds* element had changed its chromosomal location.

In summary, she says:

> It is possible, now, to reconstruct the events that gave rise to this Duplication chromosome 9 with a transposed *Ds* locus. Three assumptions regarding these events are required. (1) A *Ds* mutation occurred at its usual time – late in the development of the sporophytic tissues – in a cell of plant 4108C-1. The chromosome in which this *Ds* mutation occurred was normal in morphology and carried *I Sh Bz Wx* and *Ds* in its standard location. The *Ds* mutation resulted in breakage of the two sister chromatids at the position of the *Ds* locus in each chromatid. Evidence that a *Ds* mutation brings about breaks in sister chromatids at the locus of *Ds* is well established. This assumption is therefore legitimate. (2) The *Ds* mutation not only caused breaks to occur at the position of the *Ds* locus but resulted in the release of a submicroscopic chromatin segment that carries a *Ds* locus. This released segment carrying *Ds* has unsaturated broken ends. It could be lost from the chromosome complement if fusion with some other broken ends did not occur. . . . (3) At the same time that the events described in (1) and (2) occurred, a spontaneous chromosome break occurred just to the right of the *I* locus in this chromosome. Both sister chromatids were broken at the same locus. Evidence for frequent spontaneous breaks in maize is good (McClintock, unpublished). This assumption, therefore, is legitimately taken. These three events would give a series of broken ends as shown in A, figure 3. Fusion of broken ends could readily occur to give rise to the configuration shown in B, figure 3. The resulting chromatids are diagrammed in C, figure 3. A Duplication chromosome 9, with an inverted order of genes in the proximal duplicated segment and having a transposed *Ds* locus just to the right of the *I* locus is now formed.

"Figure 3," as referred to in the foregoing and hand-drawn by McClintock to represent the transposition event, is reproduced here as Figure 1.3. McClintock particularly drew attention to the fact that, although a chromosomal duplication accompanied the transposition, the transposition itself did not introduce any visible alteration in the appearance of the chromosome in the vicinity of the transposition. During the course of this investigation, McClintock gained the additional insight that, although the chromosome with the duplication and transposed *Ds* did not lack a telomere, *Ds* breaks occurring at subsequent cycles of chromosome replication could result in the joining of the two sister chromatids at the *Ds* insertion site. This generated a dicentric chromosome, which subsequently continued the breakage-fusion-bridge cycle, accounting for the sequential loss of markers observed in the initial exceptional kernels (Figure 1.2C).

Explaining Mutable Genes

There remained one more piece of the puzzle: the relationship between the behavior of the *Ds* element and the nature of the new *Ac*-controlled mutable genes. Recalling the earlier contributions, it is clear that McClintock's newly isolated mutations shared certain characteristics of the mutable genes studied by Emerson and Rhoades. Particularly striking was the parallel with Rhoades's *Dt*-controlled mutable a_1 allele. The solution came when McClintock realized that the *Ds* element could move, and it emerged from the analysis of one new instance of mutability at the *C* locus that surfaced in her cultures. She had crossed a plant having the chromosome 9 constitution *yg c sh Bz wx ds, ac* by a plant that was *Yg C Sh Bz wx Ds / Yg C Sh Bz wx Ds, Ac ac*. This plant, in turn, was crossed to a number of *c sh Bz ds, ac* plants in the summer of 1946. Kernels receiving *Ac* would be expected to show loss of the *C* allele, uncovering colorless *c* sectors (Figure 1.4A). She noted a

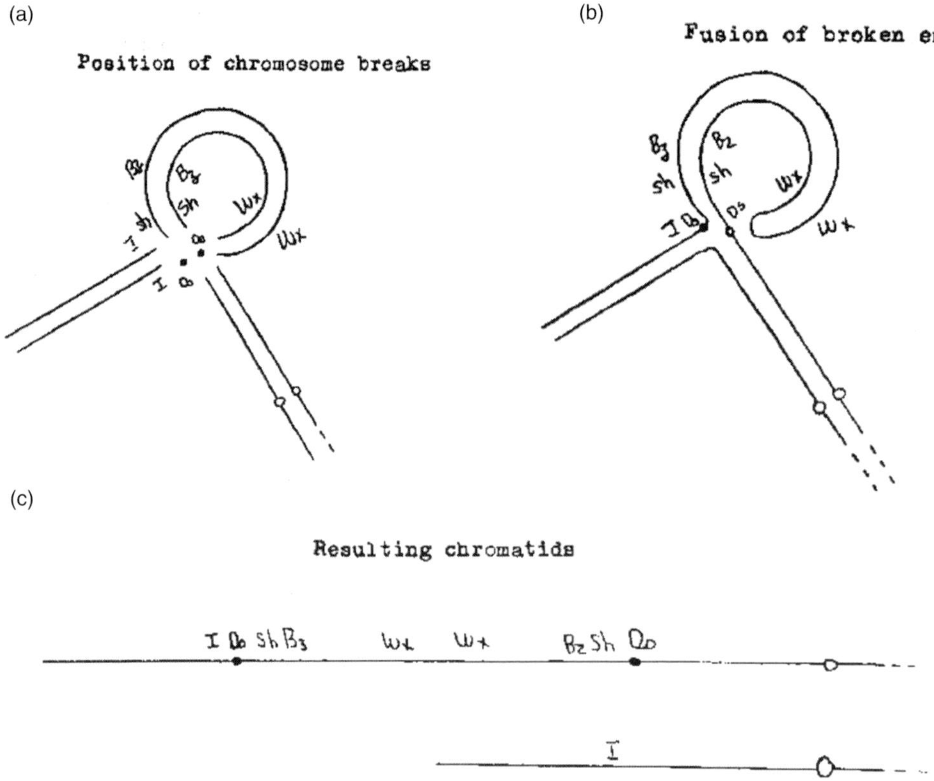

Figure 1.3 The "Figure 3" referred to in the text, hand-drawn by McClintock.

single kernel that showed variegation from the recessive colorless *c* to the dominant, pigmented *C*, unexpected for this genetic constitution (Figure 1.4B).

A plant was grown from the aberrant kernel in the summer of 1947 and designated plant 4204. Not unexpectedly, the plant showed variegation, and McClintock believed its constitution to be *yg c sh Bz wx ds / Yg c-m1 Sh Bz wx Ds, Ac ac*, wherein *c-m1* designates the first instance of a mutable allele of the *C* locus. She made a number of crosses during the summer of 1947. From the results

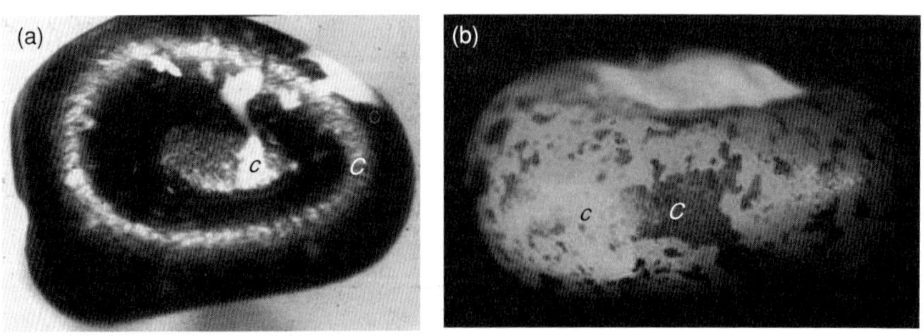

Figure 1.4 A. The phenotype of kernels losing the dominant pigmenting *C* allele to uncover the recessive null *c* due to chromosome breakage at *Ds*. B. The phenotype of a kernel in which *Ds* has inserted at the *C* locus causing loss of pigmentation except in sectors developing from cells in which the *Ds* has transposed out of the *C* locus, restoring a functional *C* allele. For color detail, see color plate section.

of a cross to a *yg c sh Bz wx ds, ac* homozygote, it was apparent that one of the chromosomes was *Yg c-m1 Bz Sh wx* and one was *yg c sh Bz wx*. However, when the plant was crossed to one that was homozygous for the chromosome 9 constitution *C sh bz wx ds, ac*, the unexpected observation was made that there was variegation that uncovered the *bz* allele, but it was almost exclusively in the *Sh* class of kernels. This was unexpected because the distance between the *Sh* locus and the *Ds* element in its original position is more than 30 cM, leading to the expectation of many kernels bearing recombinant chromosomes carrying the recessive *sh* allele and the *Ds* locus. McClintock notes parenthetically that "[t]his latter observation should have made me suspicious about the location of *Ds* but I failed to register a reaction." And she goes on to state, as noted in the opening paragraph, that she had only recently understood that *Ds* could move (McClintock, unpublished).

By the time she had finished analyzing the results of the crosses, McClintock knew something was wrong with her initial interpretation of *c-m1*, and she began an intensive study of the allele. The subsequent crosses confirmed that chromosome breakage was closely linked to the *c-m1* locus, once again implying that the expected recombination between *Ds* at its original position and *c-m1* allele was not occurring. She developed the hypothesis that the *c-m1* allele might have arisen by transposition of the *Ds* element into or nearby the *C* locus, inactivating it. In order to test this hypothesis, she selected a number of fully pigmented *C* kernels arising on *c-m1* ears. Of the 16, only 1 showed *Ds*-type chromosome breakage to reveal the *c* allele on the homolog. Moreover, she carefully examined fully pigmented (revertant *C*) sectors on kernels that contained *Ac* and were heterozygous for *c-m1* and *c* or a deficiency lacking the locus. What she observed was that these sectors almost never showed variegation from *C* to *c* or from *Wx* to *wx* as might be anticipated if *Ds* were still present on the chromosome and the *Ac* locus was causing chromosome breakage and initiating the breakage-fusion-bridge cycle. Thus, it appeared that, when the *c-m1* allele mutated either germinally or somatically to the wild-type *C* allele, all evidence of the presence of *Ds* disappeared. McClintock concludes:

> The combined observations made me suspect that the *Ds* action was associated with the presence of *Ds* in the *c-m1* locus and that the *Ds* locus was usually lost from the chromosome when a mutation to *C* occurred. The following conclusions were formulated: that (1) the *c-m1* locus arose from a *C Sh wx Ds* chromosome following removal of *Ds* from its standard location and its insertion into the normal *C* locus; (2) when a mutation of *c-m1* to *C* occurs, *Ds* disappears; (3) the insertion of *Ds* into the *C* locus brought about an inhibition in the expression of the genes at the *C* locus; consequently the modified *C* locus resembles *c* in phenotypic expression. (4) When this *Ds* element is removed from the chromosome, the normal genic action of the *C* locus can reappear. (5) Two contrasting types of *Ds* action can be recognized when *Ds* is in this new position: (a) one giving the typical dicentric chromatid through breakage and fusion at this new location and (b) one giving the *c* to *C* mutations that usually result in no visible alteration of the chromosome.

Thus, the last piece of the puzzle had fallen into place, explaining the basis of the variegation phenomena that had, by then, been under genetic scrutiny for almost half a century. Unstable mutations of the type analyzed by both Emerson and Rhoades could be understood as the result of a transposable element insertion into a locus, where it frequently underwent both somatic and germinal excision and transposition, restoring function to the affected gene. McClintock was able to make the connection between transposition of a genetic element, the *Ds* locus, and the origin of a mutable gene giving a variegated phenotype because the particular *Ds* element she first isolated had a second property, chromosome breakage, by which she was able to track it.

Many of McClintock's original *Ds* insertion mutations were not caused by chromosome-breaking *Ds* elements, which have a special structure. Nonetheless, they showed the same relationship to the

Activator element as Rhoades's a_1 mutation showed to the *Dt* locus: the mutations were unstable *only* in the presence of the second, activating locus. McClintock soon understood that *Ac* could itself transpose and cause insertion mutations that differed from those caused by a *Ds* element in being autonomously unstable. McClintock further found that the dosage of the *Ac* element, as well as more subtle changes in both the *Ds* and the *Ac* elements, affected both the timing and frequency of chromosome breaks and transposition. The *Ac* dosage effect is a negative one: the more copies, the greater the developmental delay in *Ac*-mediated breakage and transposition. Because *Ac* turned out to be the transposon causing pericarp variegation in Emerson's "calico" corn, McClintock's findings explained his puzzling observation that there is an inverse relationship between the amount of somatic reversion and the number of copies of the P^{vv} gene.

Molecular Endnote

A few explanatory notes on the underlying molecular structures might help illuminate some of the more mystifying aspects of McClintock's observations. The *Ac* element is a small transposon (4.6 kb) and encodes a single protein, its transposase (Fedoroff et al., 1983). All *Ac* elements so far isolated are virtually identical in sequence. *Ds* elements are often, although by no means always, internally deleted derivatives of an *Ac* element (Fedoroff, 1989). There are many additional, structurally different transposons that are mobilized by the *Ac* element. Some share little sequence identity with *Ac* save the 11-bp terminal inverted repetitions and some subterminal transposase binding sites (Kunze and Weil, 2002). Neither *Ac* transposons nor *Ds* transposons derived from them generally cause chromosome breakage. The chromosome-breaking *Ds* originally identified by McClintock has a unique structure: it consists of two short, identical *Ds* elements that comprise the ends of *Ac*, one inserted in inverted order almost precisely into the middle of the other (Doring et al., 1984). The ability of this *Ds* element to break chromosomes is a consequence of its transposition mechanism and the presence of both ends of the element in both orientations within its structure. The transposon causing instability of the pericarp locus in Emerson's strains, and named *Mp* by Brink and his colleagues, is the same as *Ac* (Barclay and Brink, 1954; Schwartz, 1989). Emerson's original observations that a somatic reversion event was almost always stable, except for the occasional reappearance of variegating kernels probably finds its explanation in the observation that *Ac* elements have a propensity to transpose to nearby sites, from which they can, once again, transpose back into the locus of origin. Similarly, chromosome breakage can persist after a chromosome-breaking *Ds* transposes away from a gene if it reinserts nearby on the same chromosome.

References

Barclay, P.C., and Brink, R.A. (1954) The relation between modulator and activator in maize. *Proc. Natl. Acad. Sci. USA 40*: 1118–1126.

Correns, C. (1910) Der Ubergang aus dem homozygotischen in einen heterozygotishen Zustand im selben Individuum bei buntblattrigen und gestreifbluhenden Mirabilis-Sippen. *Ber. Deutsch. Bot. Gesell. 28*: 418–434.

de Vries, H. (1905) *Species and Varieties: Their Origin by Mutation*, 2nd edn. (Chicago: Open Court Publishing Co.).

Demerec, M. (1935) Unstable genes. *Bot. Rev. 1*: 233–248.

Doring, H.P., Tillmann, E., and Starlinger, P. (1984) DNA sequence of the maize transposable element Dissociation. *Nature 307*: 127–130.

Emerson, R.A. (1914) The inheritance of a recurring somatic variation in variegated ears of maize. *Am. Naturalist 48*: 87–115.

Emerson, R.A. (1917) Genetical studies of variegated pericarp in maize. *Genetics 2*: 1–35.

Fedoroff, N. (1989) Maize transposable elements. In *Mobile DNA*, M. Howe and D. Berg, eds. (Washington, D.C.: American Society for Microbiology), pp. 375–411.

Fedoroff, N., Wessler, S., and Shure, M. (1983) Isolation of the transposable maize controlling elements *Ac* and *Ds*. *Cell 35*: 243–251.

Goldschmidt, R. (1938) *Physiological Genetics* (New York: McGraw-Hill).

Kunze, R. and Weil, C.F. (2002) The hAT and CACTA superfamilies of plant transposons. In *Mobile DNA II*, N.L. Craig, R. Craigie, M. Gellert, and A.M. Lambowitz, eds. (Washington, D. C.: ASM Press), pp. 565–610.

McClintock, B. (1938) The fusion of broken ends of sister half-chromatids following chromatid breakage at meiotic anaphases. *Missouri Ag. Exp. Sta. Res. Bul. 290*: 1–48.

McClintock, B. (1939) The behavior in successive nuclear divisions of a chromosome broken at meiosis. *Proc. Natl. Acad. Sci. USA 25*: 405–416.

McClintock, B. (1941a) The association of mutants with homozygous deficiencies in *Zea mays*. *Genetics 26*: 542–571.

McClintock, B. (1941b) The stability of broken ends of chromosomes in *Zea mays*. *Genetics 26*: 234–282.

McClintock, B. (1942) The fusion of broken ends of chromosomes following nuclear fusion. *Proc. Natl. Acad. Sci. USA 11*: 458–463.

Rhoades, M.M. (1936) The effect of varying gene dosage on aleurone colour in maize. *J. Genet. 33*: 347–354.

Rhoades, M.M. (1938) Effect of the *Dt* gene on the mutability of the a_1 allele in maize. *Genetics 23*: 377–397.

Rhoades, M.M. (1941) The genetic control of mutability in maize. *Cold Spring Harbor Symp. Quant. Biol. 9*: 138–144.

Schwartz, D. (1989) Pattern of *Ac* transposition in maize. *Genetics 121*: 125–128.

2 A Field Guide to Transposable Elements

Alan H. Schulman and Thomas Wicker

The C-value Paradox

Eukaryotic genomes can show extreme size variations (Bennett and Smith, 1976). *Amoeba dubia*, with a reported genome size of 670,000 Mbp (Gregory, 2001), and the microsporidium *Encephalitozoon cuniculi*, with a compact genome of 2.9 Mbp (Biderre et al., 1995; Katinka et al., 2001) display a 200,000-fold variation in size. However, despite the enormous variation in genome size, gene numbers in eukaryotes show only a 10-fold range – from 5,000 to 50,000 – in the basic (haploid) set of chromosomes. Vertebrates and plants are estimated to contain 25,000–40,000 genes, whereas invertebrates and fungi possess half as many (Table 2.1). This discrepancy between organismal complexity, later expressed in terms of gene number and genome size variation, came to be known as the "C-value Paradox." (Robash et al., 1974)

Within many monophyletic taxonomic groups of animals and fungi, there seems to be little paradox; genome sizes are generally in a narrow size range. Mammals, for example, which diverged approximately 70–113 million years ago (MYA; Gheerbrandt et al., 2005), have genomes with very similar sizes averaging near 3,000 Mbp (Table 2.1). Reptiles, including birds, which evolved from them about 240 MYA (Benton, 1993), have smaller genome sizes of roughly 1,000–2,000 Mbp (Krishan et al., 2005). Only among amphibians and lungfish has greater than a 100-fold genome size variation been reported (Gregory, 2001). All fungal genomes described so far are small compared to animal genomes and range from a few to less than 200 Mbp (Table 2.1). This is intriguing, considering that some of the main evolutionary lineages in fungi diverged between 700 and 1,200 MYA (Padovan et al., 2005; Taylor and Berbee, 2006).

In contrast to those of animals and fungi, plant genomes show extreme variation in size even between very closely related species. Furthermore, there is almost no correlation between genome size and phylogenetic distance in plants. For example, the grasses *Brachypodium distachyon* and bread wheat *Triticum aestivum* have genomes sizes of 273 Mbp and 5,700 Mbp, respectively, although they diverged only about 35 MYA (Bossolini et al., 2007). Wheat is a hexaploid and *B. distachyon* a diploid; nevertheless, ploidy hardly accounts for their difference in genome size. Moreover, sorghum and maize, which diverged only about 12 MYA (Swigonova et al., 2004), have genome sizes of 690 Mbp and 2,500 Mbp, respectively (Paterson et al., 2009; Schnable et al., 2009). And all of these large genomes are still dwarfed by those of some lilies, among them *Fritillaria uva-vulpis*, which has a genome size of more than 87,000 Mbp, greater than 700 times the size of the *Arabidopsis* genome (Arabidopsis Genome Initiative, 2000; Leitch et al., 2007).

Plant Transposons and Genome Dynamics in Evolution, First Edition. Edited by Nina V. Fedoroff.

Table 2.1 Genome sizes and gene numbers in publicly available genomes. Abbreviations in references: AGI, *Arabidopsis* Genome Initiative; CSC, *C. elegans* Sequencing Consortium; IBI, International *Brachypodium* Initiative; ICGSC, International Chicken Genome Sequencing Consortium; IHGSC, International Human Genome Sequencing Consortium; IRGSP, International Rice Genome Sequencing Consortium; MGSC, Mouse Genome Sequencing Consortium.

Animal genomes	Size (Mbp)	Genes	Reference
Anopheles gambiae	278	14,000	Holt et al., 2002
Caenorhabditis elegans	97	19,000	CSC, 1998
Drosophila melanogaster	120	15,200	Adams et al., 2000
Gallus gallus	1,200	20,000–23,000	ICGSC, 2004
Homo sapiens	2,850	24,000	IHGSC, 2004
Mus musculus	2,500	30,000	MGSC, 2002
Plant genomes	Size (Mbp)	Genes	Reference
Arabidopsis thaliana	120	26,200	AGI, 2000
Brachypodium distachyon	273	25,500	IBI, 2010
Fritillaria uva-vulpis	87,400	?	Leitch et al., 2007
Hordeum vulgare	5,700	38,000–48,000	Mayer et al., 2009
Oryza sativa	372	40,600	IRGSC, 2005
Physcomitrella patens	462	35,900	Rensing et al., 2008
Populus trichocarpa	410	45,500	Tuskan et al., 2006
Sorghum bicolor	659	34,500	Paterson et al., 2009
Triticum aestivum	16,000	50,000	Choulet et al., 2010
Vitis vinifera	342	30,400	Jaillon et al., 2007
Zea mays	2,061	30,000	Schnable et al., 2009
Fungal genomes	Size (Mbp)	Genes	Reference
Aspergillus nidulans	30	10,600	broadinstitute.org
Aspergillus flavus	36.8	12,600	broadinstitute.org
Fusarium verticilloides	41.8	14,200	broadinstitute.org
Magnaporthe grisea	42	11,100	Dean et al., 2005
Saccharomyces cerevisiae	11.7	5,700	broadinstitute.org
Stagonospora nodurum	37	16,600	broadinstitute.org
Tuber melanosporum	125	7,500	Martin et al., 2010
Botrytis cinerea	42.6	16,400	broadinstitute.org
Other genomes	Size (Mbp)	Genes	Reference
Encephalitozoon cuniculi	2.9	1,997	Katinka et al., 2001
Amoeba dubia	670,000	?	Gregory, 2001

The Quantity of Transposable Elements Determines Genome Size

With the acceleration of DNA sequencing, it became clear that most of the DNA of large eukaryotic genomes cannot be assigned to genes encoding a diversity of proteins but rather is comprised of various classes of repeats, primarily transposable elements (TEs). In large genomes like barley, wheat, or maize, TEs comprise as much as 80% or more of the total genomic DNA (Schnable et al., 2009; Wicker et al., 2009). Even relatively small genomes, such as that of *B. distachyon*, are populated by hundreds of different TE families; some families have hundreds of copies, whereas others have only one or a few members (International *Brachypodium* Initiative, 2010).

The most abundant TEs in plant genomes are the long terminal repeat (LTR) retrotransposons, the structure of which is described below. Most of the probably hundreds of LTR retrotransposon families in the typical plant genome are present in low or moderate copy numbers. However, the large plant genomes contain very large and successful retrotransposon families. For example, the *BARE1*, *WIS*, and *Angela* elements account for more than 10% of the Triticeae genomes (Vicient et al., 1999a; Kalendar et al., 2000; Soleimani et al., 2006; Wicker et al., 2009). A whole-genome survey in barley showed that 50% of its genome is made up of just 14 TE families, 12 of which are LTR retrotransposons (Wicker et al., 2009). It is not known what makes certain LTR retrotransposon families particularly successful at expanding to large numbers. Some abundant retrotransposons are activated by stress conditions such as drought (Kalendar et al., 2000) or UV light (Ramallo et al., 2008), but other stress-activated retrotransposons are present at low copy numbers (Grandbastien et al., 2005). Although some retrotransposons may be inherently more capable than others of achieving high copy numbers, it is also a reasonable conjecture that selective forces act to drive copy number down.

Whatever the forces driving their propagation, the LTR retrotransposons in particular are responsible for the major variation in genome size. All of the other TE superfamilies in plants constitute a minor and more or less constant fraction of the genome. *Arabidopsis thaliana* and sorghum, respectively having 120 Mbp and 700 Mbp genomes, contain a similar amount of Class II transposons (see classification below) but differ mainly in the abundance of LTR retrotransposons. A recent whole-genome survey of the barley genome using 454 sequencing showed that less than a dozen LTR retrotransposon families account for almost 50% of the genome, whereas known Class II elements contribute only slightly more than 5% (Wicker et al., 2009). The difference in genome size between two *Hordeum* species can be explained primarily by the difference in *BARE1* abundance (Vicient et al., 1999b). Additionally, analysis of *Copia* elements in rice (*Oryza sativa*) and wheat showed that different families are active at different times in "waves" lasting several hundreds of thousands of years (Wicker and Keller, 2007). The existence of epochs in which a particular family is active can be inferred from the presence of a cohort of elements of similar age, dated by applying the molecular clock to their pairs of LTRs (SanMiguel et al., 1998).

General Classification Scheme for Transposable Elements

The TEs can be thought of as small colonizing genetic units, actually "minimal genomes" within the organismal genome, containing exactly enough information to replicate, move around in the genome, or both. They are essentially intranuclear or intracellular viruses and use the DNA replication and translation machinery of their "host" to thrive within the environment of the genome. The property of self-replication led them to be designated as "selfish" DNA (Doolittle and Sapienza, 1980; Orgel and Crick, 1980). Their apparent absence of a function earned them the appellation of "junk" (Ohno, 1972). However, the twin observations that TEs contribute the bulk of the DNA of most higher organisms and that TE dynamism is important to understanding genome evolution led many to look beyond the negative monikers and focus on the TEs themselves. Moreover, it became apparent that, in order to understand the TE menagerie, it was necessary to categorize and characterize the vast variety of TEs in eukaryotic genomes.

In 2007, a group of TE experts met at the Plant and Animal Genome Conference in San Diego (California, USA) with the goal to define a consensus scheme for classifying all eukaryotic TEs. A taxonomic structure with classes, orders, superfamilies, and families was constructed, consistent criteria for them defined, and a naming system developed (Wicker et al., 2007). The proposed system was intended not only to draw together previous TE classification systems but also to provide tools for the massive job of annotating the genomes being sequenced. It groups TEs into 2 major classes, 9 orders, and 29 superfamilies (Figure 2.1). Class I comprises the TEs that replicate via an RNA intermediate in a "copy-and-paste" process, whereas the Class II TEs move as DNA primarily in

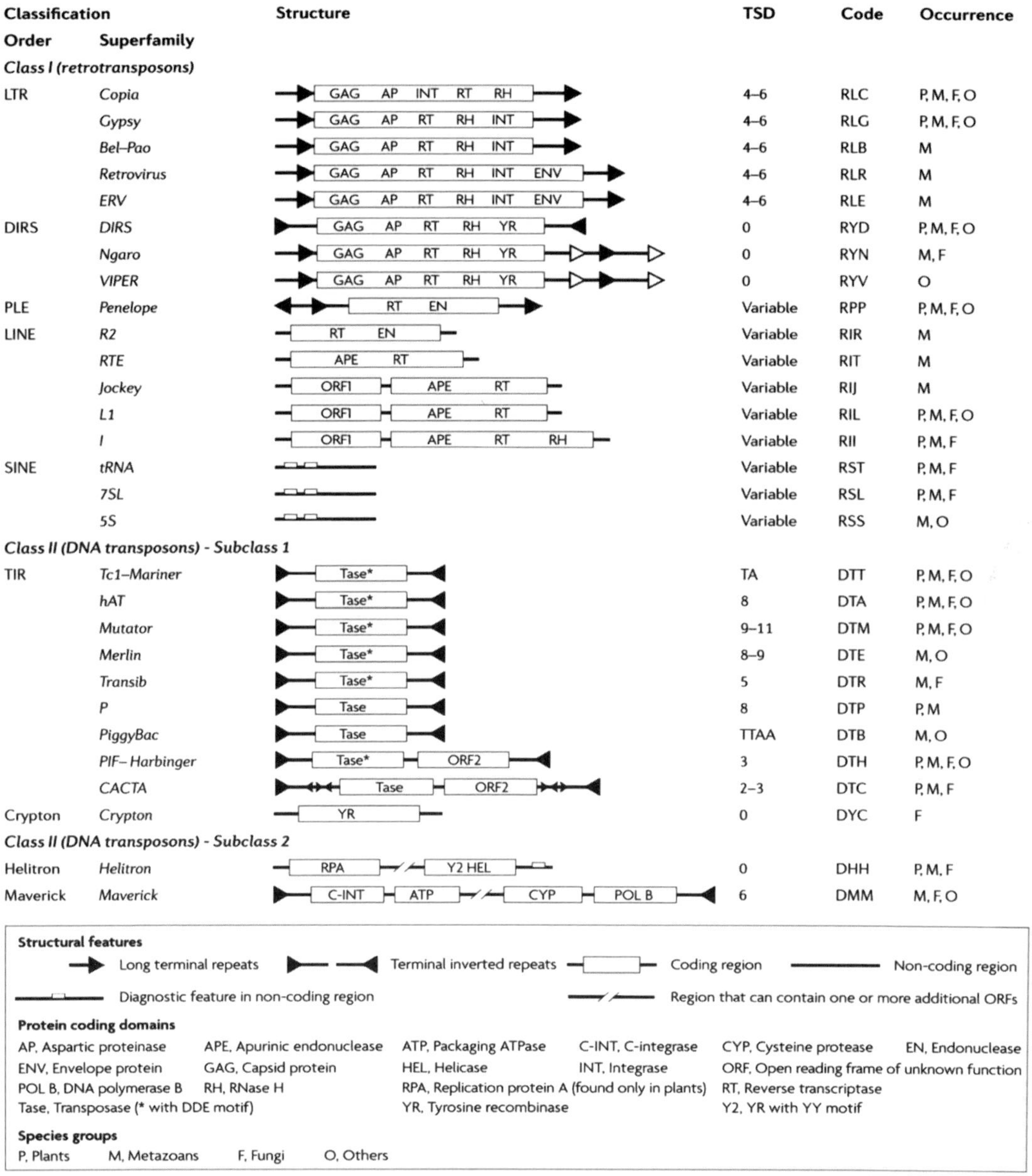

Figure 2.1 Classification system for TEs proposed by Wicker et al., (2007). The system is hierarchical and divides TEs into two main classes on the basis of the presence or absence of RNA as a transposition intermediate. They are further subdivided into orders and superfamilies. The size of the target site duplication (TSD), which is characteristic for most superfamilies, can be used as a diagnostic feature. A three-letter code precedes the family name of each TE to facilitate identification. DIRS, *Dictyostelium* intermediate repeat sequence; LINE, long interspersed nuclear element; LTR, long terminal repeat; PLE, *Penelope*-like elements; SINE, short interspersed nuclear element; TIR, terminal inverted repeat.

a "cut-and-paste" process. A novel aspect of the classification system is that the TE family name should be preceded by a three-letter code for class, order, and superfamily (Figure 2.1). Thus, the classification of the TE is contained within its name, making the type of TE immediately recognizable in genome annotations. In the following sections, the structures and proliferation mechanisms of the main TE types will be described.

Class II Elements

Arguably the most ancient class of TEs belonging to the extant DNA genomes is the DNA (Class II) transposons because these elements are found in all living organisms, both prokaryotes and eukaryotes. Most autonomous Class II elements of Subclass 1 encode a transposase, which has the ability to excise the element from the genome and integrate it elsewhere. The transposase enzyme has both an endonuclease activity for excising the transposon from the genomic DNA and a ligase activity for fusing the ends of the transposon back into the genome. DNA transposons are flanked by terminal inverted repeats that serve as recognition sites for the transposase enzyme (Figure 2.2).

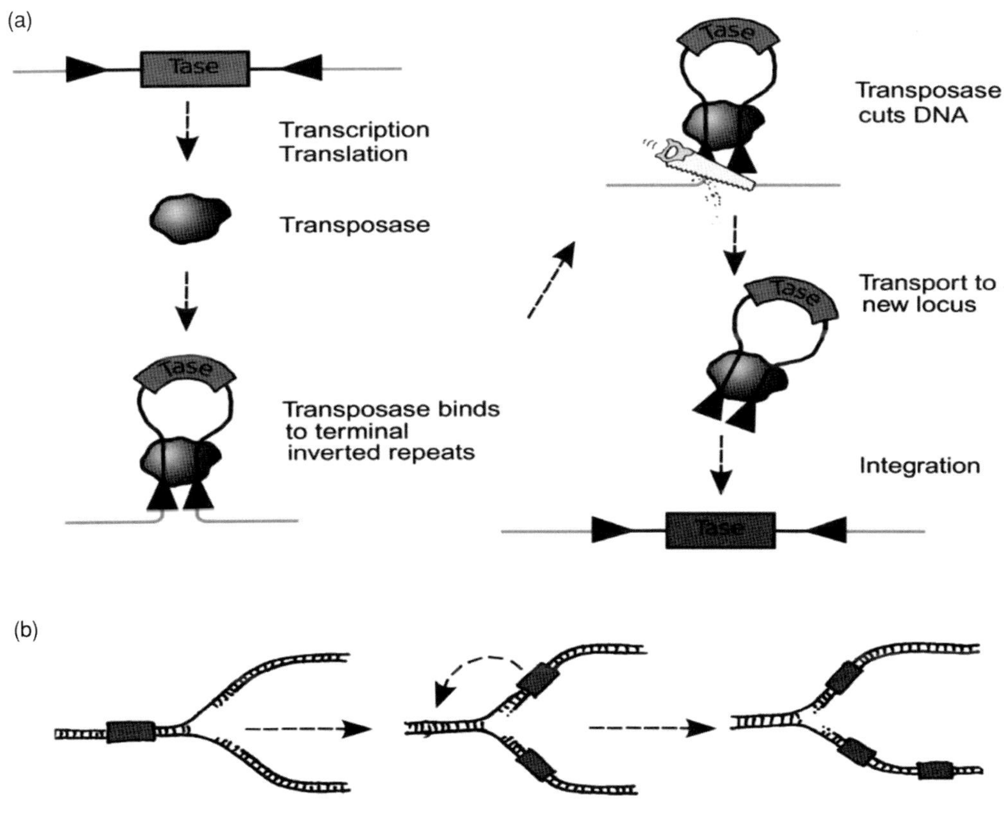

Figure 2.2 Mechanism of transposition of Class II (DNA) transposons. (a) Transposition, by "cut and paste." (b) Increasing copy number during regular DNA replication. A transposon is first replicated as part of the chromosome, then one of the newly replicated copies transposes to a locus ahead of the replication fork. Following passage of the replication fork, there is one copy on one replicated chromosome and two on the other.

Most, if not all, known transposase proteins have a so-called DDE motif. A mechanistic study showed that the aspartate and glutamate residues of the DDE motif come in close proximity to each other and form the catalytic center that enables the enzyme to excise the transposon DNA from the genome and probably also to reinsert it (Keith et al., 2008). It is widely assumed that DNA transposons multiply by moving during DNA replication from a location that has already been replicated to a location in front of the replication fork. This results in one replicated strand containing one, and the other two copies, of the element, yielding a net gain of one copy on one of the daughter chromosomes.

Class II contains, as well, a few groups that are exceptions to the rule of cut-and-paste by transposase (Figure 2.1). Among these, *Crypton* elements, which are bounded by 4- or 6-bp direct repeats, are a diverse set of eukaryotic transposons that are alone among the Subclass 1 of Class II in containing a gene for tyrosine recombinase (Goodwin et al., 2003; Kojima and Jurka, 2011). *Crypton* elements are proposed to move as circular DNA that is produced by tyrosine recombinase-driven recombination between the direct repeats and then reinserted elsewhere by recombination at a matching target site (Goodwin et al., 2003). Subclass 2 of Class II contains two groups, *Helitron* and *Maverick*, with structures and replication mechanisms that differ sharply from those of Subclass 1. The *Maverick* (also called *Polinton*) elements are a diverse group of elements that are bounded by inverted repeats, encode an integrase and five to at least nine other genes, and probably move as double-stranded DNA (Feschotte and Pritham, 2005; Kapitonov and Jurka, 2006; Pritham et al., 2007). Bioinformatic analyses indicate that *Helitrons* comprise a widespread, diverse, and sometimes abundant group of elements that are nonetheless difficult to identify due to their lack of classical terminal repeat structures (Du et al., 2006). Although some *Helitrons* encode "RepHel" proteins with replication initiator and DNA helicase domains, suggesting transposition by rolling-circle replication, their mechanism of movement has not been rigorously established and may involve excision (Kapitonov and Jurka, 2001, 2007; Li and Dooner 2009). Notably, *Helitrons* in maize appear to have moved gene fragments and pseudogenes around in the genome, greatly contributing to genome diversity (Lai et al., 2005; Morgante et al., 2005).

Class I: The Non-LTR and LTR Retrotransposons

The Class I transposable elements are united by a replication cycle involving the copying of a genomic RNA into dsDNA by reverse transcriptase. The two main orders of Class I TEs are named according to the presence or absence of an LTR at either end of the element. The long interspersed nuclear elements (LINEs; Goodier and Kazazian, 2008) often take pride of place as the representative non-LTR retrotransposons, but there are also other orders of elements besides LINEs that lack LTRs, including the DIRS (*Dictyostelium* intermediate repeat sequence), PLE (*Penelope*-like element), and SINE (short interspersed nuclear elements) retrotransposons (Wicker et al., 2007). We will return to the SINE and DIRS orders in more detail below. The non-LTR retrotransposons are found in all eukaryotes. They are the predominant TE order in vertebrates and some fungi (Spanu et al., 2010) but are generally much less abundant in plants.

The LINEs are thought to be the most ancient type of Class I element due to their simple structure, the basic forms specifying reverse transcriptase and endonuclease activities. Moreover, their use of reverse transcription can be seen as an ancient trait, linking the Class I lifecycle to the ancient transition from primordial RNA genomes to the modern DNA genomes (Brosius, 2005). Aside from lacking LTRs, the non-LTR retrotransposons do not have an integrase gene; instead, the reverse transcriptase uses the poly-A tail of the element's mRNA as a primer and starts DNA strand

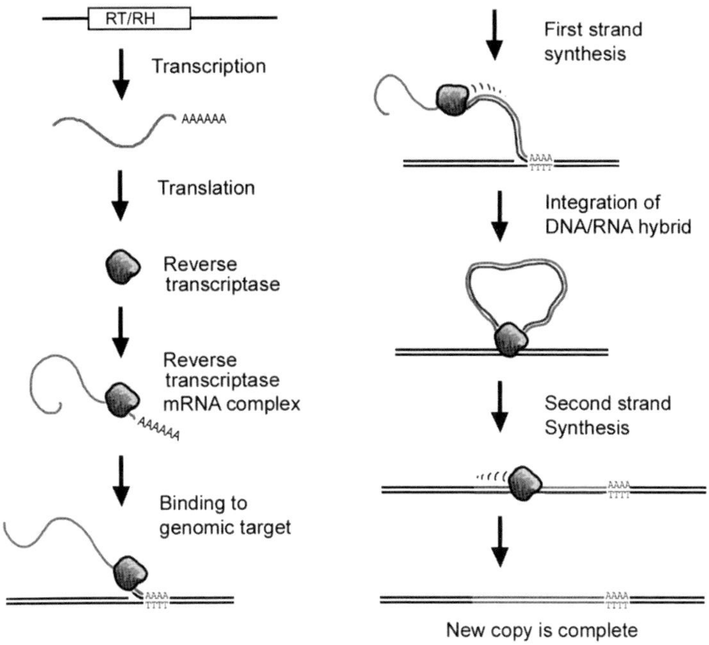

Figure 2.3 Replication mechanism of a non-LTR retrotransposon. The reverse transcriptase enzyme has the ability to transcribe mRNA into DNA and also to integrate the DNA copy into the genome.

synthesis directly at the point of insertion (Figure 2.3). The reverse transcriptase also has the ability to ligate the end of the newly synthesized DNA into the insertion point.

Like the LINEs, the LTR retrotransposons replicate via the reverse transcription of mRNA. In contrast to non-LTR retrotransposons, however, the insertion of the DNA copy is carried out through a specialized enzyme, integrase (Figure 2.4), which is encoded by the element itself. Integrase makes staggered cuts at the target site, trims extra nucleotides from the 3′ termini of the LTRs, and also joins the 3′ termini to the free 5′ ends at the staggered cut (Figure 2.4).

The LTRs direct both transcription, for which they provide a promoter at the 5′ end of the retrotransposon, and RNA termination and polyadenylation, for which signals in the LTR are operational at the 3′ end of the element. Transcription by pol II begins within the 5′ LTR, downstream from the promoter, and terminates within the 3′ LTR before its 3′ end. The result is that the transcripts lack complete LTRs at either end (Figure 2.5). The problem of restoring LTRs to both ends of the cDNA is resolved by the complex reverse transcription mechanism of LTR retrotransposons. The mechanism (Figure 2.5) achieves this through two template switches. The U5 and R domains of the LTR are reverse-transcribed from the 5′ end of the transcript. Following transfer of the minus-strand cDNA to the 3′ end of the transcript, the U3 region of the LTR is incorporated in it. The plus-strand cDNA and its LTR are then copied from the minus-strand cDNA, thereby restoring full-length LTRs to both ends of the cDNA copy and making their sequences identical.

Reverse transcription of retrotransposon RNA represents a branch point in the complex lifecycle of LTR retrotransposons (Figure 2.6). Transcripts exported from the nucleus are either translated into the polyprotein encoded by the retrotransposon or reverse transcribed as described above. During reverse transcription, the RNase H component of reverse transcriptase degrades the RNA template once it is copied. Hence, the RNA must be translated first if the same molecule is to serve

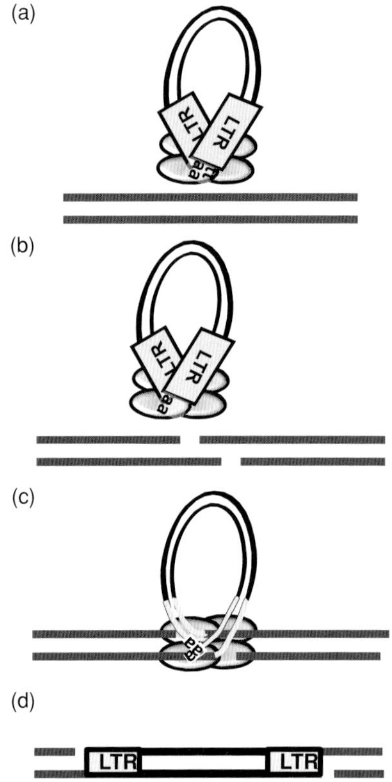

Figure 2.4 Integration mechanism of an LTR retrotransposon. The retrotransposon is represented as a loop bounded by two LTRs. Each LTR is flanked by an extra dinucleotide basepair (in this case AA/TT, as found in retrotransposon *BARE* of barley), which is copied by reverse transcriptase from the dinucleotide found between the PBS and the 3′ end of the 5′ LTR during reverse transcription. The integrase is represented bound to the LTRs as a tetramer (Dolan et al., 2009; Cherepanov et al., 2011), forming a pre-integration complex together with the retrotransposon. The genomic DNA target is shown as a pair of gray lines beneath the pre-integration complex. (a) The pre-integration complex and target site. (b) The integrase makes a 4- to 6-bp staggered cut in the genomic DNA and trims the dinucleotide from the 3′ end of each LTR, generating 5′ overhangs on both the retrotransposon and at the target site. (c) Integration of the LTR retrotransposon. The 3′ ends of the LTR are joined to the 5′ overhangs of the target. The *trans*-esterification reaction, in which the target is cleaved and retrotransposon joined, proceeds as a single-step. (d) Following the integration reaction and removal of the remaining dinucleotide from the 5′ end of each LTR, the gaps generated by the staggered cut remain. The repair of these gaps generates the TSD flanking the retrotransposon.

as a template for both cDNA and protein. The RNA template is packaged into virus-like particles (VLPs) comprised of the capsid protein Gag, together with reverse transcriptase and integrase. These proteins are derived from the polyprotein by the endoproteolytic action of aspartic proteinase, also part of the polyprotein. Following reverse transcription, the VLP is targeted to the nucleus, the cDNA enters the nucleus, and integration takes place.

The LTR retrotransposons are divided into two main superfamilies, *Gypsy* and *Copia*, which are found in almost all eukaryotic lineages and which differ both in the order and the sequence affinities of their encoded protein domains. *Gypsy* and *Copia* elements most likely originated from two independent gene fusion events, which predate the radiation of the eukaryotes (see below). Sequence and overall structural similarities indicate that the retroviruses evolved from *Gypsy* elements through the acquisition of the *env* gene encoding the envelope protein that allows them to leave the host

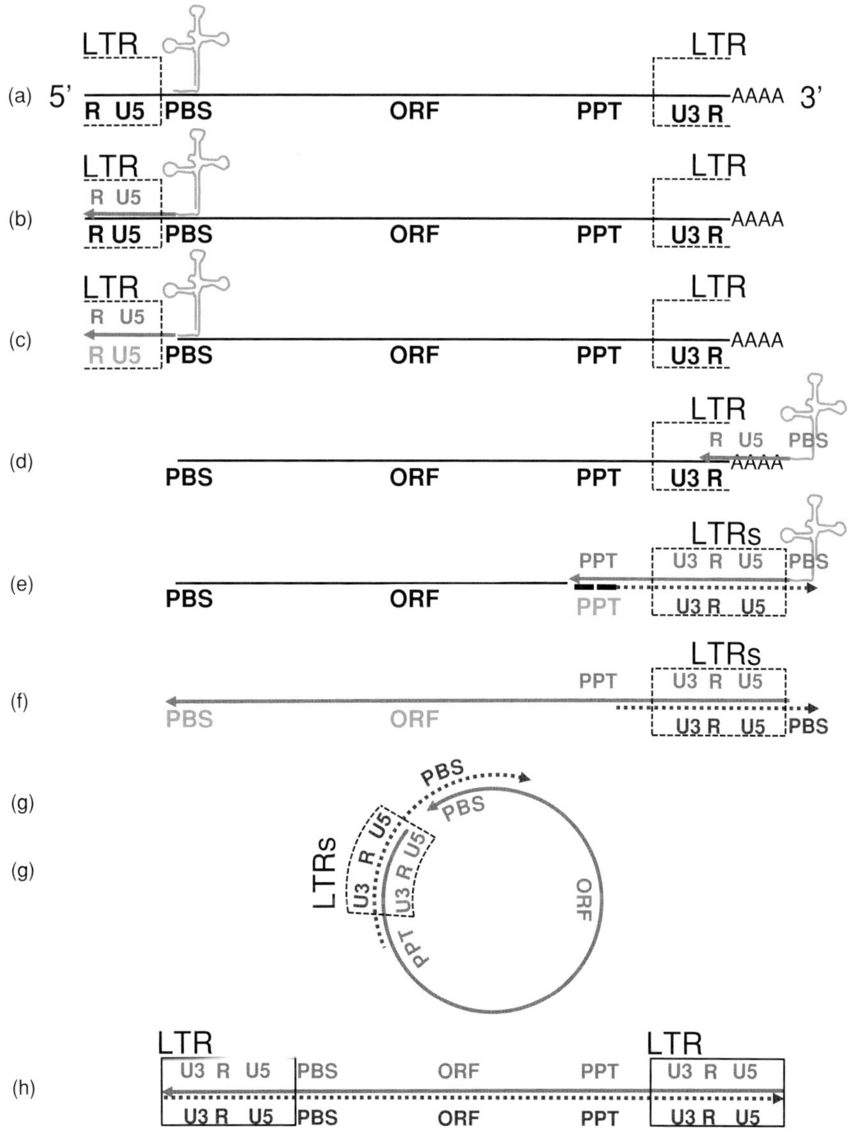

Figure 2.5 Reverse transcription of LTR retrotransposons. Diagrammatically represented are the major steps. (a) Attachment of a tRNA primer at the primer-binding site (PBS) of the retrotransposon transcript (black line), adjacent to the 3′ end of the 5′ LTR regions R and U5, to initiate reverse transcription. (b) Extension of the minus-strand cDNA (shown as a gray line) to the end of the transcript to form minus-strand strong-stop DNA (−sssDNA); (c) Degradation of the RNA from the RNA/DNA hybrid by RNaseH, exposing the repeat (R) domain that is present at both ends of the transcript. (d) Transfer of the exposed −sssDNA to the 3′ end of the transcript by hybridization of the R domain. (e) Extension of the minus-strand and concomitant degradation of the hybridized regions of the transcript by RNase H until the polypurine tract (PPT) of the cDNA is exposed, whereupon plus-strand cDNA (dotted line) synthesis is initiated from RNA fragments (short black lines) as primers. The plus strand is extended to the 5′ end of the minus-strand cDNA, and generating a complementary copy of the PBS, and forms plus-strand strong-stop DNA (+sssDNA). (f) The RNA primers are removed by RNaseH, exposing the PBS on the +sssDNA. (g) Transfer of the +sssDNA, mediated by hybridization of the PBS domain, and continuation of cDNA synthesis requiring strand displacement, each strand serving as a template for the other. (h) Completion of cDNA synthesis to generate a double-stranded linear molecular with intact LTRs at either end. The details and representation are essentially as presented earlier (Telesnitsky and Goff, 1997).

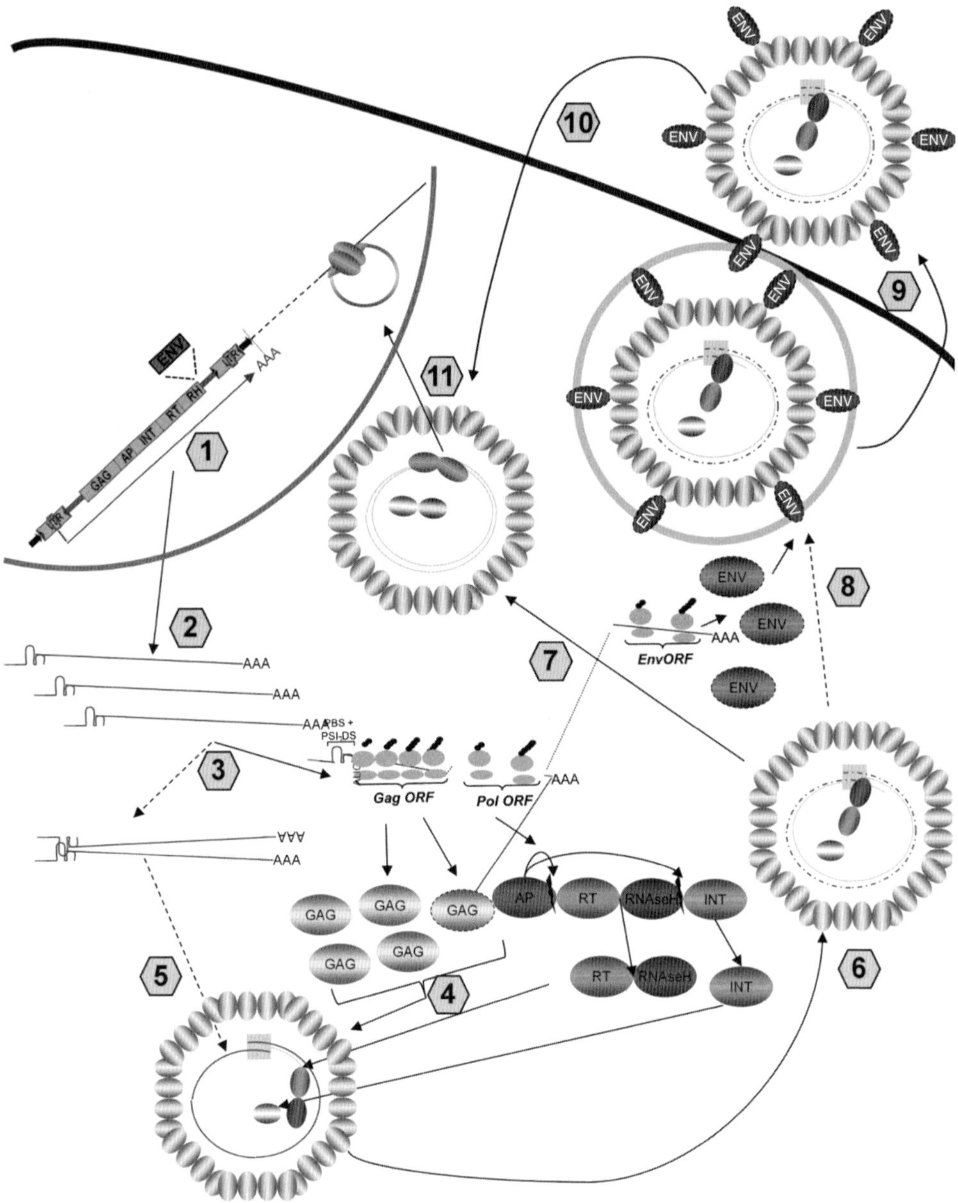

Figure 2.6 Lifecycle of LTR retrotransposons and retroviruses. The intracellular phases of both lifecycles share most steps (1–7; 11); retroviruses possess an envelope gene (env) that allows them to leave the host cell and infect other cells. The major steps of the life cycle are: (1) transcription of a copy integrated into the genome, a *Copia* element with one open reading frame (ORF) being depicted, from the promoter in the long terminal repeat (LTR); (2) nuclear export; (3) alternative translation or buckling of two transcripts destined for packaging and reverse transcription; (4)translation either of separate *gag* and *pol* ORFs to produce the capsid protein Gag and a polyprotein containing aspartic proteinase (AP), reverse transcriptase (RT), RNaseH, and integrase (INT) the order of components shown being for *Gypsy* elements, or of one common ORF; (5) assembly of a virus-like particle (VLP) from Gag containing RNA transcripts, integrase, reverse transcriptase–RNaseH; (6) reverse transcription; (7) localization of the VLP to the nucleus; (8) in addition, for retroviruses translation of the *envelope* ORF to produce the envelope protein (ENV), which then mediates packing of the capsid to a membrane envelope; (9) budding from the cell, as mediated by the envelope, and the infectious extracellular phase; (10) re-entry into the cell; and (11) passage of the cDNA–integrase complex into the nucleus and integration of the cDNA into the genome.

cell and infect other cells (Figure 2.6). In fact, the *Gypsy* family of *Drosophila*, the type element of the superfamily, can be infectious under laboratory conditions (Kim et al., 1994). Moreover, the restriction of retroviruses to some animal genomes, the widespread occurrence of an envelope gene-bearing clade of *Gypsy* elements in plants (Vicient et al., 2001), and the ubiquity of the *Gypsy* superfamily as a whole in the eukaryotes together form strong evidence that viral forms of *Gypsy* retrotransposons (retroviruses) evolved from the main line of the superfamily by gain of function.

Evolutionary Origins of Transposable Elements

As the foregoing discussion illustrates, while classification schemes help organize the massive diversity of TEs for descriptive purposes, they inevitably raise questions about their evolution and that of the genomes in which they reside. It generally held that TEs evolved from genes encoding various enzymes and structural proteins and became autonomous genetic entities through the acquisition of DNA sequences needed for regulation and replication. TE mobility requires either the physical movement of the TE itself as DNA (Class II TEs) or propagation of a copy (Class I TEs and the Helitrons in Subclass 2 of Class I). Therefore, TE mobility invariably requires either an integrase or transposase function, essentially an enzyme catalyzing recombination, to break the genomic DNA and integrate a mobile DNA segment into the chromosome.

While we tend to imagine the paste phase of the cut-and-paste or copy-and-paste mechanism as clean and dependent on the enzyme of the TE, the reality is messier. For example, integration of LINE elements involves a process called target-primed reverse transcription (TPRT) or, alternatively, endonuclease-independent insertion, both of which involve synthesis of long stretches of DNA by either host or retrotransposon enzymes to repair the duplex, an error-prone process (Sen et al., 2007). Completion of transposition is in essence a case of double-strand break repair via non-homologous end joining, which likewise is prone to errors and involves enzymes not encoded by the TE itself (Gorbunova and Levy, 1999; Puchta, 2005; Wyman and Kanaar, 2006). The evolution of TE mobility must therefore be viewed in the context of the pre-existence or co-evolution of DNA repair mechanisms for the DNA breaks they create as a consequence of their mobility.

To understand the evolutionary history of TEs themselves, however, the origins of the integrases, transposases, and reverse transcriptases need to elucidated. The transposases of Class II transposons probably evolved from an enzyme able to cleave the phosphate backbone of DNA (Figure 2.7). Moreover, recent analyses indicate that all eukaryotic transposases have a common evolutionary origin (Yuan and Wessler, 2011); their genes are arguably the most abundant and ubiquitous in nature (Aziz et al., 2010). A large range of DNA-manipulating enzymes share the DDD or DDE motif at the active site, including the V(D)J recombinases, bacterial transposases, and LTR retrotransposon integrases (Keith et al., 2008). The aspartate and glutamate residues in the active site act by providing the carboxyl groups that hold the metal ions, generally Mn and Mg, serving as cofactors. Eukaryotic transposases and integrases, as well as prokaryotic transposases such as that of bacteriophage Mu, share the DDE motif in the catalytic center, implying a common origin; recent structural studies of the enzymes strongly support this view (Hickman et al., 2010; Montaño and Rice, 2011).

The formation of terminal inverted repeats (TIRs) flanking the ancestral transposase provided recognition and binding sites for the enzyme, allowing the transposase gene to cut and paste itself into a new location (Figure 2.2). Transposases frequently recognize not only the TIRs but also additional motifs nearby, which are conserved for particular families of transposons. The sequence-specificity of particular transposases, such as that of the Class II *Mariner*-like elements (MLEs), has allowed identification of their targets in whole-genome surveys (Feschotte et al., 2003) and,

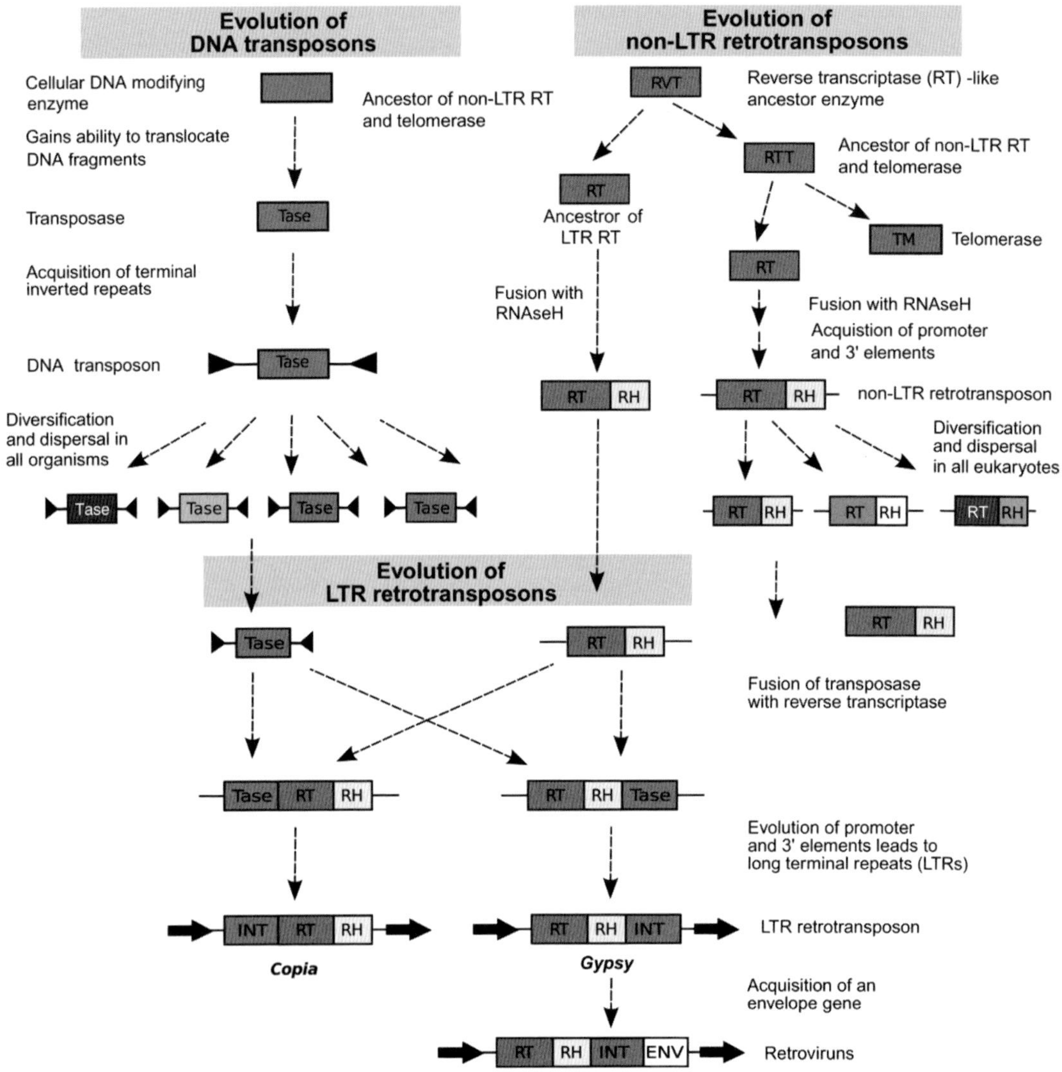

Figure 2.7 Model for the origin and evolution of the main types of TEs. The primary building blocks, transposase and reverse transcriptase–RNaseH (RT-RH), presumably evolved from genes of replication, recombination, and nucleic acid metabolism. Once the ancestor TEs evolved, differential fusions led to the rise of new superfamilies. Note that this model only includes RT-RH and integrase (INT) genes and omits additional gene products such as Gag from LTR retrotransposons, as well as additional ORFs in DNA transposons and non-LTR retrotransposons.

conversely, functional pairing of particular enzymes with given end structures (Feschotte et al., 2005; Kahlon et al., 2011).

Very early on, it was recognized that the retrotransposons, retroviruses, and bacteriophage Mu shared their terminal TG..CA ends that form the termini of the TIRs within the LTRs (Temin, 1980). Since then, the catalytic mechanisms of DDE-containing transposases and integrases have been well studied (Li et al., 2006). Progress has been made, at least for HIV-1 and a few other retroviruses, in identifying the amino acid residues of integrase that interact with the LTR (Dolan et al., 2009). However, generally speaking, surprisingly little crystallographic information is available on the interactions of integrases and transposases with DNA (Hickman et al., 2010). Hence, it is difficult at present to draw a clear picture of the co-evolution of the transposases and integrases and of their recognition sites in the mobile elements.

Not only the "cut and paste" enzymes, but also the "copy" enzyme, the reverse transcriptase of Class I TEs, appears to be derived from an ancient family of enzymes involved in nucleic acid metabolism, in this case polymerization. The evidence for this is the presence in plants, animals, fungi, protists, and bacteria of a conserved family of genes, *rvt*, encoding polymerases able to incorporate both ribonucleotides and deoxyribonucleotides (Gladyshev and Arkhipova, 2011). The catalytic centers of the reverse transcriptases of all retrotransposons contain a highly conserved motif that is, in most cases, YVDD surrounded by several small hydrophobic amino acids, together referred to as the reverse transcriptase signature. The eukaryotic telomerase enzyme, which adds telomeres to the ends of chromosomes through reverse transcription of an RNA template, contains a similar motif in its catalytic center (Lingner et al., 1997; Lue et al., 2005; Autexier and Lue, 2006). Structure-based alignments indicate that the RVT enzymes most closely resemble modern LINE reverse transcriptases, and belong with them in a larger family including the reverse transcriptases of LTR retrotransposons, retroviruses, pararetroviruses, telomerases, and the *PLE* order of Class I elements (Figure 2.1). Thus, Class I reverse transcriptase and telomerase are descendants of a common ancestral enzyme. The earliest retrotransposon reverse transcriptase probably then fused with an RNaseH gene. Subsequent acquisition of regulatory sequences gave rise to the structurally simplest known Class I elements, the non-LTR retrotransposons (Figures 2.2 and 2.7).

LTR retrotransposons seem to be the product of a gene fusion, in which the reverse transcriptase of a non-LTR retrotransposon and a Class II transposase, the primordial integrase, were combined into one element. Interestingly, such a fusion event appears to have occurred at least twice, once leading to the formation of the two main LTR superfamilies, *Gypsy* and *Copia* (Figure 2.7). Both *Gypsy* and *Copia* elements are flanked by LTRs of very similar structure. LTRs are in essence pairs of identical cassettes containing a promoter, a terminator, and a polyadenylation signal flanked by short TIRs that are recognized by the integrase, which share the almost universal terminal 5′ TG..CA 3′ terminal that forms the outer nucleotides of the TIRs. Promoters are plentiful in the genome, and terminators, polyadenylation signals, and 5- or 6-bp repeats are short enough to occur with high frequency. Hence, if the primeval *Gypsy* and *Copia* elements evolved independently, they could have acquired LTRs independently. Alternatively, both have evolved from an ancestral LTR-containing intermediate (Figure 2.7).

LTRs are, as their name implies, repeated at both ends of the retrotransposon. For most LTR retrotransposons, the minus-strand cDNA, reverse-transcribed from a tRNA primer matching the primer binding site (PBS) near the 5′ end of the LTR retrotransposon, will jump to the 3′ end of the same or different template if a sufficiently long stretch of complementary bases is found there (Figure 2.5). The process of strand jumping is repeated for the plus strand, which is primed from the polypurine tract (PPT) near the 3′ LTR. The repeat required for the second strand transfer is generated by reverse transcription of the PBS motif of tRNA primer, which remains following the first strand transfer (Figure 2.5), and therefore does not require a second evolutionary step. The two strand jumps homogenize the two ends of the final double-stranded cDNA, creating the LTRs. It is not so implausible to imagine that the acquisition of a tRNA gene near a promoter and of a purine-rich tract near a terminator, together with the presence of a stretch of a few 10s of bases of similar nucleotides at either end would have permitted reverse transcription to create two LTRs, each possessing the promoter and terminator flanking the genes. The final protein-coding domain for the aspartic proteinase became necessary only with the fusion of the other open reading frames to generate a single polyprotein, which then needs to be cleaved. The overall conceptual difficulty, as with any complex system, is to imagine fairly contemporaneous acquisition of the whole host of features needed for a functional LTR retrotransposon: the integrase, reverse transcriptase, LTRs containing promoter and terminator, and the priming sites.

A shared ancestry for TE and non-TE enzymes does not, however, require that TE enzymes have been derived from the complement of extant cellular systems. Both the ubiquity of the transposases and integrases (Aziz et al., 2010) and the pervasiveness of the retroelements suggest that TEs are of very ancient origin, predating the divergence of prokaryotes and eukaryotes 1000–1500 MYA (Javaux et al., 2001; Douzery et al., 2004). Moreover, reverse transcriptase, too, is very ancient, probably dating back to the hypothetical ancient "RNA world" (Gilbert, 1986; Cech, 2011) that predates current DNA genomes, because of its use of RNA as the template, high error rates due to lack of proofreading (Jamburuthugoda and Eickbush, 2011), likely shared ancestry with the viral RNA-directed RNA polymerases (Xiong and Eickbush, 1990; Boeke and Stoye, 1997), and much greater structural similarity to these than to DNA-directed DNA polymerases (Brautigam and Steitz, 1998). Furthermore, the prokaryotic and organellar group II introns, likely ancestors of spliceosomal introns, carry reverse transcriptase very similar to that of LINE elements (Lambowitz and Zimmerly, 2004; Gu et al., 2010). In an RNA world, RNA encoding an enzyme such as reverse transcriptase that can replicate RNA would be favored by natural selection (Boeke and Stoye, 1997).

If TEs have their origins in the very earliest stages of the evolution of life, evidence that particular modern, closely related enzymes and functions have been derived from TEs rather than vice versa is not surprising. Examples of such "domestication" of TE functions include the V(D)J recombinase from transposase, telomerase from reverse transcriptase, and various mammalian genes derived from integrases, Gag proteins, and envelope proteins (Volff, 2006). Recently, a set of widely distributed eukaryotic tyrosine kinase genes was shown to be derived from the *Crypton* transposons, Class II elements that uniquely use this enzyme as their transposase (Kojima and Jurka, 2011).

Non-autonomous Transposable Elements

A partial solution to the conundrum of the assembly of the complex functional unit that is a TE appears in the phenomenon of non-autonomous TE groups, which are abundant in modern genomes. In addition to intact TEs able to direct and catalyze their own cut-and-paste (Class II) or copy-and-paste (Class I) lifecycle, genomes are full of various sorts of deleted and mutated versions. In this regard, one might think, in retrospect, that Ohno was actually correct when he referred to the non-genic component of the genome as "junk" (Ohno, 1972). However, the idea of "fossils" in the genome, no-longer active transposable elements that have decayed but whose one-time functionality is still recognizable, is conceptually more accurate than the anthropomorphism of "junk," which suggests complete lack of value or utility.

It turns out, though, that even many of the apparently "dead" TEs, the fossils of the genome, can be brought back to life when mobilized by another element. The phenomenon was recognized right from the very beginning of the discipline, with the observation by McClintock of both autonomous and non-autonomous controlling elements, *Ac* and *Ds* (McClintock, 1948; Jones, 2005). The many mutated TEs fall into several categories. The first comprises the canonical autonomous elements, which contain intact open reading frames, promoters, structural motifs such as inverted or direct repeats that are recognized by the TE enzymes, and processing signals. The second includes autonomous elements that have been variously mutated, deleted, or recombined so that one or more of their encoded proteins or motifs are no longer functional. These tend to retain recognizable protein coding domains, even if the domains are highly degenerate. Many are inactive in that they cannot be mobilized or mobilize other elements. The third category consists of groups of non-autonomous mobile elements that have conserved structures or deletions in which one, several, or all protein coding domains are missing. The non-autonomous but active mobile elements can still be transcribed and mobilized in *trans* by proteins from autonomous elements; others may have lost the motifs required for *trans* activation and are both non-autonomous and non-mobile.

Among the Class II transposons, the term "non-autonomous element" has referred to those that cannot express transposase and catalyze their own transposition. They form binary systems with the autonomous elements able to drive their transposition. The classical examples of these include the *Ac-Ds* (McClintock, 1948; Fedoroff et al., 1983; Jones, 2005) and *Suppressor-Mutatator* (*Spm*; Fedoroff, 1999) systems, although similar ones are widespread (Hartl et al., 1992). The miniature inverted repeat transposable elements (MITEs) were first identified as insertions in the maize genome. They were named *Tourist* because they are abundant and dispersed, but have no apparent function, and were later shown through bioinformatics to be highly abundant in other plant genomes as well (Wessler et al., 1995; Feschotte et al., 2002). It has long been suspected that many families of MITEs are in fact non-autonomous elements that can be mobilized by autonomous Class II transposons (Feschotte et al., 2002; Casacuberta and Santiago, 2003). Evidence for this linkage has accumulated, starting from the circumstantial connection between a putative *Pogo*-like transposase and the MITE *Emigrant* (Feschotte and Mouchès, 2000) in *Arabidopsis* and ending with the demonstration of the activity of the *mPing-Pong* system (Jiang et al., 2003).

The MITE elements contain the minimal information needed to be mobile, which consists of TIRs that can be recognized by a transposase. Hence, it has been relatively difficult based on bioinformatic approaches to place them into families of Class II elements together with their host autonomous TE. For this reason, it is also difficult, when transitional forms are absent, to plot their evolution from autonomous elements. Moreover, more recent work leads to the view that MITEs may be generalists, able to promiscuously exploit transposase from autonomous Class II transposons that are only distantly related (Yang et al., 2009).

For Class I elements, the phenomenon of non-autonomy has several additional facets because of the complexity of their replicative life cycle (Figures 2.5 and 2.6) (Sabot and Schulman, 2006). In Class II transposons, a non-autonomous element can be mobilized as long as its termini are recognized by transposase. Class I elements must be transcribed and translated, then transcripts packaged into VLPs formed from self-encoded Gag together with integrase and reverse transcriptase (Figure 2.5). Reverse transcription, targeting and entry of the nucleus, and finally integration must occur. While any of these steps may be blocked by lack of a self-encoded protein, all potentially can be complemented in *trans* if a translationally or enzymatically defective Class I element nevertheless possesses the correct recognition signals for proteins encoded by an autonomous and competent element.

Among the non-autonomous elements, we find conserved, abundant, and insertionally polymorphic retrotransposon families such as *BARE2*, which has most of its protein-coding domains intact except for a deletion that knocks out Gag, supplied to it by *BARE1* for packaging (Tanskanen et al., 2007). At the other end of the spectrum of coding capacity are the large retrotransposon derivative (*LARD*) elements that code for no protein, but possess a long internal domain with a predicted conserved RNA structure (Kalendar et al., 2004) and the terminal repeat retrotransposons in miniature (TRIM; Witte et al., 2001; Kalendar et al., 2008), which lack protein-coding capacity and have only very short internal domains, but nevertheless are abundant and conserved in plants. Because retrotransposons encode a polyprotein, any upstream mutation generating a frameshift or stop codon will have polar effects, knocking out expression of the downstream proteins as well. Therefore, *trans* activation is expected to be very widespread among the retrotransposons, encompassing not only non-autonomous elements lacking some or all of their open reading frames, but also otherwise autonomous elements with mutations leading to loss of one or more functional proteins.

Similar to the TRIMs and MITEs in their degree of reduction are the SINEs, which constitute an order of non-autonomous Class I elements mobilized by non-LTR retrotransposons. SINEs comprise a diverse array of sequences, which can be propagated via the enzymatic machinery of the LINEs (Goodier and Kazazian, 2008). They are remarkably abundant, especially in mammalian genomes, where their numbers range from 10^4 to 10^6 (Kramerov and Vassetzky, 2005), but are also found elsewhere, including in plants (Deragon and Zhang, 2006). Despite their having been defined as an order of TEs by their mechanistic origin, SINEs are polyphyletic regarding their sequence or informational origin, including a diversity of tRNA, rRNA, and other polymerase III transcripts (Kramerov and Vassetzky, 2005). They are generally 150–200 bp; those derived from tRNA have the tRNA sequence at the 5′ end and homology at their 3′ ends to a LINE from the same genome, which is thought to provide binding sites for LINE-encoded proteins. The 3′ tails are often AT-rich, betraying origins as reverse-transcribed gene transcripts. Although the enzymology of SINE retroposition is not fully understood, at least for the *Alu* SINE element of humans, one of the LINE L1 proteins, ORF2p, is needed, whereas the other, ORF1p, may aid the movement (Kroutter et al., 2009).

A curious feature of the highly reduced, non-autonomous TEs is their exceptional abundance. It is tempting to view the much greater abundance of small non-autonomous elements compared to the lower numbers of large autonomous elements in metaphorical terms of abundant small parasites on individual larger organisms or of small, plentiful prey and large, rare predators. Whereas the metaphors imply that an ecosystem's carrying capacity for organisms of different sizes can serve as a model of the genome's carrying capacity for different types of TEs, the mechanisms driving differences in abundance between autonomous TEs and their small, non-autonomous derivatives or partners are unknown. However, the high probability of formation and low cost or the selective advantage of the symbiotic lifestyle of non-autonomous elements may be factors.

Highly reduced elements such as the SINEs and TRIMs illustrate the principal that, as long as processing and recognition signals are retained in *cis*, the needed proteins can be supplied in *trans* and non-autonomous or translationally compromised elements can continue to transpose and propagate. Hence, the non-autonomous TEs provide a window into evolution of ancestral mobile elements. Just as enzymatic and packaging functions are now supplied in *trans* to derived or partner non-autonomous elements, the proteins ancestral to those of modern TEs could have acted in *trans* to mobilize nascent Class I or Class II elements. All coding domains and signals need not have been assembled *ab initio*, but could have been captured or added sequentially. The two processes are not equally probable, however. As demonstrated by the abundance of non-autonomous elements in the genome, loss of coding capacity occurs often. Independent evolution of new types of TEs, based on the presence of only 2 classes, 9 orders, and 29 superfamilies of TEs in the eukaryotes (Wicker et al., 2007), appears to happen rarely.

Transposable Element Demography and Genome Ecology

The preceding discussion on the mechanisms and evolutionary origins of non-autonomous transposable elements begs, as well, consideration of their demographics within genomes. This is more easily done for the Class I retrotransposons than the Class II TEs for two reasons. First, the age of insertions can be estimated by comparing the LTRs of a given element, which are identical upon insertion and then diverge according to the molecular clock (SanMiguel et al., 1998; Kronmiller and Wise, 2007). Second, inserted copies do not transpose, so their age can be linked to the history of a particular region of the genome. The estimation of insertion times led to the notion of epochs, on an evolutionary timescale, of frequent insertion of new copies (SanMiguel and Bennetzen, 1998; SanMiguel et al., 2002; Wei et al., 2002), though the origins and trigger mechanisms of such "waves" of genome growth remain unclear.

Not only are TEs gained by insertion, but they are also lost by a combination of progressive small deletions and truncations (Devos et al., 2002). In addition, LTR-containing retrotransposons and endogenous retroviruses are lost by LTR–LTR recombination (Mager and Goodchild, 1989; Shirasu et al., 2000; Vitte and Panaud, 2005), which removes all of an element except one recombinant LTR. The process can also occur between nearby LTRs of sufficient sequence similarity (Vicient et al., 2005). In analyses of large sets of sequences sufficiently long to include entire retrotransposons, one can therefore infer an age structure or demographic pattern, by plotting the number of suborder (*Gypsy* or *Copia*) or family members for given age classes, estimated from the LTR pairs, and calculating an overall half-life for the group in question. Recombination into solo LTRs or truncation to delete an LTR removes a retrotransposon from consideration as a member of such a population.

In rice and *B. distachyon*, *Copia* elements have half-lives of 0.790 and 0.859 MY, respectively (International Brachypodium Initiative, 2010), whereas in wheat and barley, the half-life is too long to effectively calculate. Based on the number of solo LTRs in *B. distachyon* and on an average retrotransposon size of 10 kb, a minimum of 17.4 Mb has been lost by LTR–LTR recombination. Moreover, in *B. distachyon*, the presence of 13 *Copia* groups younger than 20 thousand years indicates that these elements are still actively propagating. These data are consistent with the notion that the small grass genomes are small due to elimination of retrotransposons primarily by LTR–LTR recombination rather than because the retrotransposons are inactive. In other plants as well, such as cotton, there is strong evidence from large-scale and comparative sequence analysis for relatively rapid recombinational loss of retrotransposon insertions through deletion (Hawkins et al., 2009).

Demographic treatments of whole genome sequences have enabled insight into the age structures and population density of retrotransposon "neighborhoods." In *B. distachyon*, large differences were found among chromosomes (International Brachypodium Initiative, 2010). The short arm of chromosome 5 (Bd5S) has a high LTR retrotransposon density and commensurately low gene density compared to the rest of the genome, as well as the youngest *Gypsy* elements and the lowest solo LTR density of all of the chromosomes. These data suggest that comparatively more retrotransposons are integrating into Bd5S and fewer are being lost by recombination than for other chromosome arms. Interestingly, the regions corresponding to Bd5S on the syntenic chromosomes of rice (Os4S) and sorghum (*Sorghum bicolor*; Sb6S) show the same overall picture, one of high retrotransposon density, low gene density, and few syntenic genes (International Brachypodium Initiative, 2010). The selective forces and mechanisms maintaining and mediating these differences remain unknown.

Differences in retrotransposon distribution and fate can be found not only between chromosomes, but also within individual chromosomes, particularly with regard to their major structural features: the telomere, centromere, nucleolar organizing regions (NORs), and the regions closely linked to these. Earlier *in situ* hybridization data showed striking differences in the distribution of various retrotransposon families, with some excluded from the aforementioned regions (Pearce et al., 1996; Suoniemi et al., 1996; Weber et al., 2010) and others, conversely, restricted to them (Presting et al., 1998; Balint-Kurti et al., 2000; Mroczek and Dawe, 2003). Recent whole-genome sequence data where pseudochromosomes have been assembled support the differences observed by *in situ* hybridization. For example, in *B. distachyon*, the centromeres are gene-poor but rich in retrotransposons, almost exclusively belonging to the *Gypsy* superfamily. One particular group of *Gypsy* elements, a subset of the chromovirus CRM clade, appears to be confined to plant centromeres (Wolfgruber et al., 2009); Neumann et al., 2011).

Although the function of centromeric retrotransposons remains unclear, their distribution may result both from negative or purifying selection against the deleterious effects of insertions into actively transcribed regions, which could inactivate genes, and from active targeting of some centromeric feature by the chromodomain of the integrase of the centromeric CRM clade. Supporting this notion, LTR retrotransposons such as *Tnt-1* and *Tos17* that do insert into genes, which makes them highly mutagenic, are normally silent and few in number (Miyao et al., 2003; Le et al., 2007). Both elements are in use as mutagenesis systems to target genes (Piffanelli et al., 2007; Tadege et al., 2008).

Negative selection may also play a role in the nesting and clustering of large retrotransposons (SanMiguel et al., 1996, 1998; Shirasu et al., 2000; Choulet et al., 2010) throughout the chromosomes, at least in the large cereal genomes. In the barley genome, concatenated *BARE* retrotransposons of the form LTR – internal domain – LTR – internal domain – LTR that result from intrachromosomal recombination between the LTRs of two different elements are present in about 5,000 copies (Vicient et al., 2005). Interspersion of genes between LTRs prone to recombination would result in loss of those genes, potentially lethal to the cell in which it occurs. The local nested, clustered, and non-genic insertional pattern of large retrotransposons stands in contrast to the MITEs (Bureau and Wessler, 1994; Sarilar et al., 2011) and *Cassandra* TRIMs (Kalendar et al., 2008), which are both small and frequently found in genes. The small MITEs and TRIMs may be less disruptive to gene function than are large Class I elements if, for example, inserted into introns, and thereby many such insertions may not have been selected against.

In contrast to negative selection for TE insertion patterns, the telomere-specific retroelements *HeT-A*, *TART*, and *TAHRE* provide classic examples of what has been called TE "domestication," or positive selection, for their activity in telomere maintenance (DeBaryshe and Pardue, 2011). Moreover, a subset of the PLE order of retrotransposons, which is distinct from the LTR, LINE, and SINE classes discussed above, has been shown to have a telomere-specific localization in protists, fungi, and plants (Gladyshev and Arkhipova, 2007). The structure of this subset, their lack of nuclease compared with most PLEs, an RT similar to that of group II introns and to telomerase, and their chromosomal position tantalizingly suggest that they may be the link between modern telomerases and ancestral non-LTR retrotransposons. Hence, the chromosomal distribution of TEs may represent a kind of current or ancestral "ecological niche" reflecting both selective pressures and competitive advantages from function.

Conclusions: Rehabilitation of Transposable Elements

Mobile genetic entities have long been viewed with suspicion by the scientific community. At the time of their discovery, McClintock's "controlling elements" (McClintock, 1948) were inscrutable in terms of classical genetics and of uncertain general importance. As molecular genetics developed, mobile DNA, assigned to the non-genic part of the genome, was impugned (Ohno, 1972) as non-functional "junk," the "silent DNA . . . serving the useful but negative function of spacing those [genes] which have succeeded."

The discovery that McClintock's controlling elements were in fact mobile segments of DNA (Fedoroff et al., 1983; Schwarz-Sommer et al., 1984) did little to improve the situation; rather than merely having no positive role, they could now be condemned as "selfish DNA" (Doolittle and Sapienza, 1980; Orgel and Crick, 1980). The negative impressions were only strengthened with the revelation that *P* elements had spread throughout *Drosophila melanogaster* populations worldwide over five decades and were phenotypically detrimental (Anxolabéhère et al., 1988); transposons were not only selfish, but also invasive parasites. The molecular biology of retroviruses was elucidated in the 1970s and 1980s. The reputation of the retrotransposons as parasites appeared sealed when the *copia* and*Ty1*–like elements on the one hand, and *gypsy* and *Ty3* on the other, were found to be structurally similar to the retroviruses and therefore entered into viral taxonomy as aspiring retroviruses, the Metaviridae and Pseudoviridae, respectively (Boeke et al., 1999).

The advent of Polymerase Chain-Reaction (PCR) and later of high-throughput and economical sequencing made it obvious that, in some cases, the "invasive parasites" actually comprised the majority of the genome, greatly outnumbering the "resident" genes, and contributed enormously to genetic diversity (Morgante, 2006). More than being agents of mutation, examples have accumulated in which TEs or genetic entities related or derived from them, including telomerase discussed above, have contributed a wide range of enzymatic functions (Volff, 2006) and patterns of epigenetic silencing (Kashkush and Khasdan, 2007; Law and Jacobsen, 2010; Lisch and Bennetzen, 2011; Rebollo et al., 2011), can move and duplicate genes or gene segments to new loci (Jiang et al., 2004; Lai et al., 2005; Morgante et al., 2005; Wicker et al., 2010), and serve as promoters and other regulatory signals (Debarry et al., 2006; Lynch et al., 2011; Studer et al., 2011; Xu et al., 2011). In the last 15 years, these observations have given rise to the concept of "domestication" of TEs for "altruistic" functions as promoters or protein-coding domains, much as insurgents might be induced to lay down their weapons and serve society (Miller et al., 1997; Pinsker et al., 2001; Hikosaka et al., 2007; Faulkner and Carninci, 2010). This amounts to granting a conditional amnesty for some TEs.

Although the TEs are more and more appreciated for their many and diverse roles, often they are seen still as late arrivals on the genomic scene, immigrants or colonists with a shady reputation who in some cases might make good citizens. It is time to reconsider this mindset. As the discussion above on the evolution of TEs shows, their protein-coding domains predate the emergence of the eukaryotes; the RNA-replicating activity of reverse transcriptase may be a link to the primordial RNA world (Poole et al., 1998), although the amino acid composition of extant forms suggests the enzyme evolved only after the transition to a DNA world (Davis, 2002). Nevertheless, TEs encoding reverse transcriptase probably gave rise to the group II introns and are the likely ancestors of spliceosomal introns; in this role they are even hypothesized to have led to the advent of the eukaryotic nucleus (Martin and Koonin, 2006; Koonin et al., 2008). Indeed, much of the DNA of modern genomes appears to be derived from transposition (Brosius, 2005). Hence, the TEs are no more invaders or parasites than any particular gene; new TE copies have the same status as do recent gene duplications that had not yet acquired an obvious selective advantage.

Transposable elements nevertheless remain distinct by their rapid mobility and replicative capacity against a background of conservation of long-term colinearity and synteny on the chromosomal level (Bolot et al., 2009). Recently, however, TEs have been shown to have no monopoly on mobility per se. Active in regulating gene expression, small RNAs are now recognized to move from cell to cell within plants (Melnyk et al., 2011). Furthermore, mRNA species as ribonucleoprotein complexes as well as many proteins have been found in phloem-conductive tissue in various species, implying that they are mobile (Lough and Lucas, 2006; Ham et al., 2009; Ma et al., 2010; Atkins et al., 2011). Hence, the mobility of TEs as information-bearing molecules is not a unique property.

If evolutionary origins, selective advantageousness, and mobility cannot sharply divide TEs from their chromosomal neighbors, what can? One might argue that because a large fraction of TEs within the genome are non-coding fossils, Ohno (1972) was ultimately right. However, evidence has accumulated that a great proportion of the DNA of eukaryotic genomes is transcribed, mostly as long noncoding RNAs (ncRNAs; Amaral et al., 2008); many of these have been shown to have regulatory roles (De Lucia and Dean, 2011; Ørom and Shiekhattar, 2011). Moreover, TEs and microRNAs functioning as promoters and regulators of genes are "noncoding" but certainly relevant to biological activity. These points suggest that even the criterion of coding capacity can no longer be used to separate TEs, particularly the non-autonomous ones, from the rest of the genome. In the end, there is nothing to do but to accept the TEs as full and rightful, and perhaps the most interesting, members of the genome community with deep roots in its earliest days. They have been built of various genomic components and continue to contribute new parts to various genes and new functions to organisms. They are (most of) the genome.

Acknowledgments

Research on which this chapter is based was carried out in part under a grant from the Academy of Finland, Decision 123074. This chapter is dedicated to the memory of Claire Shulman.

References

Adams, M.D., Celniker, S.E., Holt, R.A., Evans, C.A., Gocayne, J.D., Amanatides, P.G., Scherer, S.E., Li, P.W., Hoskins, R.A., Galle, R.F., et al. (2000) The genome sequence of *Drosophila melanogaster*. *Science 287*: 2185–2195.

Amaral, P.P., Dinger, M.E., Mercer, T.R., and Mattick, J.S. (2008) The eukaryotic genome as an RNA machine. *Science 319*: 1787–1789.

Anxolabéhère, D., Kidwell, M.G., and Periquet, G. (1988) Molecular characteristics of diverse populations are consistent with the hypothesis of a recent invasion of *Drosophila melanogaster* by mobile *P* elements. *Mol. Biol. Evol. 5*: 252–269.

Arabidopsis Genome Initiative (2000) Analysis of the genome sequence of the flowering plant *Arabidopsis thaliana*. *Nature 408*: 796–815.

Atkins, C.A., Smith, P.M., and Rodriguez-Medina, C. (2011) Macromolecules in phloem exudates – a review. *Protoplasma 248*: 165–172.

Autexier, C. and Lue, N.F. (2006) The structure and function of telomerase reverse transcriptase. *Annu. Rev. Biochem. 75*: 493–517.

Aziz, R.K, Breitbart, M., and Edwards, R.A. (2010) Transposases are the most abundant, most ubiquitous genes in nature. *Nucleic Acids Res. 38*: 4207–4217.

Balint-Kurti, P.J., Clendennen, S.K., Dolezelová, M., Valárik, M., Dolezel, J., Beetham, P.R., and May, G.D. (2000) Identification and chromosomal localization of the monkey retrotransposon in *Musa* sp. *Mol. Gen. Genet. 263*: 908–915.

Bennett, M.D. and Smith, J.B. (1976) Nuclear DNA amounts in angiosperms. *Philos. Trans. R. Soc. Lond. B Biol. Sci. 274*: 227–274.

Benton, M.J. (1993) Reptilia. In *The Fossil Record 2*, M. J. Benton, ed. (London: Chapman & Hall), pp. 681–715.

Biderre, C., Pages, M., Metenier, G., Canning, E.U., and Vivaras, C.P. (1995) Evidence for the smallest nuclear genome (2.9 Mb) in the microsporidium *Encephalitozoon cuniculi*. *Mol. Biochem. Parasitol. 74*: 229–231.

Boeke, J.D. and Stoye, J.P. (1997) Retrotransposons, endogenous retroviruses, and the evolution of retroelements. In *Retroviruses*, J.M. Coffin, S.H. Hughes, and H.E. Varmus, eds. (Cold Spring Harbor, NY: Cold Spring Harbor Laboratory Press), pp. 343–435.

Boeke, J.D., Eickbush, T.H., Sandmeyer, S.B., and Voytas, D.F. (1999) Family pseudoviridae. In *Virus Taxonomy: Classification and Nomenclature of Viruses*. Sixth Report of the International Committee on Taxonomy of Viruses, F.A. Murphy, ed. (New York: Springer-Verlag), pp. 349–357.

Bolot, S., Abrouk, M., Masood-Quraishi, U., Stein, N., Messing, J., Feuillet, C., and Salse, J. (2009) The 'inner circle' of the cereal genomes. *Curr. Opin. Plant Biol. 12*: 119–125.

Bossolini, E., Wicker, T., Knobel, P.A., and Keller, B. (2007) Comparison of orthologous loci from small grass genomes Brachypodium and rice: implications for wheat genomics and grass genome annotation. *Plant J. 49*: 704–717.

Brautigam, C.A. and Steitz, T.A. (1998) Structural and functional insights provided by crystal structures of DNA polymerases and their substrate complexes. *Curr. Opin. Struct. Biol.* *8*: 54–63.

Brosius, J. (2005) Echoes from the past – are we still in an RNP world? *Cytogenet. Genome Res.* *110*: 8–24.

Bureau, T.E. and Wessler, S.R. (1994) *Stowaway*: a new family of inverted repeat elements associated with the genes of both monocotyledonous and dicotyledonous plants. *Plant Cell* *6*: 907–916.

Casacuberta, J.M. and Santiago, N. (2003) Plant LTR-retrotransposons and MITEs: control of transposition and impact on the evolution of plant genes and genomes. *Gene* *311*: 1–11.

Cech, T.R. (2011) The RNA Worlds in Context. *Cold Spring Harb. Perspect. Biol.* doi:10.1101/cshperspect.a006742

C. elegans Sequencing Consortium. (1998) Genome sequence of the nematode *C. elegans*: a platform for investigating biology. *Science* *282*: 2012–2018.

Cherepanov, P., Maertens, G.N., and Hare, S. (2011) Structural insights into the retroviral DNA integration apparatus. *Curr. Opin. Struct. Biol.* *21*: 249–256.

Choulet, F., Wicker, T., Rustenholz, C., Paux, E., Salse, J., Leroy, P., Schlub, S., Le Paslier, M.-C., Magdelenat, G., Gonthier, C., et al. (2010) Megabase level sequencing reveals contrasted organization and evolution patterns of the wheat gene and transposable element spaces. *Plant Cell* *22*: 1686–1701.

Davis, B.K. (2002) Molecular evolution before the origin of species. *Prog. Biophys. Mol. Biol.* *79*: 77–133.

De Lucia, F. and Dean, C. (2011) Long non-coding RNAs and chromatin regulation. *Curr. Opin. Plant Biol.* *14*: 168–173.

Dean, R.A., Talbot, N.J., Ebbole, D.J., Farman, M.L., Mitchell, T.K., Orbach, M.J., Thon, M., Kulkarni, R., Xu, J.R., Pan, H., et al. (2005) The genome sequence of the rice blast fungus *Magnaporthe grisea*. *Nature* *434*: 980–986.

DeBarry, J.D., Ganko, E.W., McCarthy, E.M., and McDonald, J.F. (2006) The contribution of LTR retrotransposon sequences to gene evolution in *Mus musculus*. *Mol. Biol. Evol.* *23*: 479–481.

DeBaryshe, P.G. and Pardue, M.L. (2011) Differential maintenance of DNA sequences in telomeric and centromeric heterochromatin. *Genetics* *187*: 51–60.

Deragon, J. and Zhang, X. (2006) Short Interspersed Elements (SINEs) in plants: origin, classification, and use as phylogenetic markers. *Sys. Biol.* *55*: 949–956.

Devos, K.M., Brown, J.K.M. and Bennetzen, J.L. (2002) Genome size reduction through illegitimate recombination counteracts genome expansion in *Arabidopsis*. *Genome Res.* *12*: 1075–1079.

Dolan, J., Chen, A., Weber, I.T., Harrison, R.W., and Leis, J. (2009) Defining the DNA substrate binding sites on HIV-1 integrase. *J. Mol. Biol.* *385*: 568–579.

Doolittle, W.F. and Sapienza, C. (1980) Selfish genes, the phenotype paradigm and genome evolution. *Nature* *284*: 601–603.

Douzery, E.J., Snell, E.A., Bapteste, E., Delsuc, F., and Philippe, H. (2004) The timing of eukaryotic evolution: does a relaxed molecular clock reconcile proteins and fossils? *Proc. Natl. Acad. Sci. USA* *101*: 15386–15391.

Du, C., Fefelova, N., Caronna, J., He, L., and Dooner, H.K. (2009) The polychromatic *Helitron* landscape of the maize genome. *Proc. Natl. Acad. Sci. USA* *106*: 19916–19921.

Du, C., Swigonova, Z., and Messing, J. (2006) Retrotranspositions in orthologous regions of closely related grass species. *BMC Evol. Biol.* *6*: 62.

Faulkner, G.J. and Carninci, P. (2010) Altruistic functions for selfish DNA. *Cell Cycle* *8*: 2895–2900.

Fedoroff, N.V. (1999) The *Supressor-mutator* element and the evolutionary riddle of transposons. *Genes Cells* *4*: 11–19.

Feschotte, C. and Mouchès, C. (2000) Evidence that a family of miniature inverted-repeat transposable elements (MITEs) from the *Arabidopsis thaliana* genome has arisen from a pogo-like DNA transposon. *Mol. Biol. Evol.* *17*: 730–737.

Feschotte, C. and Pritham, E.J. (2005) Non-mammalian c-integrases are encoded by giant transposable elements. *Trends. Genet.* *21*: 551–552.

Feschotte, C., Jiang, N., and Wessler, S. (2002) Plant transposable elements: where genetics meets genomics. *Nat. Rev. Genet.* *3*: 329–341.

Feschotte, C., Swamy, L., and Wessler, S.R. (2003) Genome-wide analysis of mariner-like transposable elements in rice reveals complex relationships with stowaway miniature inverted repeat transposable elements (MITEs). *Genetics* *163*: 747–758

Fedoroff, N., Wessler, S., and Shure, M. (1983) Isolation of the transposable maize controlling elements *Ac* and *Ds*. *Cell* *35*: 235–242.

Feschotte, C., Osterlund, M.T., Peeler, R., and Wessler, S.R. (2005) DNA-binding specificity of rice mariner-like transposases and interactions with Stowaway MITEs. *Nucleic Acids Res.* *33*: 2153–2165.

Gheerbrandt, E., Domning, D.P., and Tassy, P. (2005) Paenungulata (Sirenia, Proboscidea, Hyracoidea, and relatives). In *The Rise of Placental Mammals: Origins and Relationships of the Major Extant Clades*, K.D. Rose and J.D. Archibald, eds. (Baltimore: Johns Hopkins University Press), pp. 84–105.

Gierl, A. and Saedler, H. (1986) The *En/Spm* transposable element of *Zea mays*. *Plant Mol. Biol.* *13*: 261–266.

Gilbert, W. (1986) Origin of life: the RNA world. *Nature* *319*: 618.

Gladyshev, E.A. and Arkhipova, I.R. (2007) Telomere-associated endonuclease-deficient Penelope-like retroelements in diverse eukaryotes. *Proc. Natl. Acad. Sci. USA 104*: 9352–9357.

Gladyshev, E.A. and Arkhipova, I.R. (2011) A widespread class of reverse transcriptase-related cellular genes. *Proc. Natl. Acad. Sci. USA 108*: 20311–20316.

Goodier, J.L. and Kazazian, H.H., Jr. (2008) Retrotransposons revisited: the restraint and rehabilitation of parasites. *Cell 135*: 23–35.

Goodwin, T., Butler, M.I., and Poulter, R.T. (2003) *Cryptons*: a group of tyrosine-recombinase-encoding DNA transposons from pathogenic fungi. *Microbiology 149*: 3099–3109.

Gorbunova, V.V. and Levy, A.A. (1999) How plants make ends meet: DNA double-strand break repair. *Trends Plant Sci. 4*: 263–269.

Grandbastien, M.A., Audeon, C., Bonnivard, E., Casacuberta, J.M., Chalhoub, B., Costa, A.P., Le, Q.H., Melayah, D., Petit, M., Poncet, C., et al. (2005) Stress activation and genomic impact of Tnt1 retrotransposons in Solanaceae. *Cytogenet. Genome Res. 110*: 229–241.

Gregory, T.R. (2001) Coincidence, coevolution, or causation? DNA content, cell size, and the C-value enigma. *Biol. Rev. Camb. Philos. Soc. 76*: 65–101.

Gu, S.Q., Cui, X., Mou, S., Mohr, S., Yao, J., and Lambowitz, A.M. (2010) Genetic identification of potential RNA-binding regions in a group II intron-encoded reverse transcriptase. *RNA 16*: 732–747.

Ham, B.K., Brandom, J.L., Xoconostle-Cázares, B., Ringgold, V., Lough, T.J., and Lucas, W.J. (2009) A polypyrimidine tract binding protein, pumpkin RBP50, forms the basis of a phloem-mobile ribonucleoprotein complex. *Plant Cell 21*: 197–215.

Hartl, D.L., Lozovskaya, E.R., and Lawrence, J.G. (1992) Nonautonomous transposable elements in prokaryotes and eukaryotes. *Genetica 86*: 47–53.

Hawkins, J.S., Proulx, S.R., Rapp, R.A., and Wendel, J.F. (2009) Rapid DNA loss as a counterbalance to genome expansion through retrotransposon proliferation in plants. *Proc. Natl. Acad. Sci. USA 106*: 17811–17816.

Hickman, A.B., Chandler, M., and Dyda, F. (2010) Integrating prokaryotes and eukaryotes: DNA transposases in light of structure. *Crit. Rev. Biochem. Mol. Biol. 45*: 50–69.

Hikosaka, A., Kobayashi, T., Saito, Y., and Kawahara, A. (2007) Evolution of the *Xenopus piggyBac* transposon family TxpB: domesticated and untamed strategies of transposon subfamilies. *Mol. Biol. Evol. 24*: 2648–2656.

Holt, R.A., Subramanian, G.M., Halpern, A., Sutton, G.G., Charlab, R., Nusskern, D.R., Wincker, P., Clark, A.G., Ribeiro, J.M.C., Wides, R., et al. (2002) The genome sequence of the malaria mosquito *Anopheles gambiae*. *Science 298*: 129–149.

International Brachypodium Initiative (2010) Genome sequencing and analysis of the model grass *Brachypodium distachyon*. *Nature 463*: 763–768.

International Chicken Genome Sequencing Consortium (2004) Sequence and comparative analysis of the chicken genome provide unique perspectives on vertebrate evolution. *Nature 432*: 695–716.

International Human Genome Sequencing Consortium (2004) Finishing the euchromatic sequence of the human genome. *Nature 431*: 931–945.

International Rice Genome Sequencing Project (2005) The map-based sequence of the rice genome. *Nature 436*: 793–800.

Jaillon, O., Aury, J.M., Noel, B., Policriti, A., Clepet, C., Casagrande, A., Choisne, N., Aubourg, S., Vitulo, N., Jubin, C., et al. (2007) The grapevine genome sequence suggests ancestral hexaploidization in major angiosperm phyla. *Nature 449*: 463–467.

Jamburuthugoda, V.K. and Eickbush, T.H. (2011) The reverse transcriptase encoded by the non-LTR retrotransposon R2 is as error-prone as that encoded by HIV-1. *J. Mol. Biol. 407*: 661–672.

Javaux, E.J., Knoll, A.H., and Walter, M.R. (2001) Morphological and ecological complexity in early eukaryotic ecosystems. *Nature 412*: 66–69.

Jiang, N., Bao, Z., Zhang, X., Eddy, S.R., and Wessler, S.R. (2004) Pack-MULE transposable elements mediate gene evolution in plants. *Nature 431*: 569–573.

Jiang, N., Bao, Z., Zhang, X., Hirochika, H., Eddy, S.R., McCouch, S.R., and Wessler, S.R. (2003) An active DNA transposon family in rice. *Nature 421*: 163–167.

Jones, R.N. (2005) McClintock's controlling elements: the full story. *Cytogenet. Genome Res. 109*: 90–103.

Kahlon, A.S., Hice, R.H., O'Brochta, D.A., and Atkinson, P.W. (2011) DNA binding activities of the Herves transposase from the mosquito *Anopheles gambiae*. *Mobile DNA 2*: 9.

Kalendar, R., Tanskanen, J.A., Chang, W., Antonius, K., Sela, H., Peleg, P., and Schulman, A.H. (2008) *Cassandra* retrotransposons carry independently transcribed 5S RNA. *Proc. Natl. Acad. Sci. USA 105*: 5833–5838.

Kalendar, R., Tanskanen, J., Immonen, S., Nevo, E., and Schulman, A.H. (2000) Genome evolution of wild barley (*Hordeum spontaneum*) by *BARE-1* retrotransposon dynamics in response to sharp microclimatic divergence. *Proc. Natl. Acad. Sci. USA 97*: 6603–6607.

Kalendar, R., Vicient, C.M., Peleg, O., Anamthawat-Jonsson, K., Bolshoy, A., and Schulman, A.H. (2004) LARD retroelements: novel, non-autonomous components of barley and related genomes. *Genetics 166*: 1437–1450.

Kapitonov, V. and Jurka, J. (2001) Rolling-circle transposons in eukaryotes. *Proc. Natl. Acad. Sci. USA 98*: 8714–8719.

Kapitonov, V. and Jurka, J. (2006) Self-synthesizing DNA transposons in eukaryotes. *Proc. Natl. Acad. Sci. USA 103*: 4540–4545.

Kapitonov, V.V. and Jurka, J. (2007) Helitrons on a roll: eukaryotic rolling-circle transposons. *Trends Genet. 23*: 521–529.

Kashkush, K. and Khasdan, V. (2007) Large-scale survey of cytosine methylation of retrotransposons and the impact of readout transcription from long terminal repeats on expression of adjacent rice genes. *Genetics 177*: 1975–1985.

Kawai, Y. and Otsuka, J. (2004) The deep phylogeny of land plants inferred from a full analysis of nucleotide base changes in terms of mutation and selection. *J. Mol. Evol. 58*: 479–489.

Katinka, M.D., Duprat, S., Cornillot, E., Metenier, G., Thomarat, F., Prensier, G., Barbe, V., Peyretaillade, E., Brottier, P., Wincker, P., et al. (2001) Genome sequence and gene compaction of the eukaryote parasite *Encephalitozoon cuniculi. Nature 414*: 450–453.

Keith, J.H., Schaeper, C.A., Fraser, T.S., and Fraser, M.J., Jr. (2008) Mutational analysis of highly conserved aspartate residues essential to the catalytic core of the piggyBac transposase. *BMC Mol. Biol. 9*: 73.

Kim, A., Terzian, C., Santamaria, P., Pélisson, A., Prud'homme, N., and Bucheton, A. (1994) Retroviruses in invertebrates: the *gypsy* retrotransposon is apparently an infectious retrovirus of *Drosophila melanogaster. Proc. Natl. Acad. Sci. USA 91*: 1285–1289.

Kojima, K.K. and Jurka, J. (2011) Crypton transposons: identification of new diverse families and ancient domestication events. *Mobile DNA 2*: 12.

Koonin, E.V., Wolf, Y.I., Nagasaki, K., and Dolja, V. (2008) The Big Bang of picorna-like virus evolution antedates the radiation of eukaryotic supergroups. *Nature Rev. Microbiol. 6*: 925–939.

Kramerov, D. and Vassetzky, N. (2005) Short retroposons in eukaryotic genomes. *Int. Rev. Cytol. 247*: 165–221.

Krishan, A., Dandekar, P., Nathan, N., Hamelik, R., Miller, C., and Shaw, J. (2005) DNA index, genome size, and electronic nuclear volume of vertebrates from the Miami Metro Zoo. *Cytometry A 65*: 26–34.

Kronmiller, B.A. and Wise, R.P. (2007) TE nest: automated chronological annotation and visualization of nested plant transposable elements. *Plant Physiol. 146*: 45–59.

Kroutter, E.N., Belancio, V.P., Wagstaff, B.J., and Roy-Engel, A.M. (2009) The RNA polymerase dictates ORF1 requirement and timing of LINE and SINE retrotransposition. *PLoS Genet. 5*: e1000458.

Lai, J., Li, Y., Messing, J., and Dooner, H.K. (2005) Gene movement by *Helitron* transposons contributes to the haplotype variability of maize. *Proc. Natl. Acad. Sci. USA 102*: 9068–9073.

Lambowitz, A.M. and Zimmerly, S. (2004) Mobile group II introns. *Annu. Rev. Genet. 38*: 1–35.

Law, J.A. and Jacobsen, S.E. (2010) Establishing, maintaining and modifying DNA methylation patterns in plants and animals. *Nat. Rev. Genet. 11*: 204–220.

Le, Q.H., Melayah, D., Bonnivard, E., Petit, M., and Grandbastien, M.A. (2007) Distribution dynamics of the Tnt1 retrotransposon in tobacco. *Mol. Genet. Genomics 278*: 639–651.

Leitch, I.J., Beaulieu, J.M., Cheung, K., Hanson, L., Lysak, M.A., and Fay, M.F. (2007) Punctuated genome size evolution in Liliaceae. *J. Evol. Biol. 20*: 2296–2308.

Li, Y. and Dooner, H.K. (2009) Excision of *Helitron* transposons in maize. *Genetics 182*: 399–402.

Li, M., Mizuuchi, M., Burke, T.R., Jr., and Craigie, R. (2006) Retroviral DNA integration: reaction pathway and critical intermediates. *EMBO J. 25*: 1295–1304.

Lingner, J., Hughes, T.R., Shevchenko, A., Mann, M., Lundblad, V., and Cech, T.R. (1997) Reverse transcriptase motifs in the catalytic subunit of telomerase. *Science 276*: 561–567.

Lisch, D. and Bennetzen, J.L. (2011) Transposable element origins of epigenetic gene regulation. *Curr. Opin. Plant Biol. 14*: 156–161.

Lough, T.J. and Lucas, W.J. (2006) Integrative plant biology: role of phloem long-distance macromolecular trafficking. *Annu. Rev. Plant Biol. 57*: 203–232.

Lue, N.F., Bosoy, D., Moriarty, T.J., Autexier, C., Altman, B., and Leng, S. (2005) Telomerase can act as a template- and RNA-independent terminal transferase. *Proc. Natl. Acad. Sci. USA 102*: 9778–9783.

Lynch, V.J., Leclerc, R.D., May, G., and Wagner, G.P. (2011) Transposon-mediated rewiring of gene regulatory networks contributed to the evolution of pregnancy in mammals. *Nat. Genet. 43*: 1154–1159.

Ma, Y., Miura, E., Ham, B.K., Cheng, H.W., Lee, Y.J., and Lucas, W.J. (2010) Pumpkin eIF5A isoforms interact with components of the translational machinery in the cucurbit sieve tube system. *Plant J. 64*: 536–550.

Mager, D.L. and Goodchild, N.L. (1989) Homologous recombination between the LTRs of a human retrovirus-like element causes a 5-kb deletion in two siblings. *Am. J. Hum. Genet. 45*: 848–854.

Martin, F., Kohler, A., Murat, C., Balestrini, R., Coutinho, P.M., Jaillon, O., Montanini, B., Morin, E., Noel, B., Percudani, R., et al. (2010) Périgord black truffle genome uncovers evolutionary origins and mechanisms of symbiosis. *Nature 464*: 1033–1038.

Martin, W. and Koonin, E.V. (2006) Introns and the origin of nucleus-cytosol compartmentalization. *Nature 440*: 41–45.

Mayer, K.F., Taudien, S., Martis, M., Simková, H., Suchánková, P., Gundlach, H., Wicker, T., Petzold, A., Felder, M., Steuernagel, B., et al. 2009. Gene content and virtual gene order of barley chromosome 1H. *Plant Physiol. 151,* 496–505.

McClintock, B. (1948) Mutable loci in maize. *Carnegie Inst. Wash. Yr. Bk. 47*: 155–169.

Melnyk, C.W., Molnar, A., and Baulcombe, D.C. (2011) Intercellular and systemic movement of RNA silencing signals. *EMBO J. 30*: 3553–3563.

Miller, W.J., McDonald, J.F., and Pinsker, W. (1997) Molecular domestication of mobile elements. *Genetica 100*: 261–270.

Miyao, A., Tanaka, K., Murata, K., Sawaki, H., Takeda, S., Abe, K., Shinozuka, Y., Onosato, K., and Hirochika, H. (2003) Target site specificity of the Tos17 retrotransposon shows a preference for insertion within genes and against insertion in retrotransposon-rich regions of the genome. *Plant Cell 15*: 1771–1780.

Montaño, S.P. and Rice, P.A. (2011) Moving DNA around: DNA transposition and retroviral integration. *Curr. Opin. Struct. Biol. 21*: 370–378.

Morgante, M. (2006) Plant genome organisation and diversity: the year of the junk! *Curr. Opin. Biotechnol. 17*: 168–173.

Morgante, M., Brunner, S., Pea, G., Fengler, K., Zuccolo, A., and Rafalski, A. (2005) Gene duplication and exon shuffling by *Helitron*-like transposons generate intraspecies diversity in maize. *Nat. Genet. 37*: 997–1002.

Mouse Genome Sequencing Consortium (2002) Initial sequencing and comparative analysis of the mouse genome. *Nature 420*: 520–562.

Mroczek, R.J. and Dawe, R.K. (2003) Distribution of retroelements in centromeres and neocentromeres of maize. *Genetics 165*: 809–819.

Neumann, P., Navrátilová, A., Koblížková, A., Kejnovský, E., Hřibová, E., Hobza, R., Widmer, A., Doležel, J., and Macas, J. (2011) Plant centromeric retrotransposons: a structural and cytogenetic perspective. *Mobile DNA 2*: 4.

Ohno, S. (1972) So much 'junk' in our genome. *Brookhaven Symp. Biol. 23*: 366–370.

Orgel, L.E. and Crick, F.H.C. (1980) Selfish DNA: the ultimate parasite. *Nature 284*: 604–607.

Ørom, U.A. and Shiekhattar, R. (2011) Noncoding RNAs and enhancers: complications of a long-distance relationship. *Trends Genet. 27*: 433–439.

Padovan, A.C., Sanson, G.F., Brunstein, A., and Briones, M.R. (2005) Fungi evolution revisited: application of the penalized likelihood method to a Bayesian fungal phylogeny provides a new perspective on phylogenetic relationships and divergence dates of Ascomycota groups. *J. Mol. Evol. 60*: 726–735.

Paterson, A.H., Bowers, J.E., Bruggmann, R., Dubchak, I., Grimwood, J., Gundlach, H., Haberer, G., Hellsten, U., Mitros, T., Poliakov, A., et al. (2009) The *Sorghum bicolor* genome and the diversification of grasses. *Nature 457*: 551–556.

Pearce, S.R., Pich, U., Harrison, G., Flavell, A.J., Heslop-Harrison, J.S., Schubert, I., and Kumar, A. (1996) The *Ty1-copia* group retrotransposons of *Allium cepa* are distributed throughout the chromosomes but are enriched in the terminal heterochromatin. *Chrom. Res. 4*: 357–364.

Piffanelli, P., Droc, G., Mieulet, D., Lanau, N., Bès, M., Bourgeois, E., Rouvière, C., Gavory, F., Cruaud, C., Ghesquière, A., and Guiderdoni, E. (2007) Large-scale characterization of Tos17 insertion sites in a rice T-DNA mutant library. *Plant Mol. Biol. 65*: 587–560.

Pinsker, W., Haring, E., Hagemann, S., and Miller, W.J. (2001) The evolutionary life history of P transposons: from horizontal invaders to domesticated neogenes. *Chromosoma 110*: 148–158.

Poole, A.M., Jeffares, D.C., and Penny, D. (1998) The path from the RNA world. *J. Mol. Evol. 46*: 1–17.

Presting, G.G., Malysheva, L., Fuchs, J., and Schubert, I. (1998) A Ty3/gypsy retrotransposon-like sequence localizes to the centromeric regions of cereal chromosomes. *Plant J. 16*: 721–728.

Pritham, E., Putliwala, T., and Feschotte, C. (2007) Mavericks, a novel class of giant transposable elements widespread in eukaryotes and related to DNA viruses. *Gene 390*: 3–17.

Puchta, H. (2005) The repair of double-strand breaks in plants: mechanisms and consequences for genome evolution. *J. Exp. Bot. 56*: 1–14.

Ramallo, E., Kalendar, R., Schulman, A.H., and Martínez-Izquierdo, J.A. (2008) *Reme1*, a *Copia* retrotransposon in melon, is transcriptionally induced by UV light. *Plant Mol. Biol. 66*: 137–150.

Rebollo, R., Karimi, M.M., Bilenky, M., Gagnier, L., Miceli-Royer, K., Zhang, Y., Goyal, P., Keane, T.M., Jones, S., Hirst, M., et al. (2011) Retrotransposon-induced heterochromatin spreading in the mouse revealed by insertional polymorphisms. *PLoS Genet. 7*: e1002301.

Rensing, S.A., Lang, D., Zimmer, A.D., Terry, A., Salamov, A., Shapiro, H., Nishiyama, T., Perroud, P.F., Lindquist, E.A., Kamisugi, Y., et al. (2008) The Physcomitrella genome reveals evolutionary insights into the conquest of land by plants. *Science 319*: 64–69.

Rosbash, M., Ford, P.J., and Bishop, J.O. (1974) Analysis of the *C*-value paradox by molecular hybridization. *Proc. Natl. Acad. Sci. USA 71*: 3746–3750.

Sabot, F. and Schulman, A.H. (2006) Parasitism and the retrotransposon life cycle in plants: a hitchhiker's guide to the genome. *Heredity 97*: 381–38.

SanMiguel, P. and Bennetzen, J.L. (1998) Evidence that a recent increase in maize genome size was caused by the massive amplification of intergene retrotransposons. *Ann. Bot. 82*: 37–44.
SanMiguel, P., Gaut, B.S., Tikhonov, A., Nakajima, Y., and Bennetzen, J.L. (1998) The paleontology of intergene retrotransposons of maize. *Nat. Genet. 20*: 43–45.
SanMiguel, P., Ramakrishna, W., Bennetzen, J., Busso, C., and Dubcovsky, J. (2002) Transposable elements, genes and recombination in a 215-kb contig from wheat chromosome 5Am. *Funct. Integr. Genomics 2*: 70–80.
SanMiguel, P., Tikhonov, A., Jin, Y.K., Motchoulskaia, N., Zakharov, D., Melake-Berhan, A., Springer, P.S., Edwards, K.J., Lee, M., Avramova, Z., et al. (1996) Nested retrotransposons in the intergenic regions of the maize genome. *Science 274*: 765–768.
Sarilar, V., Marmagne, A., Brabant, P., Joets, J., and Alix, K. (2011) BraSto, a Stowaway MITE from Brassica: recently active copies preferentially accumulate in the gene space. *Plant Mol. Biol. 77*: 59–75.
Schnable, P.S., Ware, D., Fulton, R.S., Stein, J.C., Wei, F., Pasternak, S., Liang, C., Zhang, J., Fulton, L., Graves, T.A., et al. (2009) The B73 maize genome: complexity, diversity, and dynamics. *Science 326*: 1112–1115.
Schwarz-Sommer, Z., Gier, A., Klösgen, R.B., Wienand, U., Peterson, P.A., and Saedler, H. (1984) The *Spm* (*En*) transposable element controls the excision of a 2-kb DNA insert at the $wx^{m\text{-}8}$ allele of *Zea mays*. *EMBO J. 3*: 1021–1028.
Sen, S.K., Huang, C.T., Han, K., and Batzer, M.A. (2007) Endonuclease-independent insertion provides an alternative pathway for L1 retrotransposition in the human genome. *Nucl. Acids Res. 35*: 3741–3751.
Shirasu, K., Schulman, A.H., Lahaye, T., and Schulze-Lefert, P. (2000) A contiguous 66-kb barley DNA sequence provides evidence for reversible genome expansion. *Genome Res. 10*: 908–915.
Soleimani, V.D., Baum, B.R., and Johnson, D.A. (2006) Quantification of the retrotransposon *BARE-1* reveals the dynamic nature of the barley genome. *Genome 49*: 389–396.
Spanu, P.D., Abbott, J.C., Amselem, J., Burgis, T.A., Soanes, D.M., Stüber, K., Ver Loren van Themaat, E., Brown, J.K., Butcher, S.A., Gurr, S.J., et al. (2010) Genome expansion and gene loss in powdery mildew fungi reveal tradeoffs in extreme parasitism. *Science 330*: 1543–1546.
Studer, A., Zhao, Q., Ross-Ibarra, J., and Doebley, J. (2011) Identification of a functional transposon insertion in the maize domestication gene *tb1*. *Nat. Genet. 43*: 1160–1163.
Suoniemi, A., Anamthawat-Jónsson, K., Arna, T., and Schulman, A.H. (1996) Retrotransposon *BARE*-1 is a major, dispersed component of the barley (*Hordeum vulgare* L.) genome. *Plant Mol. Biol. 30*: 1321–1329.
Swigonova, Z., Lai, J., Ma, J., Ramakrishna, W., Llaca, V., Bennetzen, J.L., and Messing, J. (2004) On the tetraploid origin of the maize genome. *Comp. Funct. Genomics 5*: 281–284.
Tadege, M., Wen, J., He, J., Tu, H., Kwak, Y., Eschstruth, A., Cayrel, A., Endre, G., Zhao, P.X., Chabaud, M., et al. (2008) Large-scale insertional mutagenesis using the Tnt1 retrotransposon in the model legume *Medicago truncatula*. *Plant J. 54*: 335–347.
Tanskanen, J.A., Sabot, F., Vicient, C., and Schulman, A.H. (2007) Life without GAG: the *BARE*-2 retrotransposon as a parasités parasite. *Gene 390*: 166–174.
Taylor, J.W. and Berbee, M.L. (2006) Dating divergences in the Fungal Tree of Life: review and new analyses. *Mycologia 98*: 838–849.
Telesnitsky, A. and Goff, S.P. (1997) Reverse transcriptase and the generation of retroviral DNA. In *Retroviruses*, J.M. Coffin, S.H. Hughes, and H.E. Varmus, eds. (Cold Spring Harbor, NY: Cold Spring Harbor Laboratory Press), pp. 121–160.
Temin, H.M. (1980) Origin of retroviruses from cellular moveable genetic elements. *Cell 21*: 599–600.
Tuskan, G.A., Difazio, S., Jansson, S., Bohlmann, J., Grigoriev, I., Hellsten, U., Putnam, N., Ralph, S., Rombauts, S., Salamov, A., et al. (2006) The genome of black cottonwood, *Populus trichocarpa* (Torr. & Gray). *Science 313*: 1596–1604.
Vicient, C.M., Kalendar, R., and Schulman, A.H. (2001) Envelope-containing retrovirus-like elements are widespread, transcribed and spliced, and insertionally polymorphic in plants. *Genome Res. 11*: 2041–2049.
Vicient, C.M., Kalendar, R., and Schulman, A.H. (2005) Variability, recombination, and mosaic evolution of the barley *BARE*-1 retrotransposon. *J. Mol. Evol. 61*: 275–291.
Vicient, C.M., Kalendar, R., Anamthawat-Jonsson, K., and Schulman, A.H. (1999a) Structure, functionality, and evolution of the *BARE-1* retrotransposon of barley. *Genetica 107*: 53–63.
Vicient, C.M., Suoniemi, A., Anamthawat-Jónsson, K., Tanskanen, J., Beharav, A., Nevo, E., and Schulman, A.H. (1999b) Retrotransposon *BARE*-1 and its role in genome evolution in the genus *Hordeum*. *Plant Cell 11*: 1769–1784.
Vitte, C. and Panaud, O. (2005) LTR retrotransposons and flowering plant genome size: emergence of the increase/decrease model. *Cytogenet. Genome Res. 110*: 91–107.
Volff, J.N. (2006) Turning junk into gold: domestication of transposable elements and the creation of new genes in eukaryotes. *BioEssays 28*: 913–922.
Weber, B., Wenke, T., Frömmel, U., Schmidt, T., and Heitkam, T. (2010) The Ty1-*copia* families *SALIRE* and Cotzilla populating the *Beta vulgaris* genome show remarkable differences in abundance, chromosomal distribution, and age. *Chrom. Res. 18*: 247–263.

Wei, F., Wing, R.A., and Wise, R.P. (2002) Genome dynamics and evolution of the *Mla* (powdery mildew) resistance locus in barley. *Plant Cell 14,* 1903–1917.

Wessler, S.R., Bureau, T.E., and White, S.E. (1995) LTR-retrotransposons and MITEs: important players in the evolution of plant genomes. *Curr. Opin. Genet. Dev. 5*: 814–82.

Wicker, T. and Keller, B. (2007) Genome-wide comparative analysis of *copia* retrotransposons in Triticeae, rice, and *Arabidopsis* reveals conserved ancient evolutionary lineages and distinct dynamics of individual *copia* families. *Genome Res. 17*: 1072–1081.

Wicker, T., Buchmann, J.P., and Keller, B. (2010) Patching gaps in plant genomes results in gene movement and erosion of colinearity. *Genome Res. 20*: 1229–1237.

Wicker, T., Sabot, F., Hua-Van, A., Bennetzen, J.L., Capy, P., Chalhoub, B., Flavell, A., Leroy, P., Morgante, M., Panaud, O., et al. (2007) A unified classification system for eukaryotic transposable elements. *Nat. Rev. Genet. 8*: 973–982.

Wicker, T., Taudien, S., Houben, A., Keller, B., Graner, A., Platzer, M., and Stein, N. (2009) A whole-genome snapshot of 454 sequences exposes the composition of the barley genome and provides evidence for parallel evolution of genome size in wheat and barley. *Plant J. 59*: 712–722.

Witte, C.P., Le, Q.H., Bureau, T., and Kumar, A. (2001) Terminal-repeat retrotransposons in miniature (TRIM) are involved in restructuring plant genomes. *Proc. Natl. Acad. Sci. USA 98*: 13778–13783.

Wolfgruber, T.K., Sharma, A., Schneider, K.L., Albert, P.S., Koo, D.H., Shi, J., Gao, Z., Han, F., Lee, H., Xu, R., et al. (2009) Maize centromere structure and evolution: sequence analysis of centromeres 2 and 5 reveals dynamic loci shaped primarily by retrotransposons. *PLoS Genet. 5*: e1000743.

Wyman, C. and Kanaar, R. (2006) DNA double-strand break repair: all's well that ends well. *Annu. Rev. Genet. 40*: 363–383.

Xiong, Y. and Eickbush, T.H. (1990) Origin and evolution of retroelements based upon their reverse transcriptase sequences. *EMBO J. 9*: 3353–3362.

Xu, Z., Rafi, S., and Ramakrishna, W. (2011) Polymorphisms and evolutionary history of retrotransposon insertions in rice promoters. *Genome 54*: 629–638.

Yang, G., Nagel, D.H., Feschotte, C., Hancock, C.N., and Wessler, S.R. (2009) Tuned for transposition: molecular determinants underlying the hyperactivity of a Stowaway MITE. *Science 325*: 1391–1394.

Yuan, Y.W. and Wessler, S.R. (2011) The catalytic domain of all eukaryotic cut-and-paste transposase superfamilies. *Proc. Natl. Acad. Sci. USA 108*: 7884–7889.

3 The Mechanism of *Ac/Ds* Transposition

Thomas Peterson and Jianbo Zhang

"It is now known that the Ds locus can change its position in the chromosome..." (McClintock, 1948).

With this simple declaration, McClintock overturned a major assumption implicit in genetic research: that genes maintain their relative linear orders on the chromosome. McClintock would go on to show that transposition of *Ds* required a second factor (*Ac*, *Activator*), which also transposed, and that another element, *Spm*, was also mobile. Although transposition was the subject of McClintock's 1983 Nobel award, transposition *per se* was not considered by McClintock to be the most important aspect of her work. Rather, she emphasized the ability of transposons to impact gene expression, hence her term *controlling elements*. This chapter addresses the phenomenon of transposition, as first demonstrated by the cytogenetic and genetic analyses of McClintock and others and later confirmed and extended by molecular approaches.

Transposition of *Ac/Ds* Elements

Comparative analyses of bacterial, plant, and animal transposable element systems indicate that transposition mechanisms are similar among related transposons but differ between transposon classes. This chapter will focus on transposition of *Ac/Ds,* a classical two-element transposon system in the *hAT* family. Although this system was the first identified by McClintock, most insights into *Ac/Ds* transposition were elucidated by the work of others. Interestingly, that work began more than three decades before McClintock discovered *Ac/Ds*. In the 1910s, while McClintock was still an adolescent, her future supervisor, Rollins A. Emerson, was investigating a particular variety of maize (*P-vv*) with red and white variegated kernels and cobs. Emerson and others had noted that the *P-vv* allele mutated to a stable red type (*P-rr;* red pericarp and red cob) at a high frequency; hence, *P-vv* was a member of the class of highly unstable (mutable) genes (Emerson, 1914). Moreover, Emerson showed that the *P-vv* allele exhibited an unusual dosage effect: plants homozygous for *P-vv* exhibited less kernel variegation and produced fewer germinal revertants than did *P-vv* heterozygotes (Emerson, 1917, 1929). These observations presaged McClintock's discovery of the *Ac*-negative dosage effect (see below).

Subsequent work by McClintock's contemporary R.A. Brink and his students showed that the *P-vv* allele was composed of a gene for red pericarp (*P-rr*) conjoined with a second factor (*Mp*, *Modulator of pericarp*) that could move from *P-vv* to other locations in the genome (Van Schaik

Plant Transposons and Genome Dynamics in Evolution, First Edition. Edited by Nina V. Fedoroff.

and Brink, 1959). Transposition of *Mp* from *P-vv* resulted in reversion to *P-rr*, producing both red kernel sectors and fully red (*P-rr*) progeny (Brink and Nilan, 1952). Moreover, the presence of one or more transposed *Mp* elements reduced somatic variegation of *P-vv*, i.e., *Mp* exhibited a negative dosage effect (Brink and Nilan, 1952; Brink, 1954) similar to that of *Ac* as had already been reported by McClintock (McClintock, 1951). Struck by the similarities of *Ac* and *Mp*, Barclay and Brink (1954) quickly demonstrated that *Mp* could activate *Ds*. The equivalence of *Mp* and *Ac* suggested by these early genetic experiments was later confirmed by molecular cloning (Chen et al., 1987; Lechelt et al., 1989).

While working with *P-vv*, Brink noticed frequent twinned sectors: patches of red kernels (excisions of *Ac* from *P-vv* to generate *P-rr*) adjacent to sectors of light-variegated kernels (the *P-vv* allele plus an additional transposed *Ac* (tr-*Ac*) elsewhere in the genome) (Brink and Nilan, 1952). Genetic mapping showed that different twinned sectors carried tr-*Ac* elements at different genomic locations, as would be expected for independent transposition events (Van Schaik and Brink, 1959). However, the transposed *Ac* elements present in the red and light-variegated kernels from the same twin sector mapped to the same locations (Greenblatt and Brink, 1962). This striking result was interpreted as follows: the *Ac* element transposed shortly after the *P-vv* allele had been replicated. *Ac* excised from one newly replicated chromatid and inserted into an unreplicated target site. Replication of the target site and mitotic segregation would give rise to two adjacent daughter cells, one of which would carry *P-rr*, and the other *P-vv*. Importantly, both cells would contain a copy of the transposed *Ac* element. Clonal expansion of the two daughter cells would produce the observed red/light-variegated twin sectors. These results indicated that *Ac* frequently transposed during DNA synthesis, from a replicated donor site to an unreplicated target site. This hypothesis was subsequently confirmed by Chen, Greenblatt, and Dellaporta (1987), who showed that the tr-*Ac* elements in several red/light-variegated co-twins were indeed inserted at exactly the same nucleotide position. The significant role of DNA replication in the *Ac/Ds* transposition reaction will be discussed further below.

Another important feature of *Ac* transposition is a pronounced preference for insertion into nearby sites. This was initially reported by Brink (Van Schaik and Brink, 1959) and studied in detail by the Greenblatt (Greenblatt, 1984) and Dooner laboratories (Dooner and Belachew, 1989) by classical genetic mapping of transposed *Ac* elements. More recently, molecular methods have been used to map the positions of large collections of transposed *Ac* and *Ds* elements generated by gene tagging projects (Raina et al., 2002; Kolkman et al., 2005; Vollbrecht et al., 2010). Overall, more than half of all reinsertions occur at sites linked to the donor locus; most transpositions occur into very close sites, and insertion frequency declines rapidly with increasing distance from the donor locus (Greenblatt, 1984; Vollbrecht et al., 2010). Among genetically unlinked insertion sites, intra-chromosomal transpositions are enriched relative to inter-chromosomal events (Dooner and Belachew, 1989; Vollbrecht et al., 2010). These observations suggest that, during transposition, the donor and target sites are associated; this conclusion is supported by additional results discussed further below.

The Enigmatic *Ac* Dosage Effect

> "*. . . the time and apparent frequency of Ac-controlled mutations is in large measure a function of dosage of Ac . . . the higher the Ac dosage, the later the occurrence of Ds mutations.*" (McClintock, 1948)

The effect of *Ac* dosage on transposition remains one of the most intriguing aspects of *Ac/Ds* biology. McClintock reported that increased doses of the *wx-m9::Ac* allele resulted in a delay in the timing of transposition. Kernels with a single copy of the *wx-m9* allele exhibited large revertant sectors that arose early in kernel development, whereas two or three copies resulted in progressively smaller revertant sectors arising later in kernel development. As mentioned previously, a similar effect of *Ac* dosage on kernel pericarp variegation was noted by both Emerson and Brink. Genetic experiments revealed that *Ds* elements do not contribute to the *Ac* dosage effect (Fedoroff, 1989), whereas molecular analyses indicated that increasing *Ac* dosage resulted in corresponding increases in *Ac* mRNA and TPase levels (Kunze et al., 1987; Fusswinkel et al., 1991). These results suggest that the *Ac* dosage effect occurs via post-translational mechanism(s). A potentially important caveat to this conclusion is that only the *wx-m7* allele was characterized in terms of both RNA and protein expression, and *wx-m7* actually exhibits an anomalous positive dosage effect (Schwartz, 1984; Heinlein, 1996). Experiments in petunia protoplasts showed that the frequency of *Ds* excision is proportional to the level of *Ac* TPase up to a threshold above which *Ds* excision does not increase (Becker et al., 1992; Kunze et al., 1993; Heinlein et al., 1994). In transgenic tobacco, increasing doses of *Ac* resulted in increased excision (Keller et al., 1993), although very high levels of TPase expression actually inhibited transposition (Scofield et al., 1993). Interestingly, maize endosperm cells containing three *Ac* copies exhibited large filament-like TPase complexes (Heinlein et al., 1994); similar aggregates were observed in petunia cells overexpressing *Ac* TPase (Essers et al., 2000). Taken together, these results support the hypothesis that higher *Ac* dosage leads to increased TPase levels and the formation of auto-inhibitory TPase aggregates (Scofield et al., 1993; Heinlein et al., 1994). Whether this is the sole mechanism responsible for the *Ac* negative dosage effect is not known.

cis and *trans* Effects on *Ac/Ds* Transposition

In addition to sensitivity to *Ac* dosage, *Ac/Ds* transposition frequency is also subject to *cis*-effects associated with different sites of insertion in the genome. As expected, the apparent *Ac/Ds* excision frequencies, as monitored by reversion of a reporter gene, differ depending upon the site of insertion within a gene, as excision normally leaves a remnant of the 8-bp target site duplication generated at insertion. For example, insertions in different positions of the same gene show dramatically different reversion frequencies (Athma et al., 1992). In these cases, it is assumed that excision frequency is not appreciably changed, but the probability of generating revertant alleles is diminished for open reading frame (ORF) insertions. In addition, it is reasonable to expect that expression of *Ac* can be influenced by neighboring regulatory elements to show position effects, as illustrated by the differences between the *wx-m9* and *wx-m7* alleles (Schwartz, 1984, 1986; Heinlein and Starlinger, 1991). Other kinds of differences in excision frequency still lack explanations. For example, the frequency of *Waxy* revertant sectors produced by *Ds* excision from the *wx-m9::Ds* allele was not directly proportional to the dosage of *Ds* substrates: increasing the dose of *wx-m9::Ds* from one to two copies gave less than a two-fold increase in excision sectors (Heinlein, 1996). Heinlein suggested that the *wx-m9::Ds* reporter alleles contributed from the male and female parents may be differentially competent for transposition due to imprinting, although direct evidence for imprinting of *Ds* elements is lacking. Unexpectedly, three doses of *wx-m9::Ds* produced more sectors than the sum of one and two, a result that cannot be explained by imprinting alone (Heinlein, 1996). In another case, chromosome breakage induced by *sh-m5933* regularly occurred much earlier in kernel development than did transposition of *Ac* from the *bz-m2* allele (Heinlein, 1996). This observation underscores the conclusion that different *Ac/Ds* elements can exhibit strikingly different responses to the same *Ac* transposase source.

There is little evidence that host factors play a role in modulating the transposition reaction. For example, transgenic experiments demonstrate that the *Ac/Ds* system functions not only in a wide variety of host plants, but also in yeast and zebrafish (Weil and Kunze, 2000; Emelyanov et al., 2006). It follows that any host factors essential for transposition must be conserved across the plant, animal, and fungal kingdoms. Clearly, however, *Ac/Ds* transposition is influenced by developmental signals, as indicated by the impact of *Ac* dosage on the developmental timing of transposition (McClintock, 1948). In addition, a modifier of *Ds* transposition has been identified genetically in maize (Eisses et al., 1997). Beyond this, efforts to identify *Ac* TPase interacting factors by yeast two-hybrid analyses have been largely unfruitful (Z. Li and T. Peterson, unpublished).

Molecular Characterization of Transposable Elements

Molecular cloning of *Ac* and *Ds* elements was first accomplished in the 1980s by several laboratories working independently (Courage-Tebbe et al., 1983; Fedoroff et al., 1983; Courage et al., 1984; Sutton et al., 1984). It soon became clear that *Ac* elements cloned from different maize stocks are virtually sequence-identical, whereas *Ds* elements are heterogeneous in both size and structure. *Ac* and *Ds* elements share 11-bp terminal inverted repeat sequences (TIRs) that are critical for transposition. A 1-bp deletion in the 3′ end of an *Ac* element dramatically decreases *Ac* transposition frequency (Xiao and Peterson, 2002), and deletion of 10 bp from the *Ac* 3′ end completely immobilizes the element (Conrad and Brutnell, 2005). Both *Ac* and *Ds* also contain one or more subterminal hexamer motifs (AAACGG or similar) located within 250 bp of the element termini (Figure 3.1) (Coupland et al., 1988, 1989; Bravo-Angel et al., 1995; Chatterjee and Starlinger, 1995). By progressively deleting sequences internal to the *Ac* termini, Coupland et al. (1989) found

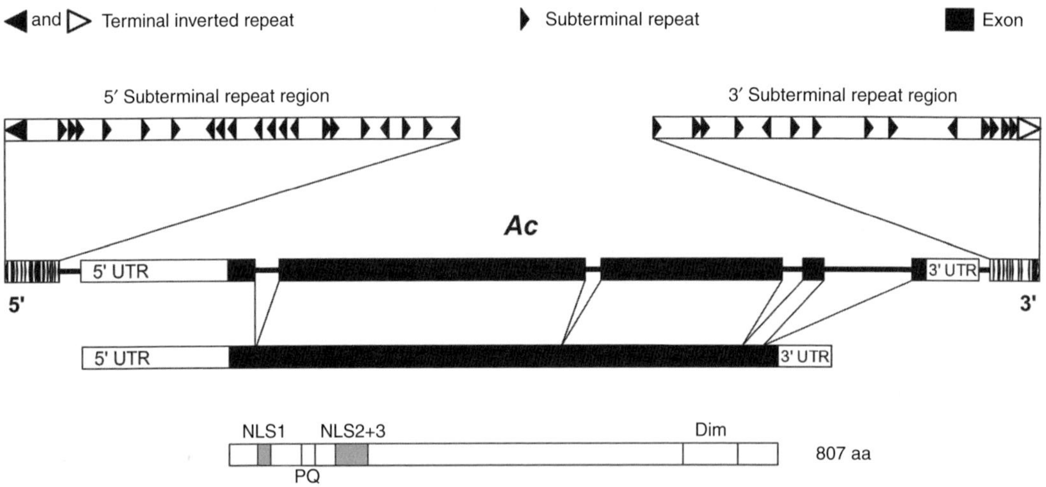

Figure 3.1 Schematic structure of *Ac* element. The upper portion shows an enlarged view of ~250-bp 5′- and 3′-terminal regions, including the 11-bp terminal inverted repeats (TIRs, larger triangles) and AAACGG-like transposase-binding motifs (smaller triangles) contained within the subterminal repetitive regions (SRRs; extended boxes). Shown in the center is the 4565-bp *Ac* element and mature spliced mRNA. The lower part of the figure shows the 807-aa transposase protein, with the nuclear localization signals (NLS1 and NLS2 + 3), the proline/glutamine repeat region (PQ), and C-terminal dimerization domains (Dim) indicated. NLS2 + 3 is involved in DNA binding. The figure was compiled from data presented in Boehm et al., 1995; Essers et al., 2000; and Kunze et al., 1993.

that the 238-bp sequence from the *Ac* 5′ end and 209-bp sequence from the *Ac* 3′ end are essential for normal levels of transposition. Moreover, engineered elements containing two *Ac* 5′ ends or two *Ac* 3′ ends are not competent for transposition, indicating that the two termini are complementary partners in the transposition reaction (Coupland et al., 1988, 1989). Interestingly, the AAACGG motifs are concentrated in these subterminal regions and appear to be the primary binding site for *Ac* TPase (see below).

The *Ac* element produces a single predominant spliced mRNA that encodes a 102-kDa transposase protein (Figure 3.1). The transposase contains subdomains including nuclear localization signals, DNA-binding domains, and a C-terminal dimerization domain. This latter region is highly conserved among *hAT* family TPases, suggesting that assembly of a functional transposition complex requires interaction of two or more TPase molecules (Kunze et al., 1993; Boehm et al., 1995; Essers et al., 2000). The *Ac* TPase binds most strongly to the AAACGG hexamer motifs that are concentrated in the ~250-bp subterminal regions discussed above (Coupland et al., 1988; Kunze and Starlinger, 1989; Becker and Kunze, 1996, 1997) and weakly to the TIR sequences.

The genetic and molecular analyses of twin sectors discussed above (Greenblatt and Brink, 1962; Greenblatt, 1984; Chen et al., 1987) showed that *Ac* transposes during or shortly after DNA replication; moreover, only one of the newly replicated *Ac* elements is competent for transposition. Molecular data indicate that the link between transposition and DNA replication can be explained by preferential binding of *Ac* TPase to hemi-methylated DNA. The *in vitro* DNA binding studies cited above revealed that binding of transposase to holomethylated sequences is much weaker than to unmethylated or hemi-methylated motifs (Wang and Kunze, 1998; Ros and Kunze, 2001). Moreover, TPase binds six-fold more strongly to 5′ termini with top-strand methylation than bottom-strand methylation. This observation explains why *Ac* transposes shortly after replication, when the element termini are still hemi-methylated, and why only one of the newly replicated elements (with top-strand methylation) transposes (Figure 3.2). The hypothesis that the pattern of strand-specific DNA methylation determines which element termini are competent for transposition has important implications for situations in which multiple elements are present at a single locus, as will be discussed below.

The Excision and Insertion Reactions

Although the exact mechanism of *Ac/Ds* transposition is still unknown, evidence to date indicates that transposition occurs through a cut-and-paste mechanism (Kunze, 1996; Gorbunova and Levy, 1997, 2000). This mechanism is conserved among *hAT* family transposons, of which the best-characterized is the insect *Hermes* element (Zhou et al., 2004). In the insertion reaction, TPase makes an 8-bp staggered cut at the target site, the transposon DNA is ligated into the cut site, and the 8-bp gaps are filled in by DNA repair synthesis to generate the characteristic 8-bp TSD flanking the newly inserted element (Figure 3.3). Excision of the *Hermes* element is initiated by transposase-induced nicking of the flanking DNA 1 bp from the transposon end; the exact site of nicking during *Ac/Ds* excision is still unknown. The exposed 3′-hydroxyl then attacks the opposite strand at the junction of host and transposon. A *trans*-esterification reaction covalently joins the flanking DNA in a hairpin structure, releasing the transposon (Figure 3.4).

For plant transposons, evidence in support of the foregoing mechanism derives primarily from analysis of sequences remaining following *Ac/Ds* excision. The excision reaction is thought to generate a chromosomal double-strand break, which is then religated to generate an excision site (Rommens et al., 1993; Rinehart et al., 1997). These excision sites almost always exhibit some

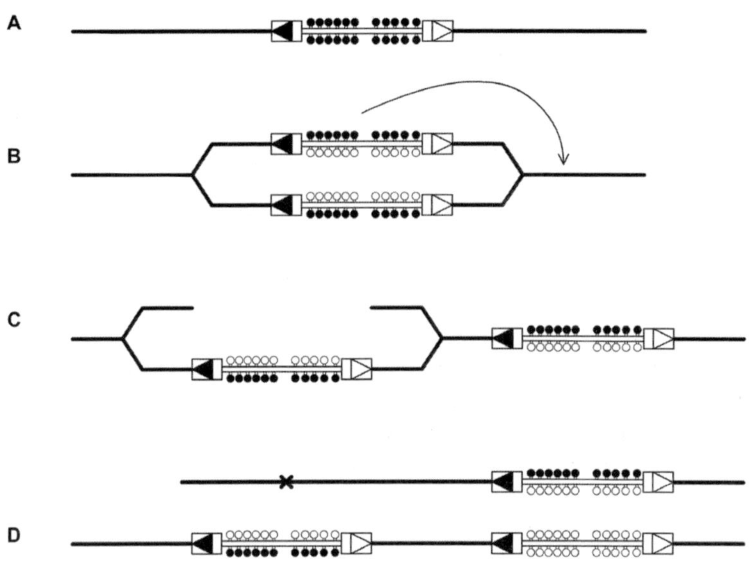

Figure 3.2 Transposition of *Ac/Ds* during DNA replication. (A) The diagrams represent an *Ac* or *Ds* element inserted into a chromosome. Solid and open triangles represent 5′ and 3′ TIRs, respectively. Filled circles represent methylation of the transposase-binding motifs in the SRRs. The 5′ SRR contains more binding motifs (20; six indicated here) than the 3′ SRR (14; five indicated here). (B) Following replication of *Ac/Ds,* the transposase-binding motifs are transiently hemi-methylated. Transposase preferentially binds the replicated *Ac/Ds* with top-strand methylation (upper element) and will initiate transposition into the unreplicated target site indicated by arrow. (C) The upper *Ac/Ds* element has excised and inserted into the flanking unreplicated target site. The broken ends left by element excision will be rejoined by host DNA repair enzymes. (D) Completion of DNA replication gives rise to two sister chromatids. Both chromatids contain the newly transposed *Ac/Ds* element at the same site. The upper chromatid contains an empty donor site with an excision footprint (X), whereas the lower chromatid retains *Ac/Ds* at the original donor site. Segregation of the two chromatids to daughter cells at a subsequent mitotic division will generate a twin sector.

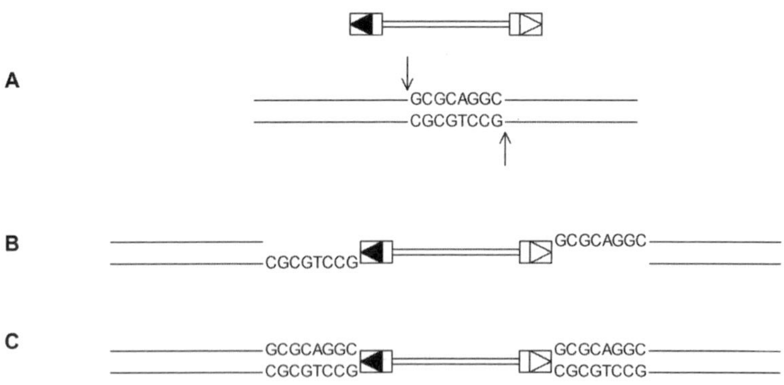

Figure 3.3 Insertion of *Ac/Ds* and formation of the target site duplication. (A) The *Ac* transposase makes 8-bp staggered cut at the target site (arrows). (B) The *Ac/Ds* element is ligated into the target site. (C) Gaps are filled to generate the 8-bp target site duplication flanking the inserted element.

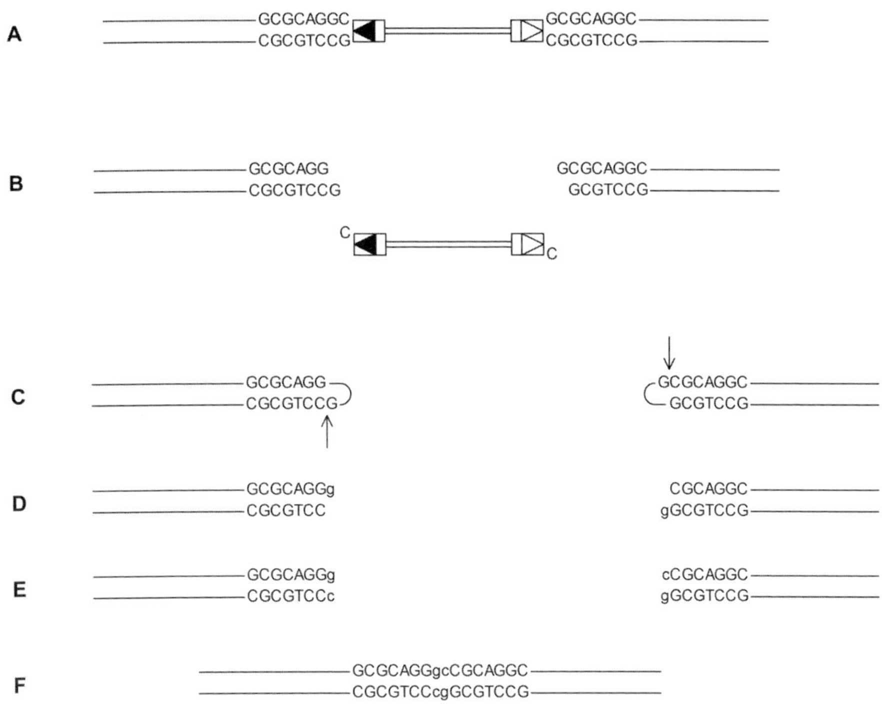

Figure 3.4 Excision of *Ac/Ds* and footprint formation. (A) A chromosomal site with an integrated *Ac/Ds* element flanked by a target site duplication. (B) The transposase makes 1-bp staggered cuts to release the transposon. The donor site has a double-strand break with 1-bp overhanging ends. (C) The ends of the cleaved donor site undergo trans-esterification to form hairpin loops, which are then nicked, usually at slightly off-center positions (arrows). (D) Bases transferred to the opposite strand are indicated by lower case letters at the ends of each donor site double-strand break. (E, F) One-bp gaps are filled in and the ends of the double-strand break are ligated to generate the empty donor site with a footprint comprising part of the initial target site duplication with a transversion of the two central nucleotides.

minor sequence changes at the new junction (footprints). Most footprints contain base substitutions and/or deletions, but palindromes of various sizes are also observed. Footprint sequences isolated from revertant alleles could be biased if the *Ac/Ds* insertion site is in an exon, because only those excisions that restore a functional open reading frame will be detected. To overcome this problem, Weil and co-workers analyzed PCR products obtained from somatic *Ac/Ds* excision events. Among these cases, the most common footprints are A↔T and G↔C transversions; deletions of 1–3 bp were also found at lower frequency (Scott et al., 1996; Rinehart et al., 1997). These results are consistent with the mechanism described above for *Hermes,* in which the target site double-strand break is repaired through a hairpin intermediate. The hairpins are thought to be nicked by a host factor, usually at sites 1 bp off center (Figure 3.4), explaining the prevalence of footprints containing A↔T and G↔C transversions. Variation in the sites of nicking can produce the observed small insertions, deletions, and occasional larger palindromes (Martin and Lister, 1989). The occurrence of large palindromes in *Ac/Ds* footprints is quite rare in plants but relatively common in yeast (Weil and Kunze, 2000), suggesting a role for host factors in the excision reaction. The footprints observed for *Ac/Ds* excision in yeast are somewhat different from those in plants. A↔T and G↔C transversions are rare, whereas short palindromes are relatively common. Based on the yeast excision footprints, Weil and Kunze (2000) proposed that a nick is made at 1 bp within *Ac/Ds* (instead of in the flanking

DNA). The transposon-flanking DNA then forms a hairpin as described above, but the hairpin is opened several base pairs from the flanking sequence–transposon junction.

Formation of *Ds* from *Ac*

A number of *Ds* elements have been isolated *de novo* from *Ac* elements inserted in different genomic locations. A majority of these new *Ds* elements carry internal deletions, with or without filler DNA (Rubin and Levy, 1997; Yan et al., 1999; Conrad et al., 2007). In all cases, the filler DNA was derived from *Ac* or nearby flanking sequences. These results indicate that *Ds* elements are derived from *Ac* excision events in which the flanking donor sequences are not ligated together but instead invade the *Ac* element in the sister chromatid and prime gap repair via a synthesis-dependent strand-annealing mechanism. Premature termination and/or template switching during gap repair synthesis can generate the observed *Ds* elements containing deletions and filler DNA (Yan et al., 1999; Conrad et al., 2007).

Standard versus Alternative Transposition

Early in her characterization of the *Ac/Ds* system, McClintock observed that *Ds* could induce not only chromosome breakage, but additional rearrangements including deletions, duplications, inversions, and translocations. As these rearrangements occurred only in the presence of *Ac*, it was believed that they were either direct or indirect products of *Ac/Ds* transposition events (McClintock, 1948, 1949b, 1950a, 1950b, 1951, 1952, 1953). *Ds*-induced chromosome breakage and associated genome rearrangements are now understood to arise from *alternative transposition* reactions: transpositions that utilize a 5′ end from one element and a 3′ end from a different, nearby element (in contrast to *standard transposition*, which utilizes the 5′ and 3′ ends of a single transposon). Standard transposition results only in a change of element position within a genome, whereas alternative transposition can induce a variety of genome rearrangements.

Because the types of alternative transposition reactions that can occur and the resulting products are highly dependent on the orientations of the participating element termini, it is important to define the terms used to indicate relative element orientations. A typical *Ac* element has 11-bp terminal inverted repeats that are usually drawn as outward-facing arrowheads in a standard element (Figure 3.5A). If the orientation of one end is flipped with respect to the other, then the two ends are now said to be in *direct* orientation; this is designated by two arrowheads pointing in the same direction (Figure 3.5B). If both ends are flipped, then the two ends are in *reversed* orientation, designated by two arrowheads pointing toward each other (Figure 3.5C). Note that these designations refer to pairwise combinations of termini (one 5′, one 3′). Loci containing two *Ac/Ds* elements will have four termini (two 5′ and two 3′) that can interact in various possible pairwise orientations (see below).

Sister Chromatid Transposition

The first type of alternative transposition reaction to be characterized at the molecular level has been termed *sister chromatid transposition*, i.e., transposition that involves *Ac/Ds* termini located on sister chromatids. Evidence for this type of transposition reaction began to emerge from molecular analysis of the structure of McClintock's *sh-m5933* allele. Early interest in this particular allele

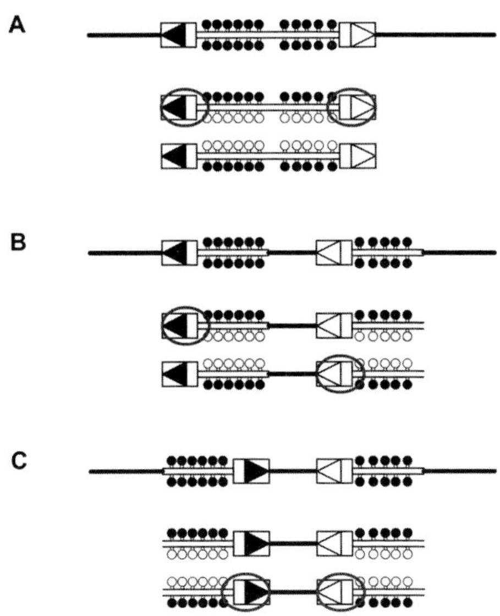

Figure 3.5 Orientations of *Ac/Ds* termini, DNA methylation, and transposition competence. Filled and open arrowheads indicate 5′ and 3′ termini, respectively. Filled circles indicate methylated transposase-binding motifs, open circles indicate non-methylated motifs, and ovals represent bound transposases. (A) A standard element with terminal inverted repeats represented by outward-facing arrowheads. Following DNA replication, the two daughter elements are transiently hemi-methylated. Transposase preferentially binds the 5′ and 3′ termini with top-strand methylation (upper element). (B) Two termini in direct orientation. Following replication, transposase binds the 5′ end on the upper chromatid and the 3′ end on the lower chromatid, leading to sister-chromatid transposition. (C) Two termini in reversed orientation. Following replication, transposase binds the 5′ and 3′ ends on the lower chromatid, leading to reversed-ends transposition.

was sparked by the fact that it contained a chromosome-breaking *Ds* element situated inside a gene for which molecular probes were available. Far from being a simple case, however, *sh-m5933* contained a complex structure with an inverted duplication (Burr and Burr, 1982) flanked by *DoubleDs* elements (Courage-Tebbe et al., 1983; Doring et al., 1989). Fedoroff (1989) proposed that the inverted duplication structure may have resulted from a transposition reaction involving *Ds* elements located on sister chromatids. Later, English et al. (1993) and Weil and Wessler (1993) proposed a model of sister-chromatid transposition to account for the phenomenon of *Ds*-induced chromosome breakage. This model predicts that transposition events involving *Ds* termini on sister chromatids should generate reciprocal deletions and duplications. Deletions were in fact reported by English et al. (1995) in a transgenic tobacco system.

More comprehensive studies of sister-chromatid transposition have been carried out using the maize *p1* (*pericarp1*) and *bz1* (*bronze 1*) genes. Both genes affect pigmentation of maize kernels, as well as other tissues, and thus serve as convenient markers for transposition events. The *p1* gene regulates the synthesis of 3-deoxyflavonoids and related compounds in kernel pericarp, whereas *bz1* affects the production of anthocyanin pigments in kernel aleurone. In the case of the *p1* gene, the allele relevant to sister chromatid transposition (*p1-vv9D9A*) traces back to Emerson's *p1-vv* allele discussed earlier. The *p1-vv9D9A* allele contains an intact *Ac* and a terminally deleted *fractured Ac* (*fAc*) inserted in the *p1* second intron (Zhang and Peterson, 1999). The *fAc* in the *p1-vv-9D9A* allele contains the 2039-bp 3′ portion of *Ac*; a similar *fAc* element was described previously at the maize *bronze1* locus (Ralston et al., 1989). In the presence of a *trans*-acting *Ac*, a single *fAc* does not undergo transposition or induce chromosome breaks (Huang and Dooner, 2008; Yu et al., 2010).

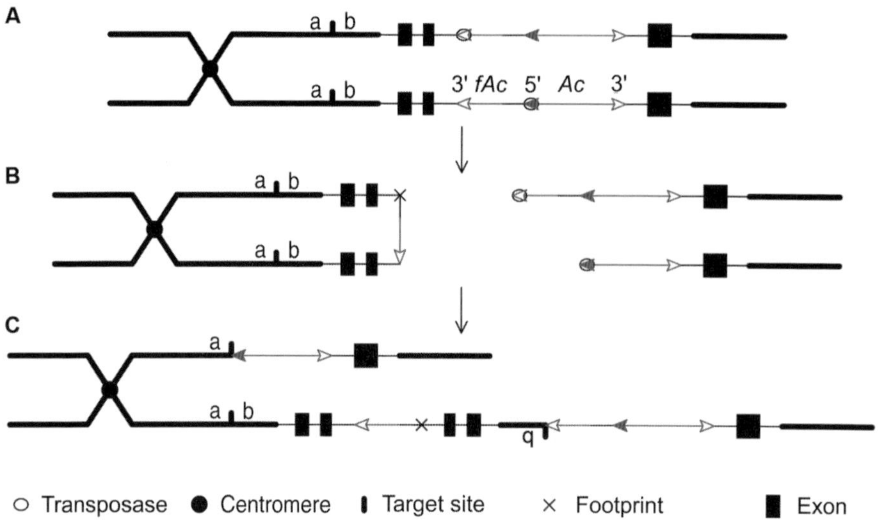

Figure 3.6 Sister chromatid transposition at the maize *p1* locus. (A) The lines represent sister chromatids joined at the centromere. Black boxes represent exons of the *P1-rr* gene. Open and filled arrowheads represent the 3′ and 5′ termini, respectively, of *Ac* and *fAc* elements. *Ac* transposase (small ovals) binds to the 5′ terminus of *Ac* in one sister chromatid and the 3′ terminus of *fAc* in the other sister chromatid. Cuts are made at the *Ac* and *fAc* termini. (B) The two non-transposon ends join together to generate a chromatid bridge, joined at the point marked by an "x." (C) Following insertion of the transposon into the target site (short vertical line between a and b), the upper chromatid contains a deletion of the b segment, and the lower chromatid contains a corresponding inverted duplication of b. Segregation of the sister chromatids to adjacent daughter cells at the subsequent mitosis will generate a potential twin sector.

However, *fAc* can participate in transposition reactions with a complementary 5′ or 3′ terminus provided by a nearby *Ac* or *Ds* element.

In the *p1-vv9D9A* allele, the 3′ end of *fAc* and the 5′ end of *Ac* are in direct orientation. The *p1-vv-9D9A* allele is unstable and spontaneously gives rise to flanking deletions and inverted duplications of various sizes. An especially informative case was presented by a twin sector on a *p1-vv9D9A* ear in which one twin contained a deletion and the other contained a duplication. Molecular analyses showed that the deletion and duplication alleles were reciprocal products of a single alternative transposition reaction (Figure 3.6) (Zhang and Peterson, 1999). Following DNA replication, the 3′ end of *fAc* in one chromatid and the 5′ end of *Ac* on the sister chromatid were the transposition substrates. These two ends were excised and then reinserted into a target site located approximately ~3600 kb toward the centromere to generate the reciprocal deletion and inverted duplication products. Importantly, the deletion and duplication chromosomes both contained the same 8-bp sequence at the rearrangement junctions, confirming that these originated in a single transposition reaction. Because the *Ac* termini that underwent transposition are located on sister chromatids, this is termed sister-chromatid transposition (Zhang and Peterson, 1999; Huang and Dooner, 2008).

The sister-chromatid transposition model predicts that reinsertion of the excised *Ac/fAc* ends into different sites in the chromatid bridge could generate a series of nested segmental deletions and segmental inverted duplications. This prediction was confirmed by the isolation of 35 interstitial deletions and 15 inverted duplications, each derived independently from the *p1-vv9D9A* allele. All of the deletions and duplications begin at the *Ac/fAc* termini in the *p1* gene and end at various sites up to 4.6 cM proximal to the *p1* gene (Zhang and Peterson, 2005).

Reversed-ends Transposition

The fact that directly oriented 5′ and 3′ termini of distinct *Ac/Ds* elements can undergo sister-chromatid transposition prompted a related question: can the reversely oriented 5′ and 3′ termini of different elements also serve as transposition substrates? This question was addressed using the *p1*-controlled kernel pigment phenotype as a convenient marker for transposition events. The particular allele studied (*P1-rr11*, which confers *r*ed pericarp, *r*ed cob) contains the same *fAc* element as in *p1-vv9D9A* and a full-length *Ac* element. The 5′ end of the *Ac* and the 3′ end of the *fAc* are oriented toward each other and separated by approximately 13 kb, a region designated the inter-transposon segment (Figure 3.7A). The *p1* gene is expressed in this allele, giving red kernels. However, there are many colorless pericarp sectors and colorless kernels, indicating that the *p1-rr11* allele is very unstable. A large number of germinal rearrangements were derived from this allele and characterized; they could be grouped into several classes, including inversions, deletions, and translocations. All of these rearrangements can be explained as products of transposition reactions involving the adjacent

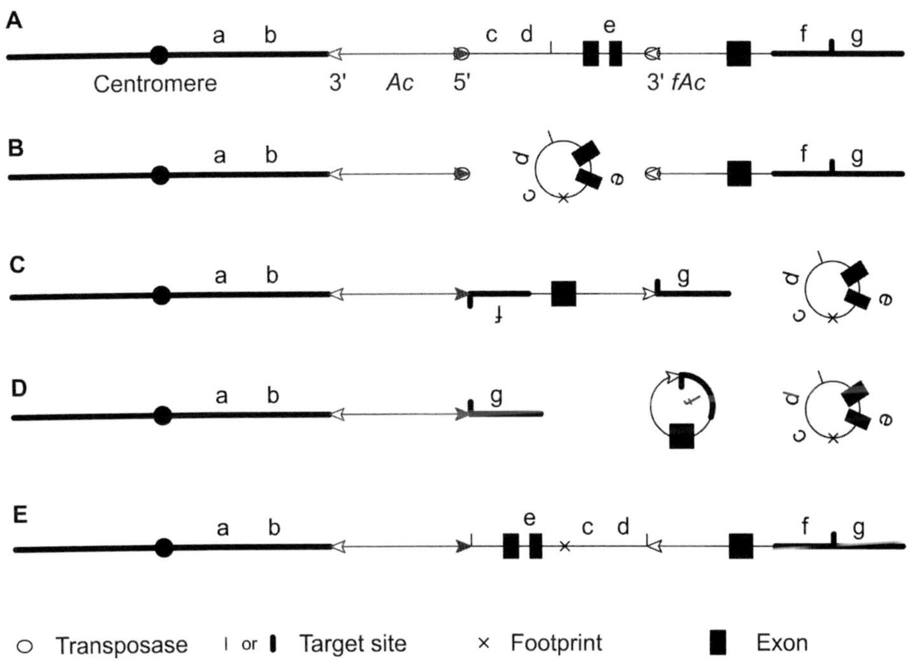

Figure 3.7 Reversed *Ac* ends transposition at the maize *p1* locus. Only one chromatid is shown because all rearrangements occur in a single chromatid. All symbols are as in Figure 3.6. (A) *Ac* transposase cleaves at the 5′ end of *Ac* and the 3′ end of *fAc*. (B) Following transposase cleavage at the junctions of *Ac/p1* and *fAc/p1*, the ends of the ITS (inter-transposon segment) are joined to form a 13-kb circle. The small x on the circle indicates a transposon footprint at the site where joining occurred. Panels C and D show the structures expected from insertion into two different target sites on the same chromatid. (C) The transposon ends insert between *f* and *g*. The *Ac* 5′ end is joined to the end adjacent to *f*, and the *fAc* 3′ end is joined to the end adjacent to *g*. The 13-kb circle is lost and segment *f* is inverted. The resulting chromosome contains an inversion of sequences from the *fAc* 3′ end to the insertion site distal to *f*. (D) The transposon ends insert between *f* and *g*. The *fAc* 3′ end is joined to the end adjacent to *f*, and the *Ac* 5′ end is joined to the end adjacent to *g*. The 13-kb circle is lost, and the segment *f* is circularized and presumed lost. The resulting chromosome contains an interstitial deletion from the *Ac* 5′ end to the insertion site distal to *f*. (E) The transposon ends insert into the 13-kb circle. The *Ac* 5′ end is joined to the end adjacent to exon 1 and the *fAc* 3′ end is joined to the other end. The 13-kb sequence is rearranged (segment *c d* and segment *e* exchange positions).

termini of the *Ac* and *fAc* elements. As shown in Figure 3.7A, the orientation of these termini (the 5′ end of *Ac* and the 3′ end of *fAc*) is reversed compared with that of a standard *Ac/Ds* element; thus, this type of transposition has been termed reversed-ends transposition. The initial step of reversed-ends transposition involves excision of the *Ac/fAc* termini and ligation of the donor site ends to generate a circular molecule composed of the inter-transposon segment (Figure 3.7B). Subsequently, insertion of the *Ac/fAc* termini into a target site on the same chromosome can generate either a flanking deletion or inversion, depending upon which side of the target site is ligated to the *Ac* or *fAc* termini (Figure 3.7C, D). Alternatively, insertion into a target site on another chromosome can generate balanced reciprocal translocation chromosomes (Figure 3.8). In all cases, one breakpoint of the rearrangement is within the *p1* locus, whereas the other breakpoints correspond to the *Ac/fAc* insertion sites. Sequence analysis confirmed this inference and also showed that the new junctions in each inversion or translocation contained identical target site duplications, indicating that these rearrangements were directly generated by the transposition reaction and not by random breakage and repair.

In some cases, the induced deletions flanking the *p1* gene extended proximally to a site within the *p2* gene, a paralog of *p1* located 70 kb upstream. Plants with these alleles produced ears with orange pericarp and cob color, indicating that these deletions somehow retained a low level of *p1* expression. Molecular analysis revealed that these alleles contained new *p2/p1* fusion genes composed of exons 1 and 2 of *p2* linked with intron 2 and exon 3 of *p1*. These examples show how alternative transposition reactions may be a source of gene creation (Zhang et al., 2006).

In addition, a few examples of a novel and particularly informative type of rearrangement were obtained from *p1-rr11* and alleles of similar structure. In this class, termed local rearrangements, there are no changes outside of the *p1* locus, but the inter-transposon segment is circularly permuted and in some cases inverted (Figure 3.7E). The new rearranged inter-transposon segment is flanked by *Ac* and *fAc* termini with identical 8-bp target site duplications, and the inter-transposon segment

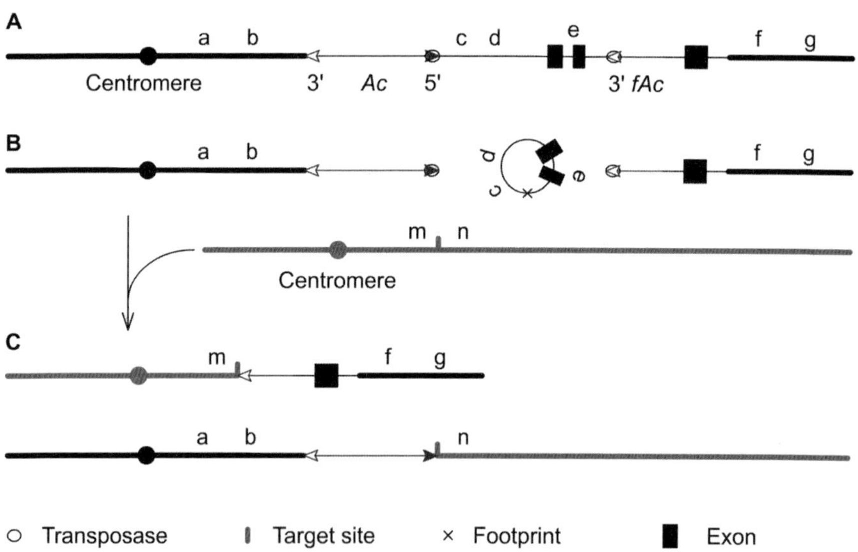

Figure 3.8 Generation of reciprocal translocations by reversed-ends transposition. All symbols are as in previous figures. (A) The *Ac* transposase binds to the *Ac* 5′ end and the reverse-oriented *fAc* 3′ end. (B) Excision of *Ac* and *fAc* termini, followed by ligation of the excision sites, results in circularization of inter-transposon segment to form a 13-kb circle, which is presumed lost. (C) Insertion of *Ac* and *fAc* termini into a target site *m/n* on a non-homologous chromosome generates a pair of balanced reciprocally translocated chromosomes.

contains the junction of the progenitor *Ac/fAc* insertion sites (Zhang and Peterson, 2004). Similar rearrangements were also isolated from a *bz1* allele with two *Ds* elements in direct orientation (Huang and Dooner, 2008). There seems to be only one possible explanation for the origin of these cases: following *Ac* and *fAc* excision, the ends of the inter-transposon segment were ligated together to form a circle, and the *Ac* and *fAc* termini inserted back into the excised inter-transposon segment circle to restore chromosome linearity. The reinsertion occurs at a new target site, resulting in permutation of the inter-transposon segment sequence. Depending upon the orientation with which the transposon ends insert into the circle, the rearranged inter-transposon segment will be in the same or inverted orientation with respect to the progenitor. Interestingly, in the case of new local rearrangement alleles derived from *p1-rr11*, their colorless pericarp exhibits very rare colored (red or orange) sectors, suggesting that a functional gene may be regenerated in subsequent events (JZ and TP, unpublished). The ability of *Ac/Ds* reversed-ends transposition to induce local rearrangements would allow researchers to repeatedly scramble *in situ* any genomic segment of interest; this may be useful for *cis*-element analysis if coupled with a functional readout.

How Does *Ds* Break Chromosomes?

McClintock identified two distinctly different "states" of *Ds* elements: State I *Ds* causes high-frequency chromosome breakage but transposes at low frequency, whereas State II *Ds* rarely causes chromosome breakage but transposes at high frequency (McClintock, 1949a). What distinguishes State I from State II elements, and how do State I elements cause chromosome breakage? Molecular cloning revealed that State II *Ds* elements are simple elements with a typical transposon structure (Kunze and Weil, 2002). Some State II *Ds* elements are derived from an *Ac* by internal deletions of various sizes, with or without the addition of "filler DNA" composed of sequences homologous to *Ac* or other genomic sequences (Yan et al., 1999; Conrad et al., 2007). Other State II *Ds* elements, such as the 405-bp *Ds1*, retain the 11-bp TIR sequences but internally have little homology to *Ac* (Kunze and Weil, 2002). In contrast, McClintock's original State I *Ds* element actually comprises two State II *Ds* elements, one inserted inside the other in opposite orientation to generate a structure termed *doubleDs*. The internal *Ds* is flanked by a typical 8-bp target site duplication, indicating that this structure arose by insertion of a State II *Ds* into a copy of itself (Doring et al., 1984). Other configurations of *Ac* and *Ds* elements can also cause chromosome breakage, including two copies of *Ds* nearby each other in either direct or inverted orientation (Weil and Wessler, 1993); *Ac* plus nearby *fAc* (Ralston et al., 1989; Yu et al., 2010); and two tightly linked *Ac/Ds* elements (Dooner and Belachew, 1991; Huang and Dooner, 2008). Together these cases might suggest that a simple excess of *Ac/Ds* termini near each other can cause breaks. However, in an elegant study, English and Jones (English et al., 1993) showed that two *Ds* termini (one 5′ and one 3′) are sufficient to induce breakage, as long as they are placed in direct orientation. By PCR and sequencing, English et al. (1995) showed that a chromatid bridge can be formed via a transposition reaction involving a *Ds* 5′ end in one chromatid and a *Ds* 3′ end from the sister chromatid; this chromatid bridge breaks in the subsequent cell division (Figure 3.9A). Weil and Wessler reached a similar conclusion by studying a pair of *Ds* elements inserted in the maize *waxy* gene (Weil and Wessler, 1993). This conclusion was confirmed by analysis of various configurations of *Ac* and *fAc* elements inserted in the maize *p1* locus: alleles in which the *fAc* 3′ end and the *Ac* 5′ end are in either direct or reversed orientations cause frequent breaks. However, alleles in which the *fAc* 3′ and *Ac* 5′ ends are in a macrotransposon configuration cause little or no breakage (Yu et al., 2010). Together, these results support a model in which breaks are induced by two specific configurations of *Ac/Ds* termini: (1) 5′ and 3′ termini in direct orientation undergo sister-chromatid transposition to form chromatid

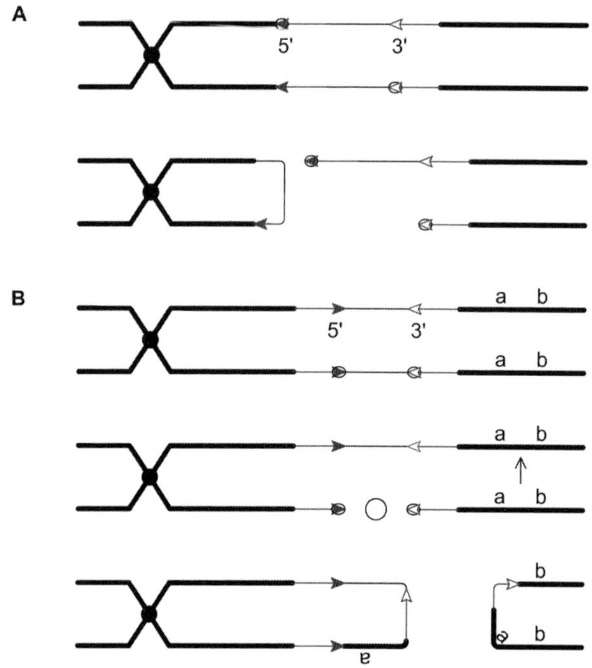

Figure 3.9 Chromosome breakage by alternative transposition. (A) In sister-chromatid transposition, excision of the 5′ and 3′ ends of the *Ds* element from sister chromatids generates a chromatid bridge that will break at a subsequent anaphase. (B) In reversed-ends transposition, excision of reverse-oriented 5′ and 3′ *Ds* termini followed by insertion into the sister chromatid generates a chromatid bridge. All symbols are as in previous figures.

bridges that subsequently break (Figure 3.9A); or (2) 5′ and 3′ termini in reversed orientation undergo reversed-ends transposition with insertion into the sister chromatid, which also generates a chromatid bridge (Figure 3.9B) (Huang and Dooner, 2008; Yu et al., 2010).

Alternative Transposition, DNA Methylation, and the Sequence of Transposition Reactions

As discussed above, Kunze and co-workers have proposed that the transposition competence of *Ac/Ds* termini is regulated by their DNA methylation state, and that termini with top-strand hemi-methylation are preferred binding sites for *Ac* transposase. This model provides simple and apparently accurate (at least in maize) rules for predicting the transposition behavior of *Ac/Ds* termini in different orientations following DNA replication. For a standard *Ac/Ds* element, the daughter element with top-strand methylation is transposition-competent (Figure 3.5A). For loci containing termini in direct orientation, the 5′ end on the upper chromatid and the 3′ end on the lower chromatid are competent, leading to sister-chromatid transposition (Figure 3.5B). For loci with termini in reversed orientation, the 5′ and 3′ termini on the same chromatid (lower) are competent, leading to reversed-ends transposition (Figure 3.5C). These rules can also be applied to cases where multiple 5′ and 3′ *Ac/Ds* termini are present at a single locus. For example, two *Ac/Ds* elements inserted nearby each other in direct orientation would have four possible transposition reactions: two standard transpositions (both of the elements on the upper chromatid); one reversed-ends transposition at the internal ends on the upper chromatid; and one macrotransposition at the two external ends, again on the upper chromatid. Two complete elements near each other in inverted orientation would also have four possible reactions: two standard transpositions (one from each chromatid) and two sister-chromatid transpositions (each utilizing one 5′ end from one chromatid and a 3′ end from the other chromatid). These rules can explain the chromosome breaking properties of various combinations of elements previously reported in the literature. For example, chromosome-breaking *doubleDs* has two sets of directly-oriented 5′ and 3′ termini, either of which can undergo sister-chromatid transposition in direct orientation; and pairs of *Ac* or *Ds* insertions at the maize *bronze1* and *waxy* loci can induce chromosome breakage whether their orientations are direct (leading to reversed-ends transposition) or inverted (leading to sister-chromatid transposition) (Dooner and Belachew, 1991; Weil and Wessler, 1993; Huang and Dooner, 2008). Whereas combinations of *Ac* and *fAc* in a macrotransposon configuration do not exhibit high-frequency chromosome breakage (Yu et al., 2010), confirming that breakage requires particular configurations of termini and not a mere multiplicity of ends.

Although these simple methylation rules are good predictors of transposition behavior in natural maize alleles, they may not always apply to elements that have been removed from maize and placed into other genomes. For example, a recent analysis of paired *Ds* elements in rice has identified a new type of alternative transposition in which directly oriented 5′ and 3′ termini *on the same chromatid* can transpose to generate flanking deletions and inversions (Xuan et al., 2011). This type of event has not been reported in maize (Zhang and Peterson, 2005); its occurrence in the rice system may reflect the fact that the particular plants assayed had been produced by a tissue culture and regeneration protocol that is known to induce extensive transposon demethylation and high transposition activity. Possibly, demethylation of the *Ds* termini may have abrogated the normal methylation-based substrate specificity of the transposase.

How far apart can two *Ac/Ds* termini be and still participate in the same transposition reaction? Not surprisingly, separation distance affects frequency: Yu et al. (2011) showed that chromosome breakage frequency increased as the distance between termini decreased. However, for very small separation distances (91 bp and 1 bp), chromosome breakage frequency was severely diminished. Interestingly, Dooner reports that chromosome breakage is still detectable for alleles with elements separated by very large physical distances (>100 kb) (Huang and Dooner, 2008). How the *Ac* transposition complex can bridge between widely spaced *Ac/Ds* termini is not known, although looping of chromosomes within the nucleus could facilitate interactions of distant elements.

Finally, it is interesting to note that the alternative transposition pathway provides some insight into the order of the cleavage and religation steps catalyzed by *Ac* transposase. For example, consider the products of sister-chromatid transposition: chromosome breakage occurs if transposon reinsertion either fails to occur or if reinsertion occurs into a target site distal to the transposon (Figure 3.9A); whereas, reciprocal deletion/duplication chromatids are formed if the transposon ends insert into a proximal target site (Figure 3.6). Note that, in order to produce a reciprocal deletion chromatid, ligation of the sequences flanking the donor site must occur *before* transposon reinsertion. If reinsertion occurred before the donor site broken ends were ligated, then one of the excised chromatid ends would be lost on an acentric fragment.

Potential Applications of Alternative Transposition

As mentioned above, the ability of sister-chromatid transposition to generate nested deletions has enabled the genetic dissection of a segment of maize chromosome 1 flanking the *p1* locus (Zhang and Peterson, 2005). In addition to deletions, alternative transposition reactions can produce a variety of rearrangements that may be useful for research and plant breeding. For example, inversions can be used to construct balancer chromosomes for genetic research (Zheng et al., 1999), whereas segmental duplications may enable increased expression of favorable genes in crop plants. The *Ac/Ds* system has already been widely applied for gene tagging in a variety of heterologous plants including *Arabidopsis* (Bancroft et al., 1992; Grevelding et al., 1992; Bancroft and Dean, 1993a, 1993b; Fedoroff and Smith, 1993; Sundaresan et al., 1995), petunia (Chuck et al., 1993), tomato (Meissner et al., 2000), and rice (Izawa et al., 1997; Chin et al., 1999). Moreover, *Ac/Ds* transposition has been demonstrated in yeast (Weil and Kunze, 2000), zebrafish, and human cells (Emelyanov et al., 2006). The fact that *Ac/Ds* elements readily undergo standard transposition in a variety of plant, animal, and fungal systems suggests that it should be feasible to reproduce alternative transposition reactions in heterologous systems. Indeed, alternative transposition and the production of genome rearrangements using transgenic constructs has already been achieved in *Arabidopsis* (Krishnaswamy et al., 2008), rice (Yu et al. 2012), and maize (TP and JZ, unpublished). In addition, existing collections of *Ac/Ds* insertion stocks can be screened for lines containing two nearby elements, and these can be an efficient source of alternative-transposition-induced rearrangements as recently demonstrated in rice (Xuan et al., 2011).

Perspective

Transposition of *Ac/Ds* elements has been studied for many years using both classical genetic and molecular approaches. More recently, transposons and the genomes they inhabit have been described with nucleotide-level precision. The results of these studies, together with insights gained from transposable element systems in other organisms, have provided a greater level of understanding of *Ac/Ds* transposition than that of any other plant transposable element. Nevertheless, major questions remain, especially regarding the auto-regulation of transposition, the composition and mechanism of assembly of transposition complexes, and the selection of insertion sites. Significant progress on these fundamental questions cannot come from computational approaches alone, but will require incisive molecular and genetic experiments that advance the investigations begun by McClintock and her colleagues more than 60 years ago.

References

Athma, P., Grotewold, E., and Peterson, T. (1992) Insertional mutagenesis of the maize *P* gene by intragenic transposition of *Ac*. *Genetics 131*: 199–209.

Bancroft, I. and Dean, C. (1993a) Transposition pattern of the maize element *Ds* in *Arabidopsis thaliana*. *Genetics 134*: 1221–1229.

Bancroft, I. and Dean, C. (1993b) Factors affecting the excision frequency of the maize transposable element *Ds* in *Arabidopsis thaliana*. *Mol. Gen. Genet. 240*: 65–72.

Bancroft, I., Bhatt, A.M., Sjodin, C., Scofield, S., Jones, J.D., and Dean, C. (1992) Development of an efficient two-element transposon tagging system in *Arabidopsis thaliana*. *Mol. Gen. Genet. 233*: 449–461.

Barclay, P.C. and Brink, R.A. (1954) The relation between *modulator and activator* in maize. *Proc. Natl. Acad. Sci. USA 40*: 1118–1126.

Becker, D., Lutticke, R., Li, M., and Starlinger, P. (1992) Control of excision frequency of maize transposable element *Ds* in Petunia protoplasts. *Proc. Natl. Acad. Sci. USA 89*: 5552–5556.

Becker, H.A. and Kunze, R. (1996) Binding sites for maize nuclear proteins in the subterminal regions of the transposable element *activator*. *Mol. Gen. Genet. 251*: 428–435.

Becker, H.A. and Kunze, R. (1997) Maize *activator* transposase has a bipartite DNA binding domain that recognizes subterminal sequences and the terminal inverted repeats. *Mol. Gen. Genet. 254*: 219–230.

Boehm, U., Heinlein, M., Behrens, U., and Kunze, R. (1995) One of three nuclear localization signals of maize *activator* (*Ac*) transposase overlaps the DNA-binding domain. *Plant J. 7*: 441–451.

Bravo-Angel, A.M., Becker, H.A., Kunze, R., Hohn, B., and Shen, W.H. (1995) The binding motifs for *Ac* transposase are absolutely required for excision of *Ds1* in maize. *Mol. Gen. Genet. 248*: 527–534.

Brink, R.A. (1954) Very light variegated pericarp in maize. *Genetics 39*: 724–740.

Brink, R.A. and Nilan, R.A. (1952) The relation between light variegated and medium variegated pericarp in maize. *Genetics 37*: 519–544.

Burr, B. and Burr, F.A. (1982) *Ds* controlling elements of maize at the *shrunken* locus are large and dissimilar insertions. *Cell 29*: 977–986.

Chatterjee, S. and Starlinger, P. (1995) The role of subterminal sites of transposable element *Ds* of *Zea mays* in excision. *Mol. Gen. Genet. 249*: 281–288.

Chen, J., Greenblatt, I.M., and Dellaporta, S.L. (1987) Transposition of *Ac* from the *P* locus of maize into unreplicated chromosomal sites. *Genetics 117*: 109–116.

Chin, H.G., Choe, M.S., Lee, S.H., Park, S.H., Koo, J.C., Kim, N.Y., Lee, J.J., Oh, B.G., Yi, G.H., Kim, S.C., et al. (1999) Molecular analysis of rice plants harboring an *Ac/Ds* transposable element-mediated gene trapping system. *Plant J. 19*: 615–623.

Chuck, G., Robbins, T., Nijjar, C., Ralston, E., Courtney-Gutterson, N., and Dooner, H.K. (1993) Tagging and cloning of a petunia flower color gene with the maize transposable element *activator*. *Plant Cell 5*: 371–378.

Conrad, L.J. and Brutnell, T.P. (2005) *Ac*-immobilized, a stable source of *activator* transposase that mediates sporophytic and gametophytic excision of *dissociation* elements in maize. *Genetics 171*: 1999–2012.

Conrad, L.J., Bai, L., Ahern, K., Dusinberre, K., Kane, D.P., and Brutnell, T.P. (2007) State II *dissociation* element formation following *activator* excision in maize. *Genetics 177*: 737–747.

Coupland, G., Baker, B., Schell, J., and Starlinger, P. (1988) Characterization of the maize transposable element *Ac* by internal deletions. *EMBO J. 7*: 3653–3659.

Coupland, G., Plum, C., Chatterjee, S., Post, A., and Starlinger, P. (1989) Sequences near the termini are required for transposition of the maize transposon *Ac* in transgenic tobacco plants. *Proc. Natl. Acad. Sci. USA 86*: 9385–9388.

Courage-Tebbe, U., Doring, H.P., Fedoroff, N., and Starlinger, P. (1983) The controlling element *Ds* at the *shrunken* locus in *Zea mays*: structure of the unstable *sh-m5933* allele and several revertants. *Cell 34*: 383–393.

Courage, U., Doring, H.P., Frommer, W.B., Kunze, R., Laird, A., Merckelbach, A., Muller-Neumann, M., Riegel, J., Starlinger, P., Tillmann, E. et al. (1984) Transposable elements *Ac* and *Ds* at the *shrunken*, *waxy*, and *alcohol dehydrogenase 1* loci in *Zea mays* L. *Cold Spring Harb. Symp. Quant. Biol. 49*: 329–338.

Dooner, H.K. and Belachew, A. (1989) Transposition pattern of the maize element *Ac* from the *bz-m2* (*Ac*) allele. *Genetics 122*: 447.

Dooner, H.K. and Belachew, A. (1991) Chromosome breakage by pairs of closely linked transposable elements of the *Ac-Ds* family in maize. *Genetics 129*: 855–862.

Doring, H.P., Tillmann, E., and Starlinger, P. (1984) DNA sequence of the maize transposable element *dissociation*. *Nature 307*: 127–130.

Doring, H.P., Nelsen-Salz, B., Garber, R., and Tillmann, E. (1989) Double *Ds* elements are involved in specific chromosome breakage. *Mol. Gen. Genet. 219*: 299–305.

Eisses, J.F., Lafoe, D., Scott, L.A., and Weil, C.F. (1997) Novel, developmentally specific control of *Ds* transposition in maize. *Mol. Gen. Genet. 256*: 158–168.

Emelyanov, A., Gao, Y., Naqvi, N.I., and Parinov, S. (2006) Trans-kingdom transposition of the maize *dissociation* element. *Genetics 174*: 1095–1104.

Emerson, R.A. (1914) The inheritance of a recurring somatic variation in variegated ears of maize. *Am. Nat. 48*: 87–115.

Emerson, R.A. (1917) Genetical studies of variegated pericarp in maize. *Genetics 2*: 1–35.

Emerson, R.A. (1929) The frequency of somatic mutation in variegated pericarp of maize. *Genetics 14*: 488–511.

English, J., Harrison, K., and Jones, J.D. (1993) A genetic analysis of DNA sequence requirements for *dissociation* state I activity in tobacco. *Plant Cell 5*: 501–514.

English, J.J., Harrison, K., and Jones, J. (1995) Aberrant transpositions of maize double *Ds*-like elements usually involve *Ds* ends on sister chromatids. *Plant Cell 7*: 1235–1247.

Essers, L., Adolphs, R.H., and Kunze, R. (2000) A highly conserved domain of the maize *activator* transposase is involved in dimerization. *Plant Cell 12*: 211.

Fedoroff, N. (1989) Maize transposable elements. In *Mobile DNA*, M. Howe and D. Berg, eds. (Washington D. C.: American Society for Microbiology), pp. 375–411.

Fedoroff, N., Wessler, S., and Shure, M. (1983) Isolation of the transposable maize controlling elements *Ac* and *Ds*. *Cell 35*: 235–242.

Fedoroff, N.V. and Smith, D.L. (1993) A versatile system for detecting transposition in *Arabidopsis*. *Plant J. 3*: 273–289.

Fusswinkel, H., Schein, S., Courage, U., Starlinger, P., and Kunze, R. (1991) Detection and abundance of mRNA and protein encoded by transposable element *activator* (*Ac*) in maize. *Mol. Gen. Genet. 225*: 186–192.

Gorbunova, V. and Levy, A.A. (1997) Circularized *Ac/Ds* transposons: formation, structure and fate. *Genetics 145*: 1161–1169.

Gorbunova, V. and Levy, A.A. (2000) Analysis of extrachromosomal *Ac/Ds* transposable elements. *Genetics 155*: 349–359.

Greenblatt, I.M. (1984) A chromosome replication pattern deduced from pericarp phenotypes resulting from movements of the transposable element, *modulator*, in maize. *Genetics 108*: 471–485.

Greenblatt, I.M. and Brink, R.A. (1962) Twin mutations in medium variegated pericarp maize. *Genetics 47*: 489–501.

Grevelding, C., Becker, D., Kunze, R., von Menges, A., Fantes, V., Schell, J., and Masterson, R. (1992) High rates of *Ac/Ds* germinal transposition in *Arabidopsis* suitable for gene isolation by insertional mutagenesis. *Proc. Natl. Acad. Sci. USA 89*: 6085–6089.

Heinlein, M. (1996) Excision patterns of *activator* (*Ac*) and *dissociation* (*Ds*) elements in *Zea mays* L.: implications for the regulation of transposition. *Genetics 144*: 1851–1869.

Heinlein, M. and Starlinger, P. (1991) Variegation patterns caused by transposable element *Ac*. *Maydica 36*: 309–316.

Heinlein, M., Brattig, T., and Kunze, R. (1994) In vivo aggregation of maize *activator* (*Ac*) transposase in nuclei of maize endosperm and Petunia protoplasts. *Plant J. 5*: 705–714.

Huang, J.T. and Dooner, H.K. (2008) Macrotransposition and other complex chromosomal restructuring in maize by closely linked transposons in direct orientation. *Plant Cell 20*: 2019–2032.

Izawa, T., Ohnishi, T., Nakano, T., Ishida, N., Enoki, H., Hashimoto, H., Itoh, K., Terada, R., Wu, C., Miyazaki, C., et al. (1997) Transposon tagging in rice. *Plant Mol. Biol. 35*: 219–229.

Keller, J., Jones, J.D., Harper, E., Lim, E., Carland, F., Ralston, E.J., and Dooner, H.K. (1993) Effects of gene dosage and sequence modification on the frequency and timing of transposition of the maize element *activator* (*Ac*) in tobacco. *Plant Mol. Biol. 21*: 157–170.

Kolkman, J.M., Conrad, L.J., Farmer, P.R., Hardeman, K., Ahern, K.R., Lewis, P.E., Sawers, R.J.H., Lebejko, S., Chomet, P., and Brutnell, T.P. (2005) Distribution of *Activator* (*Ac*) throughout the maize genome for use in regional mutagenesis. *Genetics 169*: 981–995.

Krishnaswamy, L., Zhang, J., and Peterson, T. (2008) Reversed end *Ds* element: a novel tool for chromosome engineering in Arabidopsis. *Plant Mol. Biol. 68*: 399–411.

Kunze, R. (1996) The maize transposable element *activator* (*Ac*). *Curr. Top Microbiol. Immunol. 204*: 161–194.

Kunze, R. and Starlinger, P. (1989) The putative transposase of transposable element *Ac* from *Zea mays* L. interacts with subterminal sequences of *Ac*. *EMBO J. 8*: 3177–3185.

Kunze, R. and Weil, C.F. (2002) The hAT and CACTA superfamilies of plant transposons. In *Mobile DNA II*, (Washington, D. C.: ASM Press), pp. 565–610.

Kunze, R., Stochaj, U., Laufs, J., and Starlinger, P. (1987) Transcription of transposable element *activator* (*Ac*) of *Zea mays* L. *EMBO J. 6*: 1555–1563.

Kunze, R., Behrens, U., Courage-Franzkowiak, U., Feldmar, S., Kuhn, S., and Lutticke, R. (1993) Dominant transposition-deficient mutants of maize *activator* (*Ac*) transposase. *Proc. Natl. Acad. Sci. USA 90*: 7094–7098.

Lechelt, C., Peterson, T., Laird, A., Chen, J., Dellaporta, S.L., Dennis, E., Peacock, W.J., and Starlinger, P. (1989) Isolation and molecular analysis of the maize *P* locus. *Mol. Gen. Genet. 219*: 225–234.

Martin, C. and Lister, C. (1989) Genome juggling by transposons: *Tam3*-induced rearrangements in *Antirrhinum majus*. *Dev. Genet. 10*: 438–451.

McClintock, B. (1948) Mutable loci in maize. *Carnegie Inst. Wash. Yr. Bk. 47*: 155–169.

McClintock, B. (1949a) Mutable loci in maize. *Carnegie Inst. Wash. Yr. Bk. 48*: 142–154.

McClintock, B. (1949b) Mutable loci in maize. *Carnegie Inst. Wash. Yr. Bk. 48*: 142–154.

McClintock, B. (1950a) The origin and behavior of mutable loci in maize. *Proc. Natl. Acad. Sci. USA 36*: 344–355.

McClintock, B. (1950b) Mutable loci in maize. *Carnegie Inst. Wash. Yr. Bk. 49*: 157–167.

McClintock, B. (1951) Chromosome organization and genic expression. *Cold Spring Harb. Symp. Quant. Biol. 16*: 13–47.

McClintock, B. (1952) Mutable loci in maize. *Carnegie Inst. Wash. Yr. Bk. 51*: 212–219.

McClintock, B. (1953) Induction of instability at selected loci in maize. *Genetics 38*: 579–599.

Meissner, R., Chague, V., Zhu, Q., Emmanuel, E., Elkind, Y., and Levy, A.A. (2000) Technical advance: a high throughput system for transposon tagging and promoter trapping in tomato. *Plant J. 22*: 265–274.

Raina, S., Mahalingam, R., Chen, F., and Fedoroff, N. (2002) A collection of sequenced and mapped *Ds* transposon insertion sites in *Arabidopsis thaliana*. *Plant Mol. Biol. 50*: 93–110.

Ralston, E., English, J., and Dooner, H.K. (1989) Chromosome-breaking structure in maize involving a fractured *Ac* element. *Proc. Natl. Acad. Sci. USA 86*: 9451–9455.

Rinehart, T.A., Dean, C., and Weil, C.F. (1997) Comparative analysis of non-random DNA repair following *Ac* transposon excision in maize and *Arabidopsis*. *Plant J. 12*: 1419–1427.

Rommens, C.M., van Haaren, M.J., Nijkamp, H.J., and Hille, J. (1993) Differential repair of excision gaps generated by transposable elements of the '*Ac* family'. *Bioessays 15*: 507–512.

Ros, F. and Kunze, R. (2001) Regulation of *activator/dissociation* transposition by replication and DNA methylation. *Genetics 157*: 1723–1733.

Rubin, E. and Levy, A.A. (1997) Abortive gap repair: underlying mechanism for *Ds* element formation. *Mol. Cell. Biol. 17*: 6294–6302.

Schwartz, D. (1984) Analysis of the *Ac* transposable element dosage effect in maize. *Mol. Gen. Genet. 196*: 81–84.

Schwartz, D. (1986) Analysis of the autonomous *wx-m7* transposable element mutant of maize. *Maydica 31*: 123–129.

Scofield, S.R., English, J.J., and Jones, J.D. (1993) High level expression of the *activator* transposase gene inhibits the excision of *dissociation* in tobacco cotyledons. *Cell 75*: 507–517.

Scott, L., LaFoe, D., and Weil, C.F. (1996) Adjacent sequences influence DNA repair accompanying transposon excision in maize. *Genetics 142*: 237–246.

Sundaresan, V., Springer, P., Volpe, T., Haward, S., Jones, J.D., Dean, C., Ma, H., and Martienssen, R. (1995) Patterns of gene action in plant development revealed by enhancer trap and gene trap transposable elements. *Genes. Dev. 9*: 1797–1810.

Sutton, W.D., Gerlach, W.L., Peacock, W.J., and Schwartz, D. (1984) Molecular analysis of *Ds* controlling element mutations at the *adh1* locus of maize. *Science 223*: 1265–1268.

Van Schaik, N.W. and Brink, R.A. (1959) Transpositions of *modulator*, a component of the variegated pericarp allele in maize. *Genetics 44*: 725–738.

Vollbrecht, E., Duvick, J., Schares, J.P., Ahern, K.R., Deewatthanawong, P., Xu, L., Conrad, L.J., Kikuchi, K., Kubinec, T.A., Hall, B.D., et al. (2010) Genome-wide distribution of transposed *dissociation* elements in maize. *Plant Cell 22*: 1667–1685.

Wang, L. and Kunze, R. (1998) Transposase binding site methylation in the epigenetically inactivated *Ac* derivative *Ds-cy*. *Plant J. 13*: 577–582.

Weil, C.F. and Wessler, S.R. (1993) Molecular evidence that chromosome breakage by *Ds* elements is caused by aberrant transposition. *Plant Cell 5*: 515–522.

Weil, C.F. and Kunze, R. (2000) Transposition of maize *Ac/Ds* transposable elements in the yeast *Saccharomyces cerevisiae*. *Nature Genet. 26*: 187–190.

Xiao, Y.L. and Peterson, T. (2002) *Ac* transposition is impaired by a small terminal deletion. *Mol. Genet. Genomics 266*: 720–731.

Xuan, Y.H., Piao, H.L., Je, B.I., Park, S.J., Park, S.H., Huang, J., Zhang, J.B., Peterson, T., and Han, C.-d. (2011) Transposon *Ac/Ds*-induced chromosomal rearrangements at the rice *OsRLG5* locus. *Nucleic Acids Res. 39*: e149.

Yan, X., Martinez-Ferez, I.M., Kavchok, S., and Dooner, H.K. (1999) Origination of *Ds* elements from *Ac* elements in maize: evidence for rare repair synthesis at the site of *Ac* excision. *Genetics 152*: 1733–1740.

Yu, C., Zhang, J. and Peterson, T. (2011) Genome rearrangements in maize induced by alternative transposition of reversed *Ac/Ds* termini. *Genetics 188*: 59–67.

Yu, C., Zhang, J., Han, F., Birchler, J., and Peterson, T. (2012) *Ac/Ds* transposon-induced chromosome rearrangements in transgenic rice. *Theor. Appl. Genet. 125*: 1449–1462.

Yu, C., Zhang, J., Pulletikurti, V., Weber, D.F., and Peterson, T. (2010) Spatial configuration of transposable element *Ac* termini affects their ability to induce chromosomal breakage in maize. *Plant Cell 22*: 744–754.

Zhang, J. and Peterson, T. (1999) Genome rearrangements by nonlinear transposons in maize. *Genetics 153*: 1403–1410.

Zhang, J. and Peterson, T. (2004) Transposition of reversed *Ac* element ends generates chromosome rearrangements in maize. *Genetics 167*: 1929–1937.

Zhang, J. and Peterson, T. (2005) A segmental deletion series generated by sister-chromatid transposition of *Ac* transposable elements in maize. *Genetics 171*: 333–344.

Zhang, J., Zhang, F., and Peterson, T. (2006) Transposition of reversed *Ac* element ends generates novel chimeric genes in maize. *PLoS Genet. 2*: e164.

Zheng, B., Sage, M., Cai, W.W., Thompson, D.M., Tavsanli, B.C., Cheah, Y.C., and Bradley, A. (1999) Engineering a mouse balancer chromosome. *Nat. Genet. 22*: 375–378.

Zhou, L., Mitra, R., Atkinson, P.W., Hickman, A.B., Dyda, F., and Craig, N.L. (2004) Transposition of *hAT* elements links transposable elements and V(D)J recombination. *Nature 432*: 995–1001.

4 McClintock and Epigenetics

Nina V. Fedoroff

Introduction

McClintock is widely acknowledged for her discovery of transposition, but her extraordinarily prescient work on epigenetics remains virtually unrecognized. Her work on transposon "changes of phase" (McClintock, 1955, 1956, 1957, 1958, 1971), together with that of R. A. Brink on paramutation at the maize R locus (Brink, 1956a, 1956b, 1958) and Mary Lyon on mammalian X chromosome inactivation (Lyon, 1961, 1971, 1993), presaged the now-thriving field of epigenetic regulation.

In the *Carnegie Institution Year Book 70* (McClintock, 1971), McClintock wrote:

> It is now well recognized that the genomes of most organisms have many of the same basic operational tools: the same kinds of enzymes, the same cellular structures with the same functions, and the same overall gene products required to build and maintain these structures. The differences between organisms, therefore, must reflect to a large extent differences in regulation of their genomes.

She went on to summarize her studies on what she had come to call a "change in the phase of activity" of the maize mobile elements (transposons). In her initial characterization of the *Activator* (*Ac*) element, she had noticed that the *Ac* locus, as it was then called, could undergo what she called a "change in state," which was a stable and heritable alteration in the timing and frequency of the chromosome breaks it induced at the *Dissociation* (*Ds*) locus (McClintock, 1948). It was not until her studies on the *Suppressor-mutator* element, commencing in the early 1950s, that McClintock recognized and studied a different type of heritable but reversible mechanism controlling transposon activity, today called "epigenetic" (McClintock, 1958). At the time and for many years after the importance of her discovery of transposition had been recognized, her work on epigenetics received little attention, even as it does today. In the following, I trace the genetic origins of epigenetics in McClintock's work, incorporating its molecular explication to facilitate understanding for the contemporary reader.

Spm-suppressible Alleles

McClintock named the *Suppressor-mutator* (*Spm*) transposon for its effect on certain mutant alleles of the *A1* locus, designated the *a1-m1* alleles (McClintock, 1953, 1954). Expression of the *A1* gene

Plant Transposons and Genome Dynamics in Evolution, First Edition. Edited by Nina V. Fedoroff.

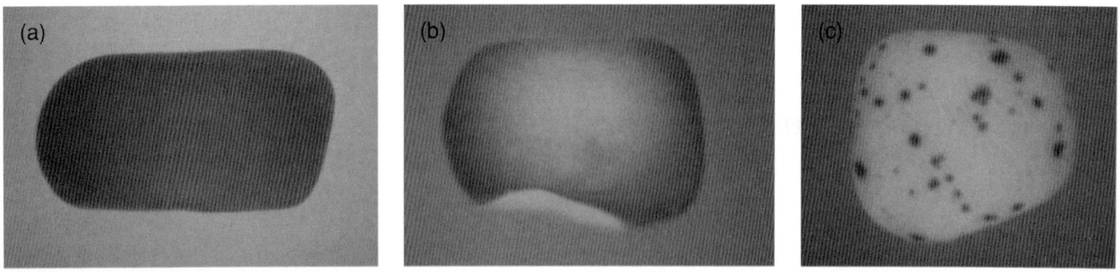

Figure 4.1 Phenotypes of kernels having the genetic constitution: (a) *a1/a1*; (b) *a1-m1/a1*, no *Spm*; (c) *a1-m1/a1*, with *Spm*. For color detail, see color plate section.

is reduced but not completely eliminated by what is now known to be an insertion of a transposition-defective *Spm* at an exon-intron boundary within the *A1* gene, one of several genes required for anthocyanin pigment production in the maize kernel's aleurone layer (Schwarz-Sommer et al., 1985; Schwarz-Sommer et al., 1987). Kernels carrying an *a1-m1* allele either homozygous or heterozygous with *a1*, a colorless null allele, but no *Spm*, are much less intensely pigmented than kernels carrying the wildtype *A1* allele (Figures 4.1a and 4.1b). Pigmentation is completely suppressed in the presence of an *Spm* element elsewhere in the genome in response to what McClintock designated as the "*suppressor*" component of *Spm*, giving a colorless kernel aleurone (Figure 4.1c). We later called this type of allele *Spm-suppressible* to distinguish it from insertions that render expression of the gene dependent on the presence of *Spm*, described below. The insertion is excised in some cells during development in response to the *mutator* component of *Spm*, restoring full *A1* gene expression in tissue sectors derived from such cells (Figure 4.1c).

Members of the series *a1-m1* alleles were derived from a single initial insertion and differed from each other in both the basal level of expression of the *A1* gene in the absence of *Spm*, as well as in the timing and frequency of reversion to the wildtype level of gene expression. The various alleles, which McClintock designated "changes of state," arose from the original isolate and were stably heritable mutations, like those she reported in the *Ac-Ds* element family (McClintock, 1956). These are now known to constitute a series of internal deletions in the transposition-defective *Spm* element (*dSpm*), itself an internally deleted *Spm* transposon (Schwarz-Sommer et al., 1985; Schwarz-Sommer et al., 1987). Subsequent studies on *Spm* revealed that the transposon-encoded TnpA protein binds strongly to the ends of the element (Masson et al., 1991; Raina et al., 1993; Raina and Fedoroff, 1995; Raina et al., 1998). Thus it appears a reasonable interpretation that the *A1* gene is transcribed, but that the transcript is either inefficiently spliced or unstable, giving a low level of *A1* gene expression in the absence of a *trans*-acting *Spm*, while transcription is blocked by TnpA binding to the insertion in the presence of an *Spm* element elsewhere in the genome.

McClintock subsequently identified similar mutant alleles of the *A2* locus (McClintock, 1957). However, her analysis of these alleles was initially confused by the occurrence of an alteration in the trans-acting *Spm* element in these cultures that changed its "phase of activity," as well as introduced a "change in state" of the transposition-defective element at the locus, rendering it unable to excise (McClintock, 1958). Both of these alterations proved extremely important in understanding the changes in expression of the *Spm* element, as well as the communications between elements (McClintock, 1958). In the *Carnegie Institution of Washington Year Book 57*, McClintock wrote:

> By means of various kinds of experiment with this state, it was first learned that the *Spm* element in the *a2-m1* cultures may undergo frequent changes in activity during the development of a plant, each

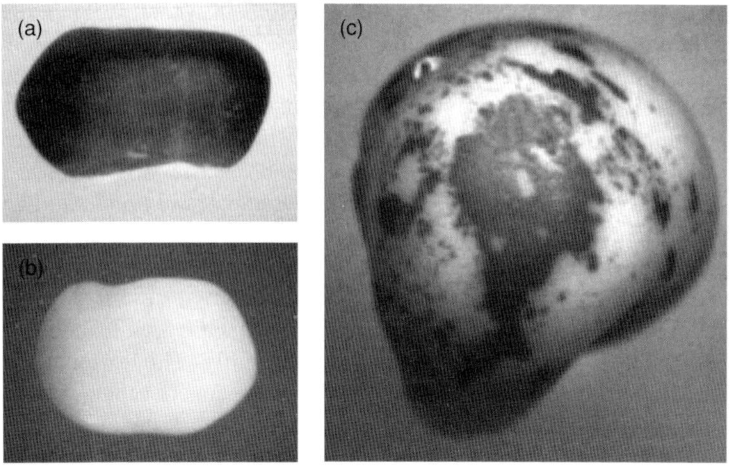

Figure 4.2 Phenotypes of kernels having the genetic constitution *a2-m1* [class II)/a2: (a) without *Spm*; (b) with a fully active *Spm*; (c) with an *Spm-c*]. For color detail, see color plate section.

> such change affecting its capacity to serve as a suppressor-mutator. Clearly, some regulatory mechanism controls the time of occurrence of such changes, although it is not yet understood. . . . The class II state of *a2-m1* readily reveals these changes in action capacity of *Spm*, for, with this state, the *Spm* element in its active phase serves only to inhibit expression of gene action at *a2-m1*.

The kernels in Figure 4.2 illustrate McClintock's observations. Kernels having the genetic constitution *a2-m1* (class II)/*a2* (*a2* is a stable recessive allele) are intensely pigmented in the absence of an *Spm* element (Figure 4.2a) and colorless in the presence of a fully active *Spm* (Figure 4.2b). The *a2-m1* allele does not exhibit the pigmented sectors observed with transposition-competent elements that result from their excision during kernel development. However, kernels of this genetic constitution vividly reveal changes in expression of the *Spm* transposon itself (Figure 4.2c). McClintock wrote:

> Changes in *Spm* action phase may alternate, and both the times and the types of change are revealed in the kernel phenotypes. In kernels having one *Spm* element, these alternating changes may be observed readily. For example, a large pigmented area may be seen in an otherwise colorless region of a kernel. Within this large pigmented area, smaller colorless areas may be observed and within these, in turn, specks of deep pigmentation. In this illustration, the sequence of changes of phase of *Spm* activity during development of the kernel was from active to inactive to active, and again to inactive.

The variegation pattern McClintock observed with a single *Spm* undergoing changes in its phase of activity during development, which I later designated a *cycling Spm* (*Spm-c*), is shown in Figure 4.2c (Fedoroff, 1983).

McClintock quickly discovered that the kernel variegation pattern depends on the number of *Spm-c* elements present. By introducing the *Spm-c* transposon through the male, the female, or both, she was able to observe the patterns produced with 1, 2, or 3 elements in the triploid endosperm. She found that the deeply pigmented areas corresponding to the element in an inactive phase were progressively smaller as the number of *Spm-c* copies increased. This implies that the deeply pigmented inactive-*Spm* phenotype requires all of the resident elements to be inactive, while the colorless active-*Spm* phenotype obtains if any one of the elements is active. This, in turn, implies that an active element produces a "suppressor" that acts in *trans* on the transposition-defective element inserted at the *A2* locus to suppress expression of the *A2* gene. It is now known that it is the TnpA protein encoded by the *Spm* transposon that binds to the element ends (Raina et al., 1998).

McClintock had by this time observed that some plants that showed no evidence of an active *Spm* in the plant or in the ear produced on the main stalk could occasionally produce a few kernels on tiller ears that have an active *Spm* phenotype. Moreover, the inactive *Spm* could be inherited in the inactive phase through many generations. She then discovered that when an *Spm-c* was introduced into a plant carrying such an inactive *Spm*, the kernels that received both the inactive *Spm* and the *Spm-c* exhibited the variegation pattern expected for multiple copies of an *Spm-c*. She inferred that the active element could reactivate the inactive element and that the *Spm-c* and the reactivated *Spm-i* then underwent inactivation during development independently. This observation suggests that a transposon-encoded protein could act *in trans* to activate the inactive one. McClintock's analysis of the *a1-m2* alleles, described below, was also consistent with the existence of such a positive, transposon-encoded regulator of transposon gene expression.

Spm-dependent Alleles

The second *Spm* insertion mutation in the *A1* locus that McClintock identified, the *a1-m2* allele, was markedly different from the *Spm-suppressible a1-m1* and *a2-m1* alleles (McClintock, 1961). McClintock reported that a fully functional *Spm* was inserted very close to the *A1* gene in the *a1-m2* allele, but that it did not completely eliminate gene expression. However, she found that the kernels were completely unpigmented when the resident *Spm* was in the inactive phase (McClintock, 1961). McClintock then identified a number of derivatives that exhibited what she called a "two-element system of control of gene action," in which a transposition-defective element capable of being mobilized *in trans* remained at the locus (McClintock, 1962). The phenotypes of one such allele, the *a1-m2 7995* allele, are shown in Figure 4.3. By contrast to the behavior of the *Spm-suppresssible* alleles, pigmentation resulting from expression of the *A* gene is not observed in the absence (Figure 4.3a), but is observed in the presence, of an active *Spm* elsewhere in the genome (Figure 4.3b). Deeply pigmented sectors signaling excision of the resident transposition-defective *Spm* appear on a uniform, less intensely pigmented background in the presence of a fully active *Spm* (Figure 4.3b).

In subsequent years, McClintock collected a number of similar derivatives differing in the level of *A1* gene expression and the pattern of somatic reversion (McClintock, 1963). We designated these alleles *Spm-dependent* in view of the fact that *A1* gene expression is observed only in the presence

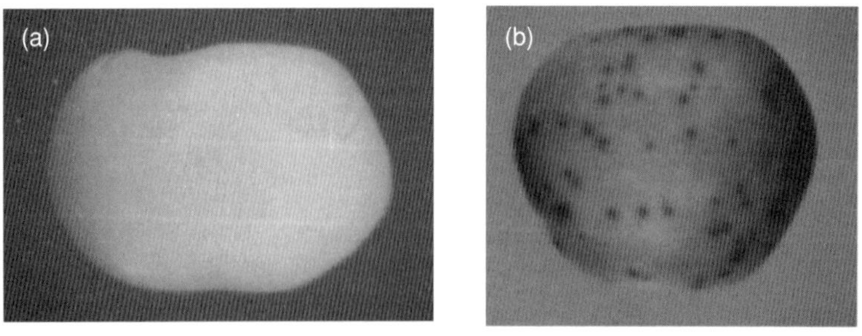

Figure 4.3 Phenotypes of the *Spm-dependent a1-m2 7995* allele in the absence (a) and presence (b) of an active *Spm*.

of an *Spm* elsewhere in the genome (Masson et al., 1987). All but one of the *Spm-dependent* alleles was found to harbor an internally deleted *Spm* element at the same insertion site in the promoter of the *A1* gene (Masson et al., 1987; Schwarz-Sommer et al., 1987). The *Spm-dependent* expression of the *A* gene in these alleles suggests that the gene has come under the control of the *Spm* element's regulatory system and that the transposon encodes a trans-acting positive regulatory factor.

Cloning and molecular analysis of the *Spm* revealed it to be an 8.3-kb sequence with a single transcription unit that gives rise to a number of alternatively spliced transcripts (Figure 4.4) (Masson et al., 1989; Masson et al., 1991; Masson et al., 1987). Two proteins, a 68-kd protein encoded by the *TnpA* transcript and a 131-kD protein encoded by the *TnpD* transcript, were found to be required for transposition based on cDNA-mediated excision of the insertion from a *dSpm*-disrupted β-glucuronidase gene in tobacco cells (Masson et al., 1991; Masson et al., 1988). The two transcripts are derived by alternative splicing from the same primary transcript and the larger protein, TnpD, is encoded entirely within the long first intron of the smaller and more abundant *TnpA* transcript (Figure 4.4). TnpD appears to be the transposase, as judged by the presence of a characteristic transposase DDE motif (Yuan and Wessler, 2011), whereas TnpA is a multifunctional regulatory protein and participates in transposition, as well.

In addition to its 13-pb terminal inverted repeats (TIR, CACTACAAGAAAA), the *Spm* has subterminal sequences of several hundred nucleotides at each end characterized by the presence of multiple copies of highly homologous direct and inverted 14-bp subterminal repeats (SR, AACC-GACACTCTTA; Figure 4.4). Subsequent studies revealed that the *Spm* promoter is coextensive with the subterminal, internally repetitive sequence immediately adjacent to the transposon's 5′ terminal repeat (Raina et al., 1993). Moreover, the 14-bp SR are the binding sites for the element-encoded TnpA protein, which is a transcriptional activator of the methylated promoter and a repressor of the unmethylated promoter (Raina et al., 1993; Schlappi et al., 1994). They are present in both orientations and in multiple copies (9 and 15 in the 5′ and 3′ ends of the transposon, respectively),

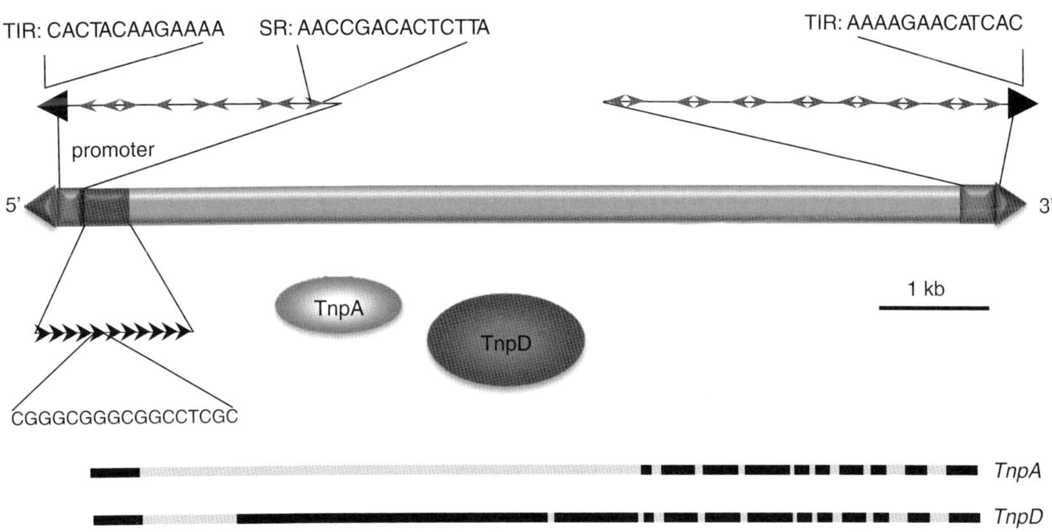

Figure 4.4 Structure of the *Spm* transposon and the *TnpA* and *TnpD* transcripts encoding the TnpA and TnpD proteins required for transposition. TnpA is a 68-kD protein with a dimerization domain and TnpD is a 131-kD protein with a DDE transposase motif and is encoded within the first intron of the TnpA transcript. The bottom diagrams represent the *Spm* transcipts, with exons represented in black and introns represented in grey.

hence the ends of the *Spm* transposon are bidirectional TnpA-dependent promoters (Raina et al., 1993). The ability of a trans-acting element to activate expression of the *A1* gene in the *a1-m2* alleles reflects the operation of the transposon's autoregulatory mechanism.

Cryptic *Spm*

Molecular analysis of the several *a1-m2* alleles in McClintock's collection identified one that appeared to have an intact element at the locus, designated *a1-m2 8167B* (Masson et al., 1987). As noted, the results of McClintock's studies on the interaction between active and inactive elements at the *A2* locus, described above, implied that an active *Spm* could *trans*-activate an inactive one. However, she reported that the activation was transient and that the inactive element reverted to the inactive form when segregated away from the active *Spm* (McClintock, 1959). Our subsequent studies showed that an active element could heritably activate an inactive one (Fedoroff, 1989), and we were able to reactivate the resident *Spm* at the *A* locus in the *a1-m2 8167B* allele by maintaining it in the presence of an active *Spm* for several generations (Banks et al., 1988). Activation was paralleled by demethylation of the promoter and GC-rich first exon sequence and was observed only in the presence of a trans-acting *Spm*. No spontaneous activation of this element was observed among several hundred thousand control kernels carrying the *a1-m2 8167B* allele produced on plants lacking an active *Spm*. We therefore designated an *Spm* in the very stably inactive state, such as the one in the *a1-m2 8167B* allele, a *cryptic Spm*. The clear implication of these studies is that an *Spm*-encoded gene product can promote the heritable activation of an epigenetically inactive element over several generations.

Presetting

Some of the *a1-m2* alleles had an odd property that McClintock called "presetting" (McClintock, 1963, 1964). Such alleles showed no *A1* gene expression when consistently maintained in the absence of an active *Spm* element, as shown in Figure 4.5a. But in some kernels on ears in which the *Spm* is newly separated from the insertion allele by meiotic segregation, some kernels exhibit continued, albeit irregular, expression of the *A1* gene (Figure 4.5b). McClintock inferred that the *Spm* element in some way "presets" the gene containing the transposition-defective element to continue expressing in its absence. The likely explanation for this phenomenon leads back to the epigenetic mechanism that underlies the "change of activity phase" of the transposon itself.

Molecular analysis of active, inactive, and cycling *Spm* elements has established that the transposon's phase of activity reflects the extent of methylation of the promoter and the internally repetitive

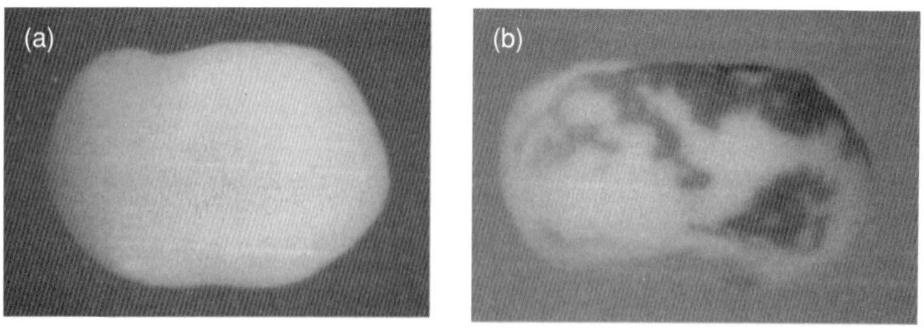

Figure 4.5 (a) The phenotype of the *Spm-dependent a1-m2 7995* allele in the long-term absence of an active *Spm*; (b) the "preset" pattern observed just after meiotic segregation of the *Spm* element. For color detail, see color plate section.

GC-rich sequence of the first exon (Banks and Fedoroff, 1989; Fedoroff and Banks, 1988). At one extreme, the deeply inactive element in the *a1-m2 8167B* allele is extensively methylated; at the other extreme, a fully active element is unmethylated in this region, although the remainder of the element is methylated irrespective of its activity phase (Banks et al., 1988). Elements exhibiting an intermediate level of methylation in the GC-rich first exon sequence can also show regular and heritable patterns in the timing or pattern of activity, although these can also readily give rise to new, heritable patterns of element expression (Banks and Fedoroff, 1989). As noted above, an element in the stably inactive state we called *cryptic*, such as the one in the *a1-m2 8167B* allele, does not undergo spontaneous activation, but can be reactivated by an active *Spm* over several generations (Banks and Fedoroff, 1989; Fedoroff and Banks, 1988). The active state is also highly stable, although less so than the *cryptic* state and heritably inactive derivatives can be selected over several generations, exhibiting increasing methylation of the GC-rich sequence in each generation (Banks and Fedoroff, 1989). *Spm* transposons with intermediate levels of methylation are inactive in the absence of a trans-acting element, but fully active in the presence of an active *Spm*. Strikingly, such elements can remain transiently active immediately after segregation of the *trans*-activating *Spm*, suggesting an explanation for presetting (Fedoroff, 1989).

The *Spm* transposon becomes reversibly inactivated in transgenic tobacco plants (Schlappi et al., 1993). Further dissection of the sequence requirements for inactivation showed that they are confined to a short sequence at the transposon's 5' end. A reporter gene expressed from the *Spm* promoter is inactivated and methylated if the promoter sequence includes the GC-rich sequence in the first exon, but not if it lacks the sequence (Schlappi et al., 1994). It follows that the first exon sequence, with its 11 closely spaced repeats of a highly GC-rich sequence (Figure 4.4), targets *Spm* for epigenetic silencing. Thus the transposon carries its own internal inactivation signal.

Of the two transposon-encoded proteins necessary for transposition, TnpA and TnpD (Masson et al., 1991), only TnpA is necessary to reactivate an inactive *Spm* in transgenic tobacco. Moreover, the methylated and inactive promoter can be reactivated and demethylated in the presence of TnpA (Schlappi et al., 1994). Subsequent studies showed that TnpA promotes active demethylation of the promoter (Cui and Fedoroff, 2002). It should be noted that the sequences that undergo methylation and demethylation are both within the promoter, as each TnpA binding site contains 3-7 C residues, and downstream from it in the GC-rich repetitive region. It is therefore a reasonable conjecture that TnpA itself recruits demethylating enzymes to the promoter.

In view of these findings, the likely explanation for presetting is that the transposon sequence inserted just upstream from the *A1* gene in the *a1-m2* alleles does not itself prevent expression of the gene, but is readily methylated, inactivating the promoter. The presence of a *trans*-acting *Spm* supplies TnpA, which recruits the demethylation machinery to the gene, allowing its expression. However, upon removal of the *Spm* by genetic segregation, the remaining transposition-defective element in the *A1* gene's promoter is not instantly remethylated, allowing its transient "preset" expression in some aleurone cells of kernels that do not receive an active *Spm*, as illustrated in Figure 4.5b.

Molecular Machinery of Epigenetic Regulation

Much progress has been made in recent years toward understanding the molecular basis of epigenetic silencing. It is now known that although histone modification and DNA methylation stabilize silencing and inactivation of genes and other genetic elements in many organisms, a genetic feedback mechanism based on small, non-coding RNAs is at the heart of epigenetic regulation (Meyer, 2011;

Simon and Meyers, 2011; Zhang and Zhu, 2011). Transposon silencing in particular is the result of RNA-directed DNA methylation (RdDM), a complex process that involves two unique plant RNA polymerases, pol IV and pol V, is mediated by 24-nt siRNAs, and results in both DNA methylation and histone modifications that suppress transcription (Haag and Pikaard, 2011; Zhang and Zhu, 2011). In brief, RdDM is initiated by pol IV-directed transcripts that are converted to RNA duplexes by RNA-DEPENDENT RNA POLYMERASE 2 (RDR2); the duplexes are then cleaved into 24-nt siRNAs by the RNAse III-family enzyme DICER-LIKE 3 (DCL3). These are stabilized by 3′-terminal methylation by HUA-ENHANCER 1 (HEN1) and the appropriate strand associates with ARGONAUTE 4 (AGO4) protein and is incorporated into an RNA-induced silencing complex (RISC). Pol V transcripts of the target locus bind to AGO4 through base-pairing with the siRNA in a large silencing complex, within which the RDM1 (REQUIRED FOR DNA METHYLATION 1) protein interacts with both AGO4 and the *de novo* cytosine methyltransferase DRM2 (DOMAINS REARRANGED METHYLTRANSFERASE 2), which in turn triggers local DNA methylation (Haag and Pikaard, 2011; Zhang and Zhu, 2011). As well, activation of RdDM promotes histone deacetylation and methylation changes that lead to the establishment of chromatin structures that repress transcription (Haag and Pikaard, 2011; Zhang and Zhu, 2011).

Demethylation is mediated by 5-methylcytosine (5-meC) DNA glycosylase/lyases, encoded by 4 genes in *Arabidopsis*: ROS1 (REPRESSOR OF SILENCING 1), DME (DEMETER); DML2 (DEMETER-LIKE 2) and DML3 (DEMETER-LIKE 3). The DNA glycosylase activity removes the 5-meC, following which the DNA backbone is cleaved at the abasic site and repaired (Zhu, 2009).

Understanding of the molecular infrastructure involved in epigenetic regulation of transposons is growing rapidly, as is awareness that epigenetic regulation is at the heart of developmental processes (Bourc'his and Voinnet, 2010; Feng et al., 2010; He et al., 2011; Henderson and Jacobsen, 2007; Mattick, 2011). The direct involvement of small RNAs in both transcriptional and post-transcriptional gene silencing explains how these gene regulatory processes are targeted to specific sequences (Almeida and Allshire, 2005; Simon and Meyers, 2011; Zhang and Zhu, 2011). But it does not explain how transposons, genes, and other genetic elements are selected for epigenetic inactivation or how the process is initiated.

Conversely, although major epigenetic reprogramming occurs in both animal and plant development, including widespread DNA demethylation (Feng et al., 2010; He et al., 2011; Johnson and Bender, 2009), little is known about the mechanisms by which specific genes are targeted. Nor is it known how specific sequences are targeted for demethylation at specific developmental stages or in a subset of cells. Recent reports suggest that DNA demethylating enzymes may themselves target subsets of genes and that there may be RNA-binding proteins that mediate targeting of demethylation (La et al., 2011; Zheng et al., 2008).

Summary

McClintock's perceptive analysis of the "changes in the phase" of *Spm* activity constitutes one of the earliest genetic investigations of an epigenetic regulatory system, only now beginning to be understood at the molecular level. An active negative feedback regulatory loop, maintained by DNA methylation, is fundamental to keeping transposon activity in check. The *Spm* transposon carries an internal sequence in its first exon that targets the element for methylation and inactivation. Extensive methylation of the *Spm* promoter and first exon sequence gives rise to a *cryptic Spm*, a deeply silenced element that neither undergoes spontaneous reactivation nor can be readily reactivated in *trans* by an active *Spm* and which can only be excised by an active element at a low frequency.

The *Spm* transposon encodes two proteins required for transposition, one of which, TnpA, is both a positive and a negative regulatory protein. TnpA is a transcriptional activator of the methylated *Spm* promoter capable of both overriding and actively reversing the DNA methylation that maintains the transposon in a silent state. It is also a repressor of the unmethylated *Spm* promoter. This property, in turn, may reflect its transpositional role, likely to be binding to the multiple TnpA binding sites at both element ends to bring them into close juxtaposition for transposition.

As a consequence of its regulatory characteristics, *Spm* transposons tend toward one of two highly heritable states, active or cryptic. Cryptic elements are extremely difficult to extricate from inactivity and even difficult to mobilize by providing a source of transposase in the form of an active *Spm* elsewhere in the genome. It is likely that the stability of the cryptic state is mediated by siRNAs and enforced by the high density of methylatable C residues in the promoter and downstream GC-rich repeats. By contrast, once activated and demethylated, *Spm* transposons stay active by virtue of a positive autoregulatory loop in which the transposon-encoded TnpA protein promotes the demethylation of its own promoter and adjacent GC-rich first exon. What remains enigmatic is how and why cryptic transposons can be activated by a variety of abiotic stresses and conditions that disrupt development, such as tissue culture and the chromosome breakage (Lisch, 2009).

References

Almeida, R., and Allshire, R.C. (2005) RNA silencing and genome regulation. *Trends Cell Biol. 15*: 251–258.

Banks, J.A., and Fedoroff, N. (1989) Patterns of developmental and heritable change in methylation of the *Suppressor-mutator* transposable element. *Dev. Genet. 10*: 425–437.

Banks, J.A., Masson, P., and Fedoroff, N. (1988) Molecular mechanisms in the developmental regulation of the maize *Suppressor-mutator* transposable element. *Genes Dev. 2*: 1364–1380.

Bourc'his, D., and Voinnet, O. (2010) A small-RNA perspective on gametogenesis, fertilization, and early zygotic development. *Science 330*: 617–622.

Brink, R.A. (1956a) A genetic change associated with the *R* locus in maize which is directed and potentially reversible. *Genetics 41*: 872–889.

Brink, R.A. (1956b) A regularly reversible change in determinative action at the *R* locus in maize. *Genetics 41*: 636.

Brink, R.A. (1958) Paramutation at the *R* locus in maize. *Cold Spring Harb. Symp. Quant. Biol. 23*: 379–391.

Cui, H., and Fedoroff, N.V. (2002) Inducible DNA demethylation mediated by the maize *Suppressor-mutator* transposon-encoded TnpA protein. *Plant Cell 14*: 2883–2899.

Fedoroff, N.V. (1983) Controlling elements in maize. In *Mobile Genetic Elements*, J. Shapiro, ed. (New York: Academic Press), pp. 1–63.

Fedoroff, N.V. (1989) The heritable activation of cryptic *Suppressor-mutator* elements by an active element. *Genetics 121*: 591–608.

Fedoroff, N.V., and Banks, J.A. (1988) Is the *Suppressor-mutator* element controlled by a basic developmental regulatory mechanism? *Genetics 120*: 559–577.

Feng, S., Jacobsen, S.E., and Reik, W. (2010) Epigenetic reprogramming in plant and animal development. *Science 330*: 622–627.

Haag, J.R., and Pikaard, C.S. (2011) Multisubunit RNA polymerases IV and V: Purveyors of non-coding RNA for plant gene silencing. Nature reviews. *Mol. Cell Biol. 12*: 483–492.

He, G., Elling, A.A., and Deng, X.W. (2011) The epigenome and plant development. *Annu. Rev. Plant. Biol. 62*: 411–435.

Henderson, I.R., and Jacobsen, S.E. (2007) Epigenetic inheritance in plants. *Nature 447*: 418–424.

Johnson, M.A., and Bender, J. (2009) Reprogramming the epigenome during germline and seed development. *Genome Biol. 10*: 232.

La, H., Ding, B., Mishra, G.P., Zhou, B., Yang, H., Bellizzi Mdel, R., Chen, S., Meyers, B.C., Peng, Z., Zhu, J.K., et al. (2011) A 5-methylcytosine DNA glycosylase/lyase demethylates the retrotransposon *Tos17* and promotes its transposition in rice. *Proc. Natl. Acad. Sci. USA 108*: 15498–15503.

Lisch, D. (2009) Epigenetic regulation of transposable elements in plants. *Annu. Rev. Plant Biol. 60*: 43–66.

Lyon, M. (1961) Gene action in the X-chromosome of the mouse. *Nature 190*: 372–373.

Lyon, M.F. (1971) Possible mechanisms of X chromosome inactivation. *Nature New Biol. 232*: 229–232.

Lyon, M.F. (1993) Epigenetic inheritance in mammals. *Trends Genet. 9*: 123–128.

Masson, P., Rutherford, G., Banks, J.A., and Fedoroff, N. (1989) Essential large transcripts of the maize *Spm* transposable element are generated by alternative splicing. *Cell 58*: 755–765.

Masson, P., Strem, M., and Fedoroff, N. (1991) The *tnpA* and *tnpD* gene products of the *Spm* element are required for transposition in tobacco. *Plant Cell 3*: 73–85.

Masson, P., Surosky, R., Kingsbury, J.A., and Fedoroff, N.V. (1987) Genetic and molecular analysis of the *Spm*-dependent *a-m2* alleles of the maize *a* locus. *Genetics 117*: 117–137.

Masson, P., Toohey, K., and Fedoroff, N. (1988) Excision of *Spm* in tobacco. *Maize Genet. Coop. Newslet. 62*: 26–27.

Mattick, J.S. (2011) The central role of RNA in human development and cognition. *FEBS Letters 585*: 1600–1616.

McClintock, B. (1948) Mutable loci in maize. *Carnegie Inst. Wash. Yr. Bk. 47*: 155–169.

McClintock, B. (1953) Mutation in maize. *Carnegie Inst. Wash. Yr. Bk. 52*: 227–237.

McClintock, B. (1954) Mutations in maize and chromosomal aberrations in *Neurospora*. *Carnegie Inst. Wash. Yr. Bk. 53*: 254–260.

McClintock, B. (1955) Controlled mutation in maize. *Carnegie Inst. Wash. Yr. Bk. 54*: 245–255.

McClintock, B. (1956) Intranuclear systems controlling gene action and mutation. *Brookhaven Symp. Biol. 8*: 58–74.

McClintock, B. (1957) Genetic and cytological studies of maize. *Carnegie Inst. Wash. Yr. Bk. 56*: 393–401.

McClintock, B. (1958) The *suppressor-mutator* system of control of gene action in maize. *Carnegie Inst. Wash. Yr. Bk. 57*: 415–429.

McClintock, B. (1959) Genetic and cytological studies of maize. *Carnegie Inst. Wash. Yr. Bk. 58*: 452–456.

McClintock, B. (1961) Further studies of the *suppressor-mutator* system of control of gene action in maize. *Carnegie Inst. Wash. Yr. Bk. 60*: 469–476.

McClintock, B. (1962) Topographical relations between elements of control systems in maize. *Carnegie Inst. Wash. Yr. Bk. 61*: 448–461.

McClintock, B. (1963) Further studies of gene-control systems in maize. *Carnegie Inst. Wash. Yr. Bk. 62*: 486–493.

McClintock, B. (1964) Aspects of gene regulation in maize. *Carnegie Inst. Wash. Yr. Bk. 63*: 592–602.

McClintock, B. (1971) The contribution of one component of a control system to versatility of gene expression. *Carnegie Inst. Wash. Yr. Bk. 70*: 5–17.

Meyer, P. (2011) DNA methylation systems and targets in plants. *FEBS Letters 585*: 2008–2015.

Raina, R., Cook, D., and Fedoroff, N. (1993) Maize *Spm* transposable element has an enhancer-insensitive promoter. *Proc. Natl. Acad. Sci. USA 90*: 6355–6359.

Raina, R., and Fedoroff, N. (1995) The role of TnpA and TnpD in transposition of *Spm*. *Maize Genet. Coop. Newslet. 69*: 13–15.

Raina, R., Schlappi, M., Karunanandaa, B., Elhofy, A., and Fedoroff, N. (1998) Concerted formation of macromolecular *Suppressor-mutator* transposition complexes. *Proc. Natl. Acad. Sci. USA 95*: 8526–8531.

Schlappi, M., Raina, R., and Fedoroff, N. (1994) Epigenetic regulation of the maize *Spm* transposable element: novel activation of a methylated promoter by TnpA. *Cell 77*: 427–437.

Schlappi, M., Smith, D., and Fedoroff, N. (1993) TnpA trans-activates methylated maize *Suppressor-mutator* transposable elements in transgenic tobacco. *Genetics 133*: 1009–1021.

Schwarz-Sommer, Z., Gierl, A., Berndtgen, R., and Saedler, H. (1985) Sequence comparison of 'states' of *a1-m1* suggest a model of *Spm* (*En*) action. *The EMBO J. 4*: 2439–2443.

Schwarz-Sommer, Z., Shepherd, N., Tacke, E., Gierl, A., Rohde, W., Leclercq, L., Mattes, M., Berndtgen, R., Peterson, P.A., and Saedler, H. (1987) Influence of transposable elements on the structure and function of the *A1* gene of *Zea mays*. *EMBO J. 6*: 287–294.

Simon, S.A., and Meyers, B.C. (2011) Small RNA-mediated epigenetic modifications in plants. *Curr. Op. Plant Biol. 14*: 148–155.

Yuan, Y.W., and Wessler, S.R. (2011) The catalytic domain of all eukaryotic cut-and-paste transposase superfamilies. *Proc. Natl. Acad. Sci. USA 108*: 7884–7889.

Zhang, H., and Zhu, J.K. (2011) RNA-directed DNA methylation. *Curr. Opin. Plant Biol. 14*: 142–147.

Zheng, X., Pontes, O., Zhu, J., Miki, D., Zhang, F., Li, W.X., Iida, K., Kapoor, A., Pikaard, C.S., and Zhu, J.K. (2008) ROS3 is an RNA-binding protein required for DNA demethylation in Arabidopsis. *Nature 455*: 1259–1262.

Zhu, J.K. (2009) Active DNA demethylation mediated by DNA glycosylases. *Annu. Rev. Genet. 43*: 143–166.

5 Molecular Mechanisms of Transposon Epigenetic Regulation

Robert A. Martienssen and Vicki L. Chandler

Introduction

In most eukaryotes, ancient transposable elements (TEs) provide chromosomes with pericentromeric and subtelomeric sequences, but recent insertions and increases in copy number are detrimental to the host. As a result, TEs are the targets of epigenetic silencing by chromatin remodeling and modification. Chromatin modification is guided by DNA replication and RNA interference, leading to symmetric and asymmetric DNA methylation, respectively. These mechanisms have overlapping targets, and can substitute for one another. RNA interference depends on the plant-specific RNA polymerases PolIV and PolV, and on the production of small interfering RNA by Argonaute, Dicer, and RNA-dependent RNA polymerases. Reprogramming of small RNA in pollen and ovules contributes to TE silencing, and mutants in RNA interference have meiotic defects and unreduced gametes. The coevolution of silencing and meiosis may reflect a requirement for recombination in the amplification of TEs in sexually reproducing organisms. An unexpected consequence of these epigenetic mechanisms is that, when TEs integrate in regulatory regions, they recruit the silencing apparatus, resulting in epigenetic regulation of genes. Paramutation is an example of transgenerational epigenetic silencing that involves these same silencing mechanisms, suggesting that paramutation and TE regulation are closely related.

Epigenetic mechanisms promote stable changes in gene expression that do not involve changes in DNA sequence. When these mechanisms act at the chromosomal level, epigenetic changes are inherited through the cell cycle along with the chromosome and are sometimes passed from one generation to the next. Epigenetic control is one of the most highly conserved properties of transposable elements and helped define transposons when they were first discovered in maize. Cycling between active and inactive states, as well as presetting of transposon-regulated gene expression, are both examples of epigenetic silencing that helped define the *Suppressor-Mutator* (*Spm*) transposon (Chapter 4 in this volume).

The molecular basis for many aspects of epigenetic regulation is now much better understood, and the first mechanism to be widely recognized was DNA methylation. For example, silencing of *Ac* and *Spm* elements (cycling) is correlated with DNA methylation (see Chapter 4 in this volume). Similarly, DNA methylation was associated with reduced transposition of nonautonomous copies of the Robertson's Mutator transposon family (Chandler and Walbot, 1986), as well as with transcriptional silencing of autonomous copies (Martienssen and Baron, 1994). In this case, loss

Plant Transposons and Genome Dynamics in Evolution, First Edition. Edited by Nina V. Fedoroff.

of autonomous Mutator activity was accompanied by simultaneous methylation of hundreds of nonautonomous elements in individual meristematic cells, visible as clones of cells with altered phenotypes (Martienssen and Baron, 1994). These phenotypes resulted from gene activation rather than repression when transposons in the promoter, untranslated leader, or intron were silenced by methylation (Chapter 6 in this volume).

This led to the idea that repressive chromatin was responsible for gene silencing, and that repression was somehow relieved when the transposons were methylated (Settles et al., 2001). More recently, histone modification has emerged as a second, more highly conserved, transposon-silencing mechanism that guides and can be guided by DNA methylation. The third broad class of mechanisms acts *in trans* as well as *in cis* and involves RNA interference (RNAi). *RNA interference* refers to the ability of double-stranded RNA to interfere with translation and transcription, usually through the production of small interfering RNA (siRNA) incorporated into Argonaute effector proteins. First found to silence transgenes and viruses, RNAi has an ancient and highly conserved role in silencing transposons and heterochromatic repeats. This is especially true in organisms that lack DNA methylation, as DNA methylation can be maintained independently of RNAi by replication in a strand-symmetric sequence context (Feng and Jacobsen, 2011; Slotkin and Martienssen, 2007).

Plants possess the full repertoire of epigenetic silencing mechanisms found in metazoans, although mechanistic differences exist and individual pathways differ. This is all the more remarkable considering that multicellularity evolved independently in plants and animals and is thought to be a *sine qua non* for polycomb silencing that occurs during cellular differentiation. In contrast, many invertebrates (including *C.elegans* and *Drosophila* spp.) lack DNA methylation, while budding yeast also lacks RNAi and many forms of histone modification, such as methylation of histone H3 lysine-9 (H3K9) and H3K27, responsible for heterochromatic and polycomb repression, respectively.

Interestingly, many of these silencing mechanisms are prominent in the germline, which in animals is a more-or-less continuous, immortal lineage set aside after the first few cell divisions in the developing embryo. In plants, the germline differentiates extremely late in development, but nonetheless possesses a richly diverse epigenetic machinery. This machinery includes germline-specific histone variants and small RNAs, although distinct classes are found in plants and animals. The germline is uniquely sensitive to transposon activation, perhaps because this is where genome reprogramming occurs to restore developmental pluripotency to the next generation. Alternatively, reprogramming itself may reflect an ancient transposon silencing mechanism. It is likely, therefore, that pathways conserved in animals and plants were responsible for silencing transposons in the first sexually reproducing eukaryotes.

Chromatin modification of DNA and histones and regulation by RNA interference act in parallel, but also influence each other, forming an interconnected network. Remarkably, these mechanisms apply equally to all classes of transposable elements, including class I transposons that move via RNA intermediates and class II retrotransposons that move via DNA intermediates, even though these transposons are only very distantly related. This is partly because these mechanisms regulate genes required for transposition, which are encoded by the elements themselves. Since they were first discovered, it was apparent that epigenetic mechanisms of gene regulation, such as imprinting and paramutation, bear a strong resemblance to transposon silencing and silencing mediated by heterochromatin. *Imprinting* refers to the expression of alleles according to their parent of origin. Gene silencing by paramutation depends on the presence of an existing silent allele *in trans*. In each case, silencing is allele specific and is chromosomally inherited. Further, silent and transcriptionally active alleles are identical in DNA sequence, leading to the concept of an "epiallele." Molecular studies have revealed that imprinting, paramutation, and transposon silencing are closely related, raising the possibility that epigenetic mechanisms of gene regulation evolved as a transposon regulatory system that was subsequently co-opted to regulate genes. Imprinting is discussed in Chapter 7 in this volume, but we will make parallels between paramutation and TE silencing at the end of this chapter.

Chromatin Remodeling, DNA and Histone Modification

Though typically heterochromatic and silent, transposable elements (TEs) can disrupt and regulate genes, and promote chromosomal rearrangements, playing a crucial role in genome structure and evolution (Biemont, 2009; Slotkin and Martienssen, 2007). To suppress this mutagenic potential, surveillance systems have evolved that target active transposons. For example, small interfering RNAs (siRNAs) target transposons for posttranscriptional silencing or for the deposition of repressive chromatin modifications via RNAi (Matzke et al., 2009; Zaratiegui et al., 2007). These modifications include DNA and histone methylation (Cheng and Blumenthal, 2010; Feng and Jacobsen, 2011; Liu et al., 2010a; Saze and Kakutani, 2011), as well as nucleosome remodeling (Clapier and Cairns, 2009).

DNA methylation is associated with TEs in vertebrates, some invertebrates, filamentous fungi, and plants (Scott and Spielman, 2006; Slotkin and Martienssen, 2007; Zemach et al., 2010; Zilberman and Henikoff, 2004). In plants, TEs are highly methylated in all contexts, while many genes have short patches of CG methylation within the transcribed region (Cokus et al., 2008; Lippman et al., 2004; Lister et al., 2008; Tran et al., 2005; Vaughn et al., 2007; X. Zhang et al., 2006; Zilberman et al., 2007). CG methylation in exons is widespread in eukaryotes, but it varies greatly in extent from 5%–10% of plant exons to more than 50% of mammalian exons, and its function is unclear. Nonetheless, DNA methylation can be maintained independently of RNAi by replication-dependent DNA methyltransferases, and can lead to transgenerational inheritance (Kankel et al., 2003; Lippman et al., 2003; Mathieu et al., 2007; Teixeira et al., 2009).

Replication-Dependent DNA Methylation

DNA METHYLTRANSFERASE1 (MET1) silences transposons by methylation of symmetric CG dinucleotides (Kato et al., 2003; Lippman et al., 2003). MET1 and its mammalian homolog Dnmt1 maintain symmetric CG methylation after replication through recognition of hemimethylated substrates. The PHD-SRA-RING domain protein VARIANT IN METHYLATION (VIM1) interacts with hemimethylated CG and CHG symmetric dinucleotide pairs via the SRA domain. VIM1 is enriched in chromocenters, and these heterochromatic regions are decondensed in natural *vim1* variants (Woo et al., 2008; Woo et al., 2007). The homolog of VIM1 in mouse cells, UHRF1/Np95, interacts directly with Dnmt1 (Bostick et al., 2007).

The chromatin remodeling ATPase DECREASE IN DNA METHYLATION1 (DDM1) was first identified because mutants of *Arabidopsis* lost DNA methylation from centromeric satellite repeats (Jeddeloh et al., 1999; Vongs et al., 1993), and lost transgene silencing (Jeddeloh et al., 1998; Mittelsten Scheid et al., 1998). DDM1 is specifically required for TE methylation (Miura et al., 2001; Singer et al., 2001), acting upstream of MET1 (Bartee and Bender, 2001; Kankel et al., 2003). Members of the Snf2 superfamily of ATP-dependent DNA and RNA helicases are capable of exchanging modified histones and of mobilizing nucleosomes along the DNA, providing access to modifying enzymes and perhaps accounting for the genetic interaction with *met1* (Kankel et al., 2003).

DDM1 is required for the silencing of the imprinted gene *FWA* (Soppe et al., 2000), presumably because the *FWA* promoter consists of a SINE element whose activity is regulated by DDM1 (Lippman et al., 2004). Mutants in *met1*, in *vim1*, and in its homolog *vim3* have similar defects in *FWA* silencing and arose in similar screens, but *met1* and *vim1vim3* mutants are late flowering, unlike *ddm1* (Jacobsen et al., 2000; Kankel et al., 2003; Woo et al., 2008). Genome-wide, *met1* and *vim1* are required for CG methylation in exons, while the effects of *ddm1* are restricted to TEs (Lippman et al., 2004).

Histone Modification and DNA Methylation

TEs are associated with specific histone modifications that account in part for the distinction between TEs and genes (Feng and Jacobsen, 2011; Saze and Kakutani, 2011). These modifications include methylation of histone H3 lysine 9 (H3K9me2) (Gendrel et al., 2002) and histone H3 lysine 27 (H3K27me1 and me2) (Jacob et al., 2009; Turck et al., 2007), as well as demethylation of H3K4 (Deleris et al., 2010). Active genes are associated with H3K4me2, promoters with H3K4me3 and H3K56Ac, and inactive genes with H3K27me3 and H3K9me3 (Tanurdzic et al., 2008; Turck et al., 2007; Zhang et al., 2009; X. Zhang et al., 2007a). DDM1 is required for the localization of heterochromatic H3K9me2 in TEs, as well as the exclusion of H3K4me2 (Lippman et al., 2004; Lippman et al., 2003). However, DDM1 is not required for overall levels of H3K9me2 or H3K4me2 (Gendrel et al., 2002), suggesting that the mechanism by which DDM1 silences transposons involves localization of these histone modifications to TEs and genes respectively (Lippman et al., 2004; Lippman et al., 2003). Redistribution of histone modifications might thus result indirectly in loss of DNA methylation.

In mammalian cells, the methyltransferase Dnmt3 has a cysteine-rich domain that binds unmodified histone H3, but is strongly inhibited by methylation of H3K4(Ooi et al., 2007). This prevents recruitment of Dnmt3 to active promoters and enhancers, which are marked with H3K4me3 and H3K4me1, respectively, in mammalian cells (Hawkins et al., 2011; Heintzman et al., 2007). Histone modifications are also predictive of DNA methylation patterns in plants (L.M. Johnson et al., 2007; X. Zhang et al., 2009). The DNA methyltransferase CHROMOMETHYLASE3 (CMT3) recognizes H3K9me2 via its chromodomain and depends on this recognition for its activity (Lindroth et al., 2004). Conversely, the H3K9 methyltransferases SUPPRESSOR OF VARIEGATION HOMOLOG4 (SUVH4)/ KRYPTONITE (KYP), SUVH5, and SUVH6 have an SRA methyl DNA-binding domain that binds preferentially to methylated CHG (Ebbs and Bender, 2006; L.M. Johnson et al., 2007). Thus, H3K9me2 and CHG methylation are mutually interdependent, and largely restricted to TEs.

Like *ddm1*, *met1* mutants lose H3K9me2 from TEs and from cytologically defined chromocenters (Rangwala and Richards, 2004; Tariq et al., 2003). This observation led to the idea that DNA methylation lies upstream of histone modification in plants (Tariq et al., 2003). Nonetheless, *met1cmt3* and *met1kyp* double mutants are much more severe phenotypically than *met1* single mutants, and it is only in double mutants that TEs are mobilized, resembling *ddm1* single mutants in this respect (Kato et al., 2003; Mirouze et al., 2009). This suggests that H3K9me2 is not lost in *met1* mutants but rather is redistributed throughout the genome (Mathieu et al., 2007), hence, *met1* mutants resemble *ddm1* mutants in this respect (Gendrel et al., 2002). The results of genome-wide bisulphite sequencing in *met1* mutants (Cokus et al., 2008; Lister et al., 2008) provide support for this idea because CG methylation in exons is replaced with CHG methylation, which is associated with H3K9me2. Similarly, a LINE element inserted in the *BONSAI* gene became hypermethylated in *ddm1* mutants, resulting in rare dwarf plants (Saze and Kakutani, 2007).

Enzymatic Loss of DNA and Histone Methylation

The foregoing observations indicate that CG and non-CG methylation interact to some extent, and can play redundant roles. An important clue to the mechanism of this interaction came from mutations that inhibited the gain of non-CG methylation at the *BONSAI* gene (*ibm* mutants). The H3K9 demethylase, IBM1, is required to prevent accumulation of ectopic H3K9me2 and the associated CHG methylation in the transcribed region of *BONSAI* and of thousands of other genes (Inagaki et al., 2010). One idea is that IBM1 and gene transcription form an autocatalytic loop that keeps transcribed genes free of marks normally associated with transposons. Conversely, CMT3 and KYP form another autocatalytic loop that maintains such marks over silent transposons. Interestingly, *ddm1* enhances the phenotype of *ibm1* mutants very substantially (Saze and Kakutani, 2007; Saze et al., 2008), indicating perhaps that DDM1 maintains these autocatalytic loops by facilitating the distribution of modified histones between genes and TEs (Lippman et al., 2004).

Consistent with this idea, histone modifications normally associated with genes, such as histone acetylation and H3K27me3, interact strongly with DNA methylation. HISTONE DEACETYLASE6 (HDA6) helps maintain CpG methylation (Aufsatz et al., 2007; Earley et al., 2010), while H3K27me3 and H3K4me2 are excluded from methylated gene bodies (Bender, 2004; Fuks, 2005; X. Zhang et al., 2009). On the other hand, H3K4me1 is found in most, if not all, methylated exons in plants (X. Zhang et al., 2009). Only some of these interactions are common to mammalian cells, reflecting perhaps the differing compositions of the modification complexes involved (Feng et al., 2010).

In plants, the homologous DNA glycosylases DEMETER (DME) and REPRESSOR OF SILENCING1 (ROS1) are responsible for demethylation via base excision repair (Morales-Ruiz et al., 2006). DME removes DNA methylation from the promoters of *MEDEA (MEA), FWA* and *FERTILIZATION INDEPENDENT SEEDS2 (FIS2)*, which are only expressed from the maternal chromosome in the central cell of the female gametophyte (Gehring et al., 2009; Gehring and Henikoff, 2007; Hsieh et al., 2009). Recently, a similar mechanism, involving deamination via APOBEC (apolipoprotein B mRNA editing enzyme, catalytic polypeptide-like) proteins, has been uncovered in vertebrates (Rai et al., 2008). Demethylation of histone H3K4me3, H3K9me3, and H3K27me3 is accomplished by related oxidoreductases that carry the jumonjiC domain, and mutations in these oxidoreductases can also result indirectly in ectopic DNA methylation (Deleris et al., 2010; Saze and Kakutani, 2007; Saze et al., 2008). The function of histone demethylation in normal development is still not clear.

RNA Interference (RNAi) and RNA-Directed DNA Methylation (RdDM)

In plants, worms, and in fission yeast, siRNA precursors are made by RNA-dependent RNA polymerase and then cleaved into 23-24nt duplexes by Dicer. siRNA duplexes are loaded onto Argonaute proteins that bind siRNA at the 5′ end, and use it to guide endonucleolytic cleavage of the passenger strand, as well as to guide complementary transcripts from TEs and repeats. siRNA guides histone methylation in plants and animals and in the fission yeast, *S. pombe* (Grewal and Elgin, 2007; Verdel and Moazed, 2005; Zaratiegui et al., 2007). In plants, siRNA can also guide DNA methylation, by a related mechanism called RNA-directed DNA methylation (RdDM). We will first discuss RNAi-guided histone modification in fission yeast, where the mechanism is best understood, and then make parallels with RdDM in plants.

RNAi-Guided Histone Modification

In *S.pombe*, Argonaute is found in the RITS (RNA initiator of transcriptional silencing) complex (Verdel et al., 2004), and binds to the GW (glycine tryptophan) repeat protein Tas3, which interacts with the chromodomain protein Chp1 (Schalch et al., 2011). The RITS complex is recruited via its chromodomain to heterochromatic histone modifications like H3K9me2, but also to the PolII accessory factor Mlo3, which is also methylated by the histone methyltransferase Clr4 (Zhang et al., 2011). Catalytic activity of Ago1 is required for both posttranscriptional and transcriptional silencing, which is mediated by the CLRC (Cryptic Locus Regulatory Complex). CLRC contains the histone-modifying enzymes Clr4 and Lid2 (which demethylates H3K4) and proteins involved in DNA repair, and is translocated along the chromosome during DNA replication by association with DNA polymerase epsilon (Li et al., 2011). The role of RNAi is to release PolII from the chromosome in order to allow progression of the replication fork (Zaratiegui et al., 2011).

At least three different factors required for DNA replication, including DNA polymerase epsilon, are also required for transgene silencing, H3K9 methylation, and suppression of homologous recombination in *Arabidopsis* (J. Liu et al., 2010b; Q. Liu et al., 2010c; Yin et al., 2009). DNA replication mutants have little effect on DNA methylation, possibly because viable alleles are weak. Also reminiscent of *S.pombe*, dsRNA from hairpins induces DNA and H3K9 methylation of target sequences via the chromodomain protein CMT3 and KRYPTONITE (SUVH4), which may play similar roles to Chp1 and Clr4 in *S.pombe* (L.M. Johnson et al., 2007). Two homologs of SUVH4 (SUVH5 and SUVH6) are further required for silencing of TEs (Ebbs and Bender, 2006), and influence DNA methylation via CMT3.

RNA-Dependent DNA Methylation

DNA methylation induced by homologs of mammalian Dnmt3, DOMAINS REARRANGED METHYASE1 (DRM1), and DRM2 is also guided by RNAi, and screens for mutants in RdDM recovered several additional genes, including *POLIV, POLV, RNA-DEPENDENT RNA POLYMERASE2 (RDR2)*, and *DEFECTIVE IN RNA DEPENDENT DNA METHYLATION1 (DRD1)* (Herr et al., 2005; Kanno et al., 2004; Onodera et al., 2005). The mechanism by which these and other factors mediate RdDM is rapidly being unveiled (Figure 5.1).

Pol IV and Pol V are plant-specific RNA polymerases required for siRNA biogenesis and RdDM, respectively, and may play the same role as Pol II in *S.pombe* (Pikaard et al., 2008; Ream et al., 2009; Wierzbicki et al., 2008). Although distinguishable by various accessory factors (Gao et al., 2010; He et al., 2009a; He et al., 2009b; Law et al., 2010; X. Zheng et al., 2008; Z. Zheng et al., 2010), PolIV and PolV share most subunits with each other and with PolII, but the two largest subunits are plant specific, and presumably confer substrate specificity to each enzyme. PolV is required for the accumulation of intergenic transcripts found in heterochromatin, indicating that heterochromatic DNA is the likely template (Haag et al., 2009; Pikaard et al., 2008; Pontes et al., 2009; Wierzbicki et al., 2008).

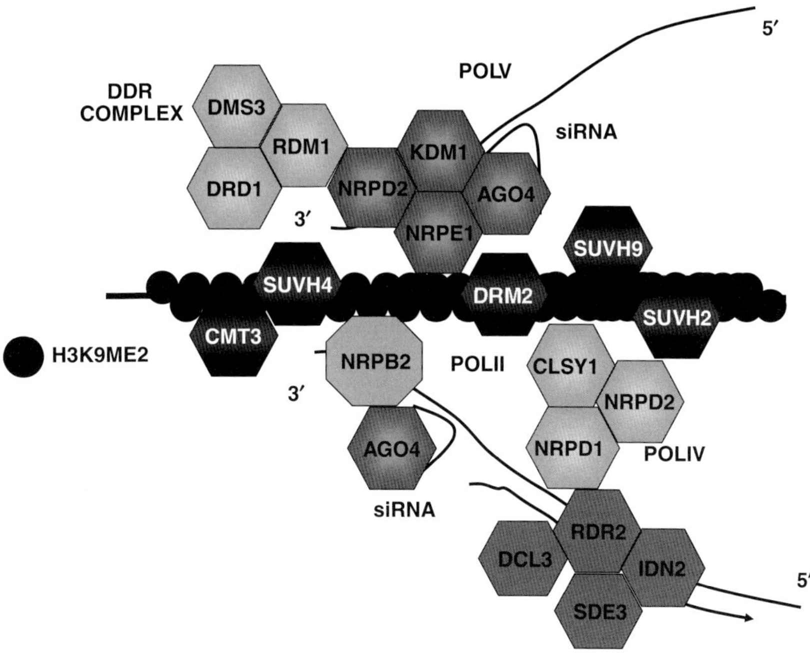

Figure 5.1 A model for RNA-dependent DNA methylation (RdDM) in *Arabidopsis*. Transposable elements, centromeric repeats, and intergenic noncoding repeats are transcribed by PolV, which is recruited to chromatin by the DDR chromatin remodeling complex. AGO4 interacts with GW/WG repeats in the C-terminal domains of the largest subunit of PolV (NRPE1) and the associated Spt5-like protein KDM1. AGO4binds 24nt siRNA and cleaves PolV transcripts. PolII shares many subunits with and can substitute for PolV, and also recruits AGO4 via RDM1. POLIV shares its second subunit (NRPD2) with PolV, but unlike PolII and PolV, PolIV is essential for siRNA biogenesis. The template for RNA synthesis is not known, but PolIV is recruited to chromatin by CLSY1, a homolog of DRD1. PolIV transcripts are converted to double-stranded RNA by RDR2. IDN2, DCL3, and SDE3 are shown associated with Rdr2 by analogy with homologs in *Arabidopsis* (SGS3) and fission yeast (DCR1and HRR1), respectively, and are similarly required for biogenesis of 24nt siRNA. RdDM requires the histone methyltransferases SUVH2 and SUVH9, as well as the DNA methyltransferases DRM2 and CMT3. SUVH2, SUVH9, and SUVH4 have SRA domains that bind methylated cytosines in CHH, CG, and CHG contexts, respectively, while CMT3 has a chromodomain that binds H3K9me2. However, the mechanism by which these methyltransferases are recruited by RNAi is still unclear. By analogy with fission yeast, this mechanism may involve the DNA replication machinery, as mutations in DNA polymerase reduce silencing in *Arabidopsis* and fission yeast. For color detail, see color plate section.

PolIV, on the other hand, is essential for siRNA production and may play a role in promoting RNA-dependent RNA polymerase activity (Vaughn and Martienssen, 2005) via direct interaction with PolIV and the associated chromatin remodeler, CLSY1 (Law et al., 2011). Similarly, PolV and PolII subunits associate with the DDR chromatin remodeling complex, which comprises DMS3 (DEFECTIVE IN MERISTEM SILENCING 3), the chromatin remodeler DRD1, and RDM1 (Law et al., 2010). PolII can substitute for PolV to some extent, and can also contribute to heterochromatic silencing (Zheng et al., 2009).

AGO4 specifically binds 24nt siRNAs that begin with a 5′A or G (Mi et al., 2008), and guides non-CG DNA methylation in a way that requires catalytic (cleavage) activity (Qi et al., 2006). Noncatalytic mutants of AGO4 can substitute for catalytic forms at some TEs, but the presence of the close homolog AGO6 may supply this activity in these cases (Havecker et al., 2010; Zheng et al., 2007). AGO4 forms a complex with PolV by binding to the GW-WG (glycine tryptophan) repeats found in the C-terminal domain (CTD) of the largest subunit NRPE1 and in the CTD of the accessory factor KDM1 (Figure 5.1), which resembles the transcription factor Spt5. AGO4, PolV, and the DDR complex colocalize with DRM2 in the nucleus (Gao et al., 2010), but a direct connection between RNAi and DRM2 has not yet been established.

Interaction Between RNA-Dependent and Replication-Dependent DNA Methylation

Just as DNA methylation is maintained in the absence of RNAi by *DDM1*, *MET1*, and by *HDA6* (Matzke and Birchler, 2005), RNAi maintains transposon silencing in the absence of MET1 or DDM1. The 24nt siRNAs silence some classes of TEs via RdDM, but others lose their 24nt siRNA, and novel 21nt "epigenetically activated" siRNAs (easiRNAs) accumulate instead, especially from Gypsy class (Athila) retrotransposons (Lippman et al., 2003; Slotkin et al., 2009; Teixeira et al., 2009). The easiRNAs closely resemble trans-acting siRNAs (tasiRNA), which are derived from noncoding and coding RNA templates, first targeted by matching 22nt miRNA (Allen et al., 2005; Cuperus et al., 2010; Howell et al., 2007; Ronemus et al., 2006). *RDR6* and *DICER-LIKE 4 (DCL4)* are also required for tasiRNA biogenesis, and have been recovered in screens for mobile silencing signals generated by artificial hairpins (Voinnet, 2005). The tasiRNA can spread from cell to cell (Adenot et al., 2006; Chitwood et al., 2009; Fahlgren et al., 2006; Garcia et al., 2006; Schwab et al., 2009), and can trigger both posttranscriptional and transcriptional silencing. However, transcriptional silencing also requires the 24nt pathway in a mechanism that is still unclear (Brosnan et al., 2007; Eamens et al., 2008; Smith et al., 2007).

Why are some 24nt siRNAs lost in *ddm1* and *met1* mutants? The 24nt siRNAs are not lost in *cmt3*, *kyp*, or *hda6* mutants, suggesting that histone and DNA methylation are not responsible (Lippman et al., 2003). Based on similar phenotypes, interaction with PolV and the DDR complex seems a plausible explanation for the selective reduction of 24nt siRNA observed in *ddm1* and *met1* mutants (Pontes et al., 2009). By contrast, the accumulation of 21nt easiRNAs in *ddm1* and *met1* mutants seems to be a consequence of transposon expression. easiRNA are also found in tissue culture cells, in which Athila retrotransposons are neither silenced nor methylated (Tanurdzic et al., 2008). The easiRNA resembles tasiRNA in many respects, indicating that target specificity may be conferred by miRNA. miRNAs are encoded by many transposons (Piriyapongsa and Jordan, 2008), including Athila, and could therefore target antisense RNA from the transposon, should it arise. Some miRNAs actually target the DNA methylation machinery itself: miR820/miR583 targets DRM in rice (Luo et al., 2006; Wu et al., 2010), while mir773 targets MET2 (Howell et al., 2007). It is not yet clear if these miRNA and tasiRNA affect levels of methylation.

Functional Consequences of Epigenetic Regulation

For some transposons, retrotransposition in *met1* mutants is regulated by a combination of H3K9 methylation and RNAi, although spontaneous activation can also occur during inbreeding (Mirouze et al., 2009). The consequences of transposon activation include transposition and genome damage, as well as gene regulation: for example, *ddm1* mutants accumulate a "syndrome" of epigenetic defects on inbreeding, and many of these defects are caused by TEs found next to or within a gene (Kakutani et al., 1996; Kinoshita et al., 2007; Lippman et al., 2004; Miura et al., 2001; Saze and Kakutani, 2007; Stokes et al., 2002; Stokes and Richards, 2002; Vongs et al., 1993). Genetic background has a pronounced effect, and *ddm1* mutants, for example, flower earlier in the L*er* background than in a Col background (Singer et al., 2001). This variability may well be due to polymorphic TE insertions, which are responsible for the majority of indels detected between accessions (Borevitz et al., 2007; Liu et al., 2004; Vaughn et al., 2007; Zhang et al., 2007b; Zhang et al., 2008). Transgenerational epigenetic regulation of transposons could account for aspects of hybrid vigor, hybrid sterility, and inbreeding depression (Mathieu et al., 2007; Slotkin and Martienssen, 2007). Genes come under TE regulation when they integrate nearby and become subject to epigenetic regulation (Slotkin and Martienssen, 2007). If deleterious alleles were caused by such insertions, their phenotypic effects could be different in hybrids than in inbreds if transposons were activated by hybrid dysgenesis. However, these phenotypic effects would only be revealed after subsequent inbreeding if they were complemented by wild type in the hybrids.

Alternatively, activation of TEs may result in hybrid lethality, as it does in hybrid dysgenesis in *Drosophila*. RNAi may play a role in transposon activation, as siRNA in the seed parent may not recognize TE from the pollen, and vice versa (Martienssen, 2010b). piRNAs play a similar role in *Drosophila* hybrid dysgenesis, which results after fertilization of eggs from the female parent that lack small RNA corresponding to TEs from the male parent (Brennecke et al., 2008; Chambeyron et al., 2008; Jensen et al., 2008). In *Arabidopsis*, both interspecific and interploidy crosses give rise to seed collapse in F1 hybrid endosperm, either due to overproliferation and failure to cellularize or underproliferation and premature cellularization (Adams et al., 2000; Bushell et al., 2003; Comai, 2000). While lethality has generally been ascribed to defects in imprinting and dosage, it is possible that TE activation and small RNA contribute to at least some aspects of these phenomena.

Heterochromatin Reprogramming and Germ Cell Fate

Epigenetic changes are reversible, making them amenable to short-term environmental as well as developmental modulation (Rando and Verstrepen, 2007). In each generation, germline reprogramming must reset epigenetic marks without losing control of TEs. We have previously proposed that controlled germline transposon activity can reveal and then silence TEs via reprogramming and RNAi (Johnson and Bender, 2009; Slotkin et al., 2009). Reprogramming in animals may serve a similar purpose (Bourc'his and Voinnet, 2010), as transient expression of transposons is induced in the germline and during early development (Dupressoir and Heidmann, 1996; Ostertag et al., 2002; Pasyukova et al., 1997). In animals, transposon transcripts are processed into piRNAs (Aravin, Hannon, and Brennecke, 2007; Lau et al., 2006) that may direct methylation in sperm (Aravin et al., 2008; Carmell et al., 2007) and into endogenous siRNA that may have a similar function in oocytes (Malone et al., 2009; Watanabe et al., 2008). In plants, TE transcripts are converted into siRNAs, which accumulate in the male and female germline (Slotkin et al., 2009). Imprinting, hybrid dysgenesis, and epigenetic inheritance are potential outcomes in the next generation (Gehring et al., 2009; Lippman et al., 2004; Peaston et al., 2004). Potential mechanisms and parallels with animal systems are described below.

Reprogramming in the Male Germline

Pollen development in *Arabidopsis* starts with a diploid meiocyte in the anther, which undergoes meiosis, forming a tetrad of haploid microspores. This is followed by the first round of pollen mitosis (PMI), resulting in a bicellular grain consisting of a vegetative cell and a generative cell (Borg et al., 2009). The generative cell undergoes a second mitotic division (PMII), resulting in two haploid sperm cells (SC) in the tricellular mature pollen grain. During fertilization, one sperm cell fuses with the egg cell, while the other fuses with the central cell, initiating the development of the embryo and endosperm, respectively.

The vegetative nucleus (VN) undergoes a dramatic loss of heterochromatin in the binuclear stage, and in mature pollen (Figure 5.2). Transposons are derepressed in the VN, resulting in the reactivation and mobilization of transposable elements (Slotkin et al., 2009). The activated transposons generate pollen-specific 21nt siRNAs that accumulate in sperm cells (Slotkin et al., 2009). Thus, active transposons are "unmasked" in the VN, resulting in the production of small RNAs, which reenforce silencing in sperm.

In somatic cells, 21nt siRNA accumulation occurs in *met1* and *ddm1* mutants as well as in cultured cells. DDM1 accumulates in sperm cells but not in the VN, so it follows that transposon derepression occurs in the vegetative nucleus (Slotkin et al., 2009). *MET1* is also expressed in sperm cells and not the VN (Honys and Twell, 2004), but both genes are expressed in uninucleate microspores. Because these cells only divide once to give rise to the generative and vegetative cells (Figure 5.2), methylation is likely to occur during replication. Indeed, most symmetric CG sites are fully methylated in both SC and VN, while some sites lose methylation in the VN. These sites are found in *AtMu1* transposons (Slotkin et al., 2009), but not in centromeric satellite repeats (Schoft et al., 2009), suggesting selective DNA demethylation in the VN. Interestingly, 24nt siRNA from

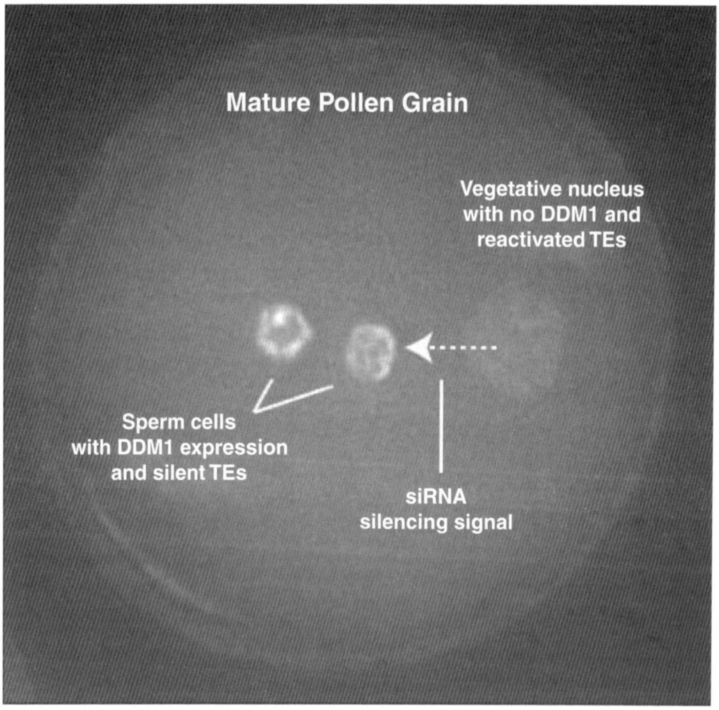

Figure 5.2 Small RNA Silencing in Pollen. Mature pollen grains comprise two cell types—a companion vegetative cell that does not contribute its genome to the next generation, and sperm cells that participate in double fertilization of female gametes, namely the egg and the central cell. Loss of heterochromatin in the vegetative nucleus is accompanied by loss of the chromatin remodeler DDM1, and generates small RNA signals that silence TE in sperm (Slotkin et al., 2009). Reprinted, with permission from Creasey and Martienssen, 2010.

AtMu1 is not found in pollen (Slotkin et al., 2009), but centromeric siRNA is retained, suggesting VN methylation may reflect changes in siRNA composition (Matzke et al., 2009). Pollen mitosis I and II are classical models for plant stem cell division (Borg et al., 2009), and regions of asymmetric CHH methylation would follow the inheritance of DNA strands in the absence of RdDM (Klar, 2007). Mammals also lose heterochromatin transiently in primordial germ cells (PGC), and orthologs of both DDM1 (Lsh1) and MET1 (Dnmt1) are expressed in mouse sperm, but a link between reprogramming and small RNA biogenesis has yet to be established (Aravin et al., 2009; Branco et al., 2008; Carmell et al., 2007; Hajkova et al., 2008).

In both plants and animals, sperm cells undergo reprogramming via histone exchange. For example, *Arabidopsis* MGH3 (AtMGH3 or HTR10) is a Histone 3.3- (H3.3) like protein that is expressed specifically in sperm (Ingouff et al., 2007; Okada et al., 2005). One possibility is that reprogramming of the VN and of the SC is coupled, in that genome reprogramming in sperm cells destabilizes gene silencing and thus makes TE movement more likely. TE-derived siRNAs from the VN may act as a surveillance mechanism by accumulating in sperm cells and silencing TEs during the short time window when the epigenome is reorganized (Jacob and Martienssen, 2011).

Reprogramming in the Female Germline

Female gamete development in *Arabidopsis* begins with the differentiation of a diploid megaspore mother cell (MMC) that undergoes meiosis to form four haploid megaspores. Three of these megaspores die, and the surviving megaspore undergoes three rounds of mitosis, resulting in a multinucleate syncytium that subsequently cellularizes. The mature female gametophyte consists of three antipodal cells, one (diploid) central cell, two synergid cells, and one egg cell. Double fertilization of the egg and central cells results in a diploid zygote and triploid endosperm, respectively. PolIV-dependent siRNA in the developing ovule are 24nt in length, maternal in origin, and enriched in the female gametophyte (Mosher et al., 2009). The surrounding maternal integuments (seed coat) express AGO9, one of ten siRNA-binding Argonaute homologs in *Arabidopsis* (Olmedo-Monfil et al., 2010). It is possible that maternal siRNA signals travel into the female gamete from the ovule, as AGO9 is necessary to silence TEs in the egg cell (Olmedo-Monfil et al., 2010). AGO9 is also expressed in the pollen grain cytoplasm (Olmedo-Monfil et al., 2010), where it may play a similar role to the one it plays in the ovule.

The maternal endosperm genome is partially demethylated before fertilization by the DNA glycosylase DEMETER (see above), and also after fertilization by replication-dependent passive demethylation, mediated by the Retinoblastoma (Rb) pathway (Jullien and Berger, 2010; Jullien et al., 2008). This is because a complex involving RB-RELATED (RBR) blocks MET1 expression. DNA demethylation in the endosperm is accompanied by DNA remethylation that may be mediated by small RNAs during early embryo development (Gehring et al., 2009; Hsieh et al., 2009). In animals, deamination by cytosine deaminase (Bhutani et al., 2010; Popp et al., 2010) or by conversion of methylcytosine to hydroxymethylcytosine (Kriaucionis and Heintz, 2009; Tahiliani et al., 2009), in combination with DNA repair (Wossidlo et al., 2010), contributes to active DNA demethylation.

Remarkably, Argonaute and siRNA biogenesis mutations have a dramatic effect on female germ-cell fate in both maize (Singh et al., 2011) and *Arabidopsis* (Olmedo-Monfil et al., 2010). In *Arabidopsis*, one or more somatic cells in mutant ovules differentiate as diploid functional megaspores without meiosis (Olmedo-Monfil et al., 2010). These diploid eggs are fertile and can be pollinated to give rise to triploid progeny. In some plant species, the differentiation of diploid somatic egg cells is an important step in asexual development (apomixis). One possibility is that diploid germ-cell differentiation results from failure of normal meiosis. The Argonaute MEL1/AGO5 is required for normal meiosis in rice (Nonomura et al., 2007), while AGO9 is required for meiosis II in maize (Singh et al., 2011), raising the possibility that failure of meiosis in RNAi mutants triggers asexual germ-cell fate in surrounding cells. As TE siRNAs appear to move from cell to cell, both in the pollen grain (Slotkin et al., 2009) and in the ovule (Olmedo-Monfil et al., 2010), they could provide an inhibitory signal once the MMC is correctly specified.

TE are propagated much more efficiently in sexual rather than asexual species (Hickey, 1993), providing a biological rationale for such a signal. Phylogenomic analysis of more than 20,000 genes in 150 taxa has revealed that amino acid substitutions in the shared second subunit of PolIV and PolV contribute more significantly to the partition of gymnosperms from angiosperms than changes in other proteins. Gymnosperms have maternal endosperm and no 24nt siRNA at all, providing a rationale for such a radical change in this protein (Morin et al., 2008). Gymnosperms have very large genomes, in which methylation of TEs appears to be suppressed, perhaps as a result of the loss of RNA-dependent DNA methylation (Rabinowicz et al., 2005). Thus, TE regulation has had a profound impact on reproductive strategy (apospory) and on genome organization in land plants.

Transgenerational Inheritance of Transposon Silencing

Loss of silencing in homozygous *ddm1* and in homozygous *met1* mutants and loss of DNA methylation from TEs are epigenetically inherited in backcrosses to wild-type plants (Kakutani et al., 1999; Lippman et al., 2003; Saze et al., 2003). That is, many TEs that lose methylation and silencing in homozygous mutants remain unmethylated and transcribed in heterozygous progeny, and even in homozygous wild types following additional backcrosses. No other mutations that affect DNA and histone modification in *Arabidopsis* display this marked epigenetic inheritance (Lippman et al., 2003), not even mutations in *vim1*, which is thought to guide MET1 to hemimethylated DNA (Woo et al., 2008; Woo et al., 2007).

The inheritance of unmethylated DNA makes it possible to observe the consequences of the loss of MET1 and DDM1 function in the germline. This is because unmethylated TEs are inherited from the gametophyte after fertilization. For example, the loss of MET1 methyltransferase activity in the haploid megaspore results in so-called passive loss of DNA methylation in the three subsequent cell divisions, as unmethylated nucleotides are incorporated during DNA replication (Saze et al., 2003). As a result, methylation is reduced in *met1* megagametophytes (Saze et al., 2003) when they arise following meiosis in *met1/+* heterozygotes. In contrast, *ddm1* gametophytes from *ddm1/+* heterozygotes have normal levels of DNA methylation (Kakutani et al., 1999; Kankel et al., 2003). This is further evidence that loss of DNA methylation is an indirect consequence of the chromatin-remodeling defect in *ddm1* mutants and that loss takes longer than two to three cell divisions to become apparent.

While epigenetic inheritance of unmethylated TEs is widespread (Lippman et al., 2003), some TEs retain some methylation in *ddm1* mutants and in heterozygous backcross progeny (Teixeira et al., 2009). Typically, these TEs have large numbers of matching siRNAs, consistent with the idea that siRNAs can guide DNA methylation to matching sequences in the absence of DDM1 (Teixeira et al., 2009). Consistent with this observation, such methylation is lost in double mutants also defective in genes coding for enzymes in the 24nt siRNA pathway, such as *rdr2* (Teixeira et al., 2009). In wild-type backgrounds, methylation levels at these remethylatable TEs rise stepwise in each subsequent generation of inbreeding, once DDM1 activity is restored. DDM1 is expressed predominantly in the germline, and it seems likely that methylation initially guided by small RNAs is maintained in the germline by DDM1. Some of these siRNAs appear to act in trans, as they are absent from *ddm1* homozygous mutants and are only restored in backcross heterozygotes (Lippman et al., 2003; Teixeira et al., 2009).

The inheritance of transposon silencing from one generation to the next would be expected to impact the patterns of methylation observed in natural populations, as well as in artificial populations, such as recombinant inbred lines. Profiling of DNA methylation in different accessions of inbreeding *Arabidopsis* has revealed that TE methylation is highly conserved, while exon methylation is much more variable (Vaughn et al., 2007). This is true even at genes such as FWA, where the first two exons comprise the remnants of a SINE transposon (Lippman et al., 2004; Saze and Kakutani, 2007). In this case, methylation is conserved across a panel of 96 different inbred accessions, perhaps reflecting selection for gene silencing and early flowering (Vaughn et al., 2007). In contrast, methylated exons that are not derived from TEs were only methylated in a subset of the accessions examined, typically about half of them. These observations suggested that exon methylation was relatively unstable, and it was subsequently observed that exon methylation was lost or gained from about 5%–10% of the parental alleles in F2 progeny between the accessions (Vaughn et al., 2007).

More recently, whole genome bisulphite sequencing has been used to investigate the stability of epigenetic variation over 30 generations of inbreeding in *Arabidopsis* (Becker et al., 2011; Schmitz et al., 2011). About 10% of all methylated cytosines varied in at least one strain, but reversion was very frequent, so that overall patterns were remarkably stable. Methylation differences with functional consequences are generally clustered together, and when differentially methylated regions were examined (DMR), only 30 such regions were found in each strain, similar to the number of spontaneous mutations (nucleotide substitutions). These DMR were nonrandom, appearing in several strains independently, and most if not all of them were found in TEs. Sequencing of independent lines of *ddm1* mutants, as well as additional natural accessions, has revealed that widespread transposition is limited to only a handful of transposon families and occurs sporadically in individual lines (Tsukahara et al., 2009).

The transgenerational inheritance of methylation patterns in *Arabidopsis* bears a strong resemblance to classical phenomena in maize, now thought of as "epigenetic." For example, the recurrence of epialleles over many generations is highly reminiscent of "cycling" of transposons from active to inactive states, first observed by McClintock and subsequently shown to be associated with methylation (Martienssen et al., 1990; McClintock, 1956) (see Chapter 4 in this volume). In another example, the loss and gain of methylation from one allele in trans is strongly reminiscent of the classical phenomenon of paramutation. In maize, both cycling and paramutation have been found to involve the RNAi pathway. The role of RNAi in cycling of transposons betweeen active and inactive states is discussed in Chapter 6 in this volume. We discuss paramutation below.

Paramutation

Paramutation is a form of allele-specific gene silencing that was first described at several loci that regulate anthocyanin pigmentation in maize. At the *R* (red color) locus, for example, the complex *R-st* allele can silence *R-r* alleles in *R-st/R-r* heterozygotes, resulting in mottled aleurone pigmentation in the seed (Brink, 1956; Chandler et al., 2000). This modified *R-r′* allele remains silent when the heterozygotes are used to pollinate silks from plants carrying neutral (null) alleles, such as *r-g*. Remarkably, the silent *R-r′* allele can silence other *R-r* alleles when such heterozygotes are used to pollinate silks from plants carrying *R-r*. However, while the silencing is heritable, it is not necessarily permanent, as *R-r′* will occasionally lose silencing spontaneously, especially when present in the hemizygous condition. Silencing is also lost transiently in the endosperm, when the R-r′ allele is transmitted through the central cell, but this is due to a parent-of-origin effect (imprinting) described in Chapter 7 in this volume.

Paramutation can also be visualized directly in the plant, when it occurs at the *B* (booster) (Coe, 1959) and *Pl* (plant color) (Hollick et al., 1995) loci, and it is through studies with this sporophytic silencing system that the mechanism is best understood (Arteaga-Vazquez and Chandler, 2010). The *B-I* allele conditions intense purple color in the leaves, but spontaneously converts to *B′*, which undergoes a permanent form of gene silencing. In heterozygotes between *B′* and *B-I*, *B-I* is effectively silenced early in plant development, leading to F1 plants that have the *B′* phenotype. Progeny tests confirm that the *B-I* allele has been efficiently converted to *B′* in these heterozygotes and behaves as a *B′* allele in subsequent crosses, such that in wild-type genetic backgrounds *B-I* is never recovered. This behavior includes efficient silencing of additional *B-I* alleles in heterozygotes, leading to an inexorable loss of gene activity in successive generations.

The *Pl*-Rhoades allele undergoes a very similar silencing phenomenon and is identical in nucleotide sequence to the silent allele *Pl′*-mahogany. Again, heterozygotes are efficiently silenced, reducing pigmentation in the seedling and anthers (Hollick et al., 1995). Because of the stability of the silent *B′* and *Pl′* alleles, mutagenesis and forward genetic screens can be used to recover rare trans-acting mutations in which silencing is lost. The first of these to be characterized at the molecular level were termed *mop* genes for *mediator of paramutation*, and revealed that silencing was lost in mutants that lost 24nt "heterochromatic" small RNA, such as *RDR2* (*mop1*) (Alleman et al., 2006; Dorweiler et al., 2000; Nobuta et al., 2008). Subsequently, *rmr* mutations, which lose silencing of *Pl′*, and other *mop* mutants were found in the first (Erhard et al., 2009)and second largest subunits of PolIV (Sidorenko et al., 2009; Stonaker et al., 2009), both of which are also required for accumulation of 24nt siRNAs. Aspects of paramutation were prevented in trans in these mutants depending on the distinct paramutation systems.

Molecular and genetic analysis of the *B-I* upstream regions revealed that the key regulatory region required for paramutation and high expression in the *B-I* state was comprised of 7 copies of an 853-bp repeat only found 100 kb upstream of the transcription start site (Stam et al., 2002b). Five copies of the repeat could still participate in paramutation, but one copy could not (Stam et al., 2002a). Plants containing three repeat units were still able to carry out paramutation, but the frequency and stability was reduced, suggesting repeat numbers were "counted," possibly by RdRP (Martienssen, 2003). In leaf tissue, a low level of 24nt siRNA corresponding to these repeats depend on *mop1* for their accumulation, and a transgene that produces dsRNA and siRNAs homologous to the repeats is capable of reproducing paramutation (Arteaga-Vazquez et al., 2010).

At the Pl and R loci, full-length and nonautonomous *Doppia* transposable elements occupy each promoter, which has led to the speculation that transposons may play a key role in paramutation within these systems. It is tempting to make the analogy with genes regulated by the Robertson's Mutator family of TEs, which are also regulated in trans by dominant factors that encode hairpin RNA (Slotkin et al., 2005). In each case, silencing requires *mop1*, is progressive over generations, and is ultimately irreversible. Furthermore, silencing occurs in the embryo and in successive dividing cells, giving rise to large clonal sectors that grow in size with each successive node (Martienssen and Baron, 1994). These sectors are very similar to those that occur in *B-I* homozygous plants undergoing spontaneous paramutation (Coe, 1968). Finally, in *Arabidopsis*, silencing of the PHOSHPHOANTHRANILATE ISOMERASE (PAI) gene is also induced by a hairpin, depends on small RNA (Mathieu and Bender, 2004), and occurs in large sectors that progress from leaf to leaf (Bender and Fink, 1995). Unlike paramutation, the silenced PAI2 locus does not itself become paramutagenic, and silencing does not require RDR2. However, the 24nt siRNAs responsible for silencing PAI have been shown to move from shoots into grafted roots, indicating sectoring in each case could be because of translocation of the siRNA from surrounding cells into the meristem (Martienssen, 2010a).

Conclusions

Epigenetic mechanisms of transposon silencing in plants fall into three major classes, namely DNA methylation, histone modification, and RNAi. Chromatin remodeling links all three, as does the transcriptional machinery, including the plant-specific RNA polymerases PolIV and PolV. Transgenerational effects of these epigenetic mechanisms involve RNAi, and RNAi controls both transposon silencing and germ-cell fate during reproduction. Transposon silencing is reminiscent of paramutation, a classical gene-silencing mechanism in maize, and paramutation requires the same siRNA-silencing pathway that controls heterochromatic histone and DNA modification in plants. Finally, coevolution of epigenetics with reproductive strategy may reflect a requirement for meiosis in transposon propagation. The presence of unreduced maternal gametes depends on these mechanisms, and may have contributed to the evolution of some of the defining properties of flowering plants, such as double fertilization of both egg and central cell and polyploidy, which results from fusion of unreduced gametes (Martienssen, 2010b).

References

Adams, S., Vinkenoog, R., Spielman, M., Dickinson, H.G., and Scott, R.J. (2000) Parent-of-origin effects on seed development in *Arabidopsis thaliana* require DNA methylation. *Development 127*: 2493–2502.

Adenot, X., Elmayan, T., Lauressergues, D., Boutet, S., Bouche, N., Gasciolli, V., and Vaucheret, H. (2006) DRB4-dependent TAS3 trans-acting siRNAs control leaf morphology through AGO7. *Curr. Biol. 16*: 927–932.

Alleman, M., Sidorenko, L., McGinnis, K., Seshadri, V., Dorweiler, J.E., White, J., Sikkink, K., and Chandler, V.L. (2006) An RNA-dependent RNA polymerase is required for paramutation in maize. *Nature 442*: 295–298.

Allen, E., Xie, Z., Gustafson, A.M., and Carrington, J.C. (2005) microRNA-directed phasing during trans-acting siRNA biogenesis in plants. *Cell 121*: 207–221.

Aravin, A.A., Hannon, G.J., and Brennecke, J. (2007) The Piwi-piRNA pathway provides an adaptive defense in the transposon arms race. *Science 318*: 761–764.

Aravin, A.A., Sachidanandam, R., Bourc'his, D., Schaefer, C., Pezic, D., Toth, K.F., Bestor, T., and Hannon, G.J. (2008) A piRNA pathway primed by individual transposons is linked to *de novo* DNA methylation in mice. *Mol. Cell 31*: 785–799.

Aravin, A.A., van der Heijden, G.W., Castaneda, J., Vagin, V.V., Hannon, G.J., and Bortvin, A. (2009) Cytoplasmic compartmentalization of the fetal piRNA pathway in mice. *PLoS Genet. 5*: e1000764.

Arteaga-Vazquez, M.A., and Chandler, V.L. (2010) Paramutation in maize: RNA mediated trans-generational gene silencing. *Curr. Opin. Genet. Dev. 20*: 156–163.

Arteaga-Vazquez, M., Sidorenko, L., Rabanal, F.A., Shrivistava, R., Nobuta, K., Green, P.J., Meyers, B.C., and Chandler, V.L. (2010) RNA-mediated trans-communication can establish paramutation at the *b1* locus in maize. *Proc. Natl. Acad. Sci. USA 107*: 12986–12991.

Aufsatz, W., Stoiber, T., Rakic, B., and Naumann, K. (2007) *Arabidopsis* histone deacetylase 6: a green link to RNA silencing. *Oncogene 26*: 5477–5488.

Bartee, L., and Bender, J. (2001) Two *Arabidopsis* methylation-deficiency mutations confer only partial effects on a methylated endogenous gene family. *Nucleic Acids Res. 29*: 2127–2134.

Becker, C., Hagmann, J., Muller, J., Koenig, D., Stegle, O., Borgwardt, K., and Weigel, D. (2011) Spontaneous epigenetic variation in the *Arabidopsis thaliana* methylome. *Nature 480*: 245–249.

Bender, J. (2004) DNA methylation and epigenetics. *Annu. Rev. Plant Biol. 55*: 41–68.

Bender, J., and Fink, G.R. (1995) Epigenetic control of an endogenous gene family is revealed by a novel blue fluorescent mutant of *Arabidopsis*. *Cell 83*: 725–734.

Bhutani, N., Brady, J.J., Damian, M., Sacco, A., Corbel, S.Y., and Blau, H.M. (2010) Reprogramming towards pluripotency requires AID-dependent DNA demethylation. *Nature 463*: 1042–1047.

Biemont, C. (2009) Are transposable elements simply silenced or are they under house arrest? *Trends Genet. 25*: 333–334.

Borevitz, J.O., Hazen, S.P., Michael, T.P., Morris, G.P., Baxter, I.R., Hu, T.T., Chen, H., Werner, J.D., Nordborg, M., Salf, D.E., et al. (2007) Genome-wide patterns of single-feature polymorphism in *Arabidopsis thaliana*. *Proc. Natl. Acad. Sci. USA 104*: 12057–12062.

Borg, M., Brownfield, L., and Twell, D. (2009) Male gametophyte development: a molecular perspective. *J. Exp. Bot. 60*: 1465–1478.

Bostick, M., Kim, J.K., Esteve, P.O., Clark, A., Pradhan, S., and Jacobsen, S.E. (2007) UHRF1 plays a role in maintaining DNA methylation in mammalian cells. *Science 317*: 1760–1764.

Bourc'his, D., and Voinnet, O. (2010) A small-RNA perspective on gametogenesis, fertilization, and early zygotic development. *Science 330*: 617–622.

Branco, M.R., Oda, M., and Reik, W. (2008) Safeguarding parental identity: Dnmt1 maintains imprints during epigenetic reprogramming in early embryogenesis. *Genes Dev. 22*: 1567–1571.

Brennecke, J., Malone, C.D., Aravin, A.A., Sachidanandam, R., Stark, A., and Hannon, G.J. (2008) An epigenetic role for maternally inherited piRNAs in transposon silencing. *Science 322*: 1387–1392.

Brink, R.A. (1956) A genetic change associated with the *R* locus in maize which is directed and potentially reversible. *Genetics 41*: 872–889.

Brosnan, C.A., Mitter, N., Christie, M., Smith, N.A., Waterhouse, P.M., and Carroll, B.J. (2007) Nuclear gene silencing directs reception of long-distance mRNA silencing in *Arabidopsis*. *Proc. Natl. Acad. Sci. USA 104*: 14741–14746.

Bushell, C., Spielman, M., and Scott, R.J. (2003) The basis of natural and artificial postzygotic hybridization barriers in *Arabidopsis* species. *Plant Cell 15*: 1430–1442.

Carmell, M.A., Girard, A., van de Kant, H.J., Bourc'his, D., Bestor, T.H., de Rooij, D.G., and Hannon, G.J. (2007) MIWI2 is essential for spermatogenesis and repression of transposons in the mouse male germline. *Dev. Cell 12*: 503–514.

Chambeyron, S., Popkova, A., Payen-Groschene, G., Brun, C., Laouini, D., Pelisson, A., and Bucheton, A. (2008) piRNA-mediated nuclear accumulation of retrotransposon transcripts in the *Drosophila* female germline. *Proc. Natl. Acad. Sci. USA 105*: 14964–14969.

Chandler, V.L., Eggleston, W.B., and Dorweiler, J.E. (2000) Paramutation in maize. *Plant Mol. Biol. 43*: 121–145.

Chandler, V.L., and Walbot, V. (1986) DNA modification of a maize transposable element correlates with loss of activity. *Proc. Natl. Acad. Sci. USA 83*: 1767–1771.

Cheng, X., and Blumenthal, R.M. (2010) Coordinated chromatin control: structural and functional linkage of DNA and histone methylation. *Biochemistry 49*: 2999–3008.

Chitwood, D.H., Nogueira, F.T., Howell, M.D., Montgomery, T.A., Carrington, J.C., and Timmermans, M.C. (2009) Pattern formation via small RNA mobility. *Genes Dev. 23*: 549–554.

Clapier, C.R., and Cairns, B.R. (2009) The biology of chromatin remodeling complexes. *Annu. Rev. Biochem. 78*: 273–304.

Coe, E.H. (1959) A regular and continuing conversion-type phenomenon at the *B* locus in maize. *Proc. Natl. Acad. Sci. USA 45*: 828–832.

Coe, E.H., Jr. (1968) Heritable repression due to paramutation in maize. *Science 162*: 925.

Cokus, S.J., Feng, S., Zhang, X., Chen, Z., Merriman, B., Haudenschild, C.D., Pradhan, S., Nelson, S.F., Pellegrini, M., and Jacobsen, S.E. (2008) Shotgun bisulphite sequencing of the *Arabidopsis* genome reveals DNA methylation patterning. *Nature 452*: 215–219.

Comai, L. (2000) Genetic and epigenetic interactions in allopolyploid plants. *Plant Mol. Biol. 43*: 387–399.

Creasey, K.M., and Martienssen, R.A. (2010) Germline reprogramming of heterochromatin in plants. *Cold Spring Harb. Symp. Quant. Biol. 75*: 269–274.

Cuperus, J.T., Carbonell, A., Fahlgren, N., Garcia-Ruiz, H., Burke, R.T., Takeda, A., Sullivan, C.M., Gilbert, S.D., Montgomery, T.A., and Carrington, J.C. (2010) Unique functionality of 22-nt miRNAs in triggering RDR6-dependent siRNA biogenesis from target transcripts in *Arabidopsis*. *Nature Struct. Mol. Biol. 17*: 997–1003.

Daxinger, L., Kanno, T., Bucher, E., van der Winden, J., Naumann, U., Matzke, A.J., and Matzke, M. (2009) A stepwise pathway for biogenesis of 24-nt secondary siRNAs and spreading of DNA methylation. *EMBO J. 28*: 48–57.

Deleris, A., Greenberg, M.V., Ausin, I., Law, R.W., Moissiard, G., Schubert, D., and Jacobsen, S.E. (2010) Involvement of a Jumonji-C domain-containing histone demethylase in DRM2-mediated maintenance of DNA methylation. *EMBO Rep. 11*: 950–955.

Dorweiler, J.E., Carey, C.C., Kubo, K.M., Hollick, J.B., Kermicle, J.L., and Chandler, V.L. (2000) *Mediator of paramutation1* is required for establishment and maintenance of paramutation at multiple maize loci. *Plant Cell 12*: 2101–2118.

Dupressoir, A., and Heidmann, T. (1996) Germ line-specific expression of intracisternal A-particle retrotransposons in transgenic mice. *Mol. Cell Biol. 16*: 4495–4503.

Eamens, A., Vaistij, F.E., and Jones, L. (2008) NRPD1a and NRPD1b are required to maintain post-transcriptional RNA silencing and RNA-directed DNA methylation in *Arabidopsis*. *Plant J. 55*: 596–606.

Earley, K.W., Pontvianne, F., Wierzbicki, A.T., Blevins, T., Tucker, S., Costa-Nunes, P., Pontes, O., and Pikaard, C.S. (2010) Mechanisms of HDA6-mediated rRNA gene silencing: suppression of intergenic Pol II transcription and differential effects on maintenance versus siRNA-directed cytosine methylation. *Genes Dev. 24*: 1119–1132.

Ebbs, M.L., and Bender, J. (2006) Locus-specific control of DNA methylation by the *Arabidopsis* SUVH5 histone methyltransferase. *Plant Cell 18*: 1166–1176.

Erhard, K.F., Jr., Stonaker, J.L., Parkinson, S.E., Lim, J.P., Hale, C.J., and Hollick, J.B. (2009) RNA polymerase IV functions in paramutation in *Zea mays*. *Science 323*: 1201–1205.

Fahlgren, N., Montgomery, T.A., Howell, M.D., Allen, E., Dvorak, S.K., Alexander, A.L., and Carrington, J.C. (2006) Regulation of *AUXIN RESPONSE FACTOR3* by TAS3 ta-siRNA affects developmental timing and patterning in *Arabidopsis*. *Curr. Biol. 16*: 939–944.

Feng, S., Cokus, S.J., Zhang, X., Chen, P.Y., Bostick, M., Goll, M.G., et al. (2010) Conservation and divergence of methylation patterning in plants and animals. *Proc. Natl. Acad. Sci. USA 107*: 8689–8694.

Feng, S., and Jacobsen, S.E. (2011) Epigenetic modifications in plants: an evolutionary perspective. *Curr. Opin. Plant. Biol. 14*: 179–186.

Fuks, F. (2005) DNA methylation and histone modifications: teaming up to silence genes. *Curr. Opin. Genet. Dev. 15*: 490–495.

Gao, Z., Liu, H.L., Daxinger, L., Pontes, O., He, X., Qian, W., et al. (2010) An RNA polymerase II- and AGO4-associated protein acts in RNA-directed DNA methylation. *Nature 465*: 106–109.

Garcia, D., Collier, S.A., Byrne, M.E., and Martienssen, R.A. (2006) Specification of leaf polarity in *Arabidopsis* via the trans-acting siRNA pathway. *Curr. Biol. 16*: 933–938.

Gehring, M., Bubb, K.L., and Henikoff, S. (2009) Extensive demethylation of repetitive elements during seed development underlies gene imprinting. *Science 324*: 1447–1451.

Gehring, M., and Henikoff, S. (2007) DNA methylation dynamics in plant genomes. *Biochim. Biophys. Acta. 1769*: 276–286.

Gendrel, A.V., Lippman, Z., Yordan, C., Colot, V., and Martienssen, R.A. (2002) Dependence of heterochromatic histone H3 methylation patterns on the *Arabidopsis* gene *DDM1*. *Science 297*: 1871–1873.

Grewal, S.I., and Elgin, S.C. (2007) Transcription and RNA interference in the formation of heterochromatin. *Nature 447*: 399–406.

Haag, J.R., Pontes, O., and Pikaard, C.S. (2009) Metal A and metal B sites of nuclear RNA polymerases Pol IV and Pol V are required for siRNA-dependent DNA methylation and gene silencing. *PLoS One 4*: e4110.

Hajkova, P., Ancelin, K., Waldmann, T., Lacoste, N., Lange, U.C., Cesari, F., Lee, C., Almouzni, G., Schneider, R., and Surani, M.A. (2008) Chromatin dynamics during epigenetic reprogramming in the mouse germ line. *Nature 452*: 877–881.

Havecker, E.R., Wallbridge, L.M., Hardcastle, T.J., Bush, M.S., Kelly, K.A., Dunn, R.M., Schwach, F., Doonan, J.H., and Baulcombe, D.C. (2010) The *Arabidopsis* RNA-directed DNA methylation Argonautes functionally diverge based on their expression and interaction with target loci. *Plant Cell 22*: 321–334.

Hawkins, R.D., Hon, G.C., Yang, C., Antosiewicz-Bourget, J.E., Lee, L.K., Ngo, Q.M., et al. (2011) Dynamic chromatin states in human ES cells reveal potential regulatory sequences and genes involved in pluripotency. *Cell Res. 21*: 1393–1409.

He, X.J., Hsu, Y.F., Pontes, O., Zhu, J., Lu, J., Bressan, R.A., Pikaard, C., Wang, C.S., and Zhu, J.K. (2009a) NRPD4, a protein related to the RPB4 subunit of RNA polymerase II, is a component of RNA polymerases IV and V and is required for RNA-directed DNA methylation. *Genes Dev. 23*: 318–330.

He, X.J., Hsu, Y.F., Zhu, S., Wierzbicki, A.T., Pontes, O., Pikaard, C.S., Liu, H.L., Wang, C.S., Jin, H., and Zhu, J.K. (2009b) An effector of RNA-directed DNA methylation in *Arabidopsis* is an ARGONAUTE 4- and RNA-binding protein. *Cell 137*: 498–508.

Heintzman, N.D., Stuart, R.K., Hon, G., Fu, Y., Ching, C.W., Hawkins, R.D., et al. (2007) Distinct and predictive chromatin signatures of transcriptional promoters and enhancers in the human genome. *Nat. Genet. 39*: 311–318.

Herr, A.J., Jensen, M.B., Dalmay, T., and Baulcombe, D.C. (2005) RNA polymerase IV directs silencing of endogenous DNA. *Science 308*: 118–120.

Hickey, D.A. (1993) Molecular symbionts and the evolution of sex. *J. Hered 84*: 410–414.

Hollick, J.B., Patterson, G.I., Coe, E.H., Jr., Cone, K.C., and Chandler, V.L. (1995) Allelic interactions heritably alter the activity of a metastable maize *pl* allele. *Genetics 141*: 709–719.

Honys, D., and Twell, D. (2004) Transcriptome analysis of haploid male gametophyte development in *Arabidopsis*. *Genome Biol. 5*: R85.

Howell, M.D., Fahlgren, N., Chapman, E.J., Cumbie, J.S., Sullivan, C.M., Givan, S.A., Kasschau, K.D., and Carrington, J.C. (2007) Genome-wide analysis of the RNA-DEPENDENT RNA POLYMERASE6/DICER-LIKE4 pathway in *Arabidopsis* reveals dependency on miRNA- and tasiRNA-directed targeting. *Plant Cell 19*: 926–942.

Hsieh, T.F., Ibarra, C.A., Silva, P., Zemach, A., Eshed-Williams, L., Fischer, R.L., and Zilberman, D. (2009) Genome-wide demethylation of *Arabidopsis* endosperm. *Science 324*: 1451–1454.

Inagaki, S., Miura-Kamio, A., Nakamura, Y., Lu, F., Cui, X., Cao, X., Kimura, H., Saze, H., and Kakutani, T. (2010) Autocatalytic differentiation of epigenetic modifications within the *Arabidopsis* genome. *EMBO J. 29*: 3496–3506.

Ingouff, M., Hamamura, Y., Gourgues, M., Higashiyama, T., and Berger, F. (2007) Distinct dynamics of HISTONE3 variants between the two fertilization products in plants. *Curr. Biol. 17*: 1032–1037.

Jacob, Y., Feng, S., LeBlanc, C.A., Bernatavichute, Y.V., Stroud, H., Cokus, S., Johnson, L.M., Pellegrini, M., Jacobsen, S.E., and Michaels, S.D. (2009) ATXR5 and ATXR6 are H3K27 monomethyltransferases required for chromatin structure and gene silencing. *Nat. Struct. Mol. Biol. 16*: 763–768.

Jacob, Y., and Martienssen, R.A. (2011) Chromatin reprogramming: gender equality during *Arabidopsis* germline differentiation. *Curr. Biol. 21*: R20–22.

Jacobsen, S.E., Sakai, H., Finnegan, E.J., Cao, X., and Meyerowitz, E.M. (2000) Ectopic hypermethylation of flower-specific genes in *Arabidopsis*. *Curr. Biol. 10*: 179–186.

Jeddeloh, J.A., Bender, J., and Richards, E.J. (1998) The DNA methylation locus *DDM1* is required for maintenance of gene silencing in *Arabidopsis*. *Genes Dev. 12*: 1714–1725.

Jeddeloh, J.A., Stokes, T.L., and Richards, E.J. (1999) Maintenance of genomic methylation requires a SWI2/SNF2-like protein. *Nat. Genet. 22*: 94–97.

Jensen, P.A., Stuart, J.R., Goodpaster, M.P., Goodman, J.W., and Simmons, M.J. (2008) Cytotype regulation of *P* transposable elements in *Drosophila melanogaster*: Repressor polypeptides or piRNAs? *Genetics 179*: 1785–1793.

Johnson, L.M., Bostick, M., Zhang, X., Kraft, E., Henderson, I., Callis, J., and Jacobsen, S.E. (2007) The SRA methyl-cytosine-binding domain links DNA and histone methylation. *Curr. Biol. 17*: 379–384.

Johnson, M.A., and Bender, J. (2009) Reprogramming the epigenome during germline and seed development. *Genome Biol. 10*: 232.

Jullien, P.E., and Berger, F. (2010) DNA methylation reprogramming during plant sexual reproduction? *Trends Genet. 26*: 394–399.

Jullien, P.E., Mosquna, A., Ingouff, M., Sakata, T., Ohad, N., and Berger, F. (2008) Retinoblastoma and its binding partner MSI1 control imprinting in *Arabidopsis*. *PLoS Biol. 6*: e194.

Kakutani, T., Jeddeloh, J.A., Flowers, S.K., Munakata, K., and Richards, E.J. (1996) Developmental abnormalities and epimutations associated with DNA hypomethylation mutations. *Proc. Natl. Acad. Sci. USA 93*: 12406–12411.

Kakutani, T., Munakata, K., Richards, E.J., and Hirochika, H. (1999) Meiotically and mitotically stable inheritance of DNA hypomethylation induced by *ddm1* mutation of *Arabidopsis thaliana*. *Genetics 151*: 831–838.

Kankel, M.W., Ramsey, D.E., Stokes, T.L., Flowers, S.K., Haag, J.R., Jeddeloh, J.A., Riddle, N.C., Verbsky, M.L., and Richards, E.J. (2003) *Arabidopsis MET1* cytosine methyltransferase mutants. *Genetics 163*: 1109–1122.

Kanno, T., Bucher, E., Daxinger, L., Huettel, B., Kreil, D.P., Breinig, F., et al. (2010) RNA-directed DNA methylation and plant development require an IWR1-type transcription factor. *EMBO Rep. 11*: 65–71.

Kanno, T., Mette, M.F., Kreil, D.P., Aufsatz, W., Matzke, M., and Matzke, A.J. (2004) Involvement of putative SNF2 chromatin remodeling protein DRD1 in RNA-directed DNA methylation. *Curr. Biol. 14*: 801–805.

Kato, M., Miura, A., Bender, J., Jacobsen, S.E., and Kakutani, T. (2003) Role of CG and non-CG methylation in immobilization of transposons in *Arabidopsis*. *Curr. Biol. 13*: 421–426.

Kinoshita, Y., Saze, H., Kinoshita, T., Miura, A., Soppe, W.J., Koornneef, M., and Kakutani, T. (2007) Control of *FWA* gene silencing in *Arabidopsis thaliana* by *SINE*-related direct repeats. *Plant J. 49*: 38–45.

Klar, A.J. (2007) Lessons learned from studies of fission yeast mating-type switching and silencing. *Annu. Rev. Genet. 41*: 213–236.

Kriaucionis, S., and Heintz, N. (2009) The nuclear DNA base 5-hydroxymethylcytosine is present in Purkinje neurons and the brain. *Science 324*: 929–930.

Lau, N.C., Seto, A.G., Kim, J., Kuramochi-Miyagawa, S., Nakano, T., Bartel, D.P., and Kingston, R.E. (2006) Characterization of the piRNA complex from rat testes. *Science 313*: 363–367.

Law, J.A., Ausin, I., Johnson, L.M., Vashisht, A.A., Zhu, J.K., Wohlschlegel, J.A., and Jacobsen, S.E. (2010) A protein complex required for polymerase V transcripts and RNA-directed DNA methylation in *Arabidopsis*. *Curr. Biol. 20*: 951–956.

Law, J.A., Vashisht, A.A., Wohlschlegel, J.A., and Jacobsen, S.E. (2011) SHH1, a homeodomain protein required for DNA methylation, as well as RDR2, RDM4, and chromatin remodeling factors, associate with RNA polymerase IV. *PLoS Genet. 7*: e1002195.

Li, F., Martienssen, R., and Cande, W.Z. (2011) Coordination of DNA replication and histone modification by the Rik1-Dos2 complex. *Nature 475*: 244–248.

Lindroth, A.M., Shultis, D., Jasencakova, Z., Fuchs, J., Johnson, L., Schubert, D., et al. (2004) Dual histone H3 methylation marks at lysines 9 and 27 required for interaction with CHROMOMETHYLASE3. *EMBO J. 23*: 4286–4296.

Lippman, Z., Gendrel, A.V., Black, M., Vaughn, M.W., Dedhia, N., McCombie, W.R., et al. (2004) Role of transposable elements in heterochromatin and epigenetic control. *Nature 430*: 471–476.

Lippman, Z., May, B., Yordan, C., Singer, T., and Martienssen, R. (2003) Distinct mechanisms determine transposon inheritance and methylation via small interfering RNA and histone modification. *PLoS Biol. 1*: E67.

Lister, R., O'Malley, R.C., Tonti-Filippini, J., Gregory, B.D., Berry, C.C., Millar, A.H., and Ecker, J.R. (2008) Highly integrated single-base resolution maps of the epigenome in *Arabidopsis*. *Cell 133*: 523–536.

Liu, C., Lu, F., Cui, X., and Cao, X. (2010a) Histone methylation in higher plants. *Annu. Rev. Plant Biol. 61*: 395–420.

Liu, J., He, Y., Amasino, R., and Chen, X. (2004) siRNAs targeting an intronic transposon in the regulation of natural flowering behavior in *Arabidopsis*. *Genes Dev. 18*: 2873–2878.

Liu, J., Ren, X., Yin, H., Wang, Y., Xia, R., and Gong, Z. (2010b) Mutation in the catalytic subunit of DNA polymerase alpha influences transcriptional gene silencing and homologous recombination in *Arabidopsis*. *Plant J. 61*: 36–45.

Liu, Q., Wang, J., Miki, D., Xia, R., Yu, W., He, J., Zheng, Z., Zhu, J.K., and Gong, Z. (2010c) DNA replication factor C1 mediates genomic stability and transcriptional gene silencing in *Arabidopsis*. *Plant Cell 22*: 2336–2352.

Luo, Y.C., Zhou, H., Li, Y., Chen, J.Y., Yang, J.H., Chen, Y.Q., and Qu, L.H. (2006) Rice embryogenic calli express a unique set of microRNAs, suggesting regulatory roles of microRNAs in plant post-embryogenic development. *FEBS Lett. 580*: 5111–5116.

Malone, C.D., Brennecke, J., Dus, M., Stark, A., McCombie, W.R., Sachidanandam, R., and Hannon, G.J. (2009) Specialized piRNA pathways act in germline and somatic tissues of the *Drosophila* ovary. *Cell 137*: 522–535.

Martienssen, R.A. (2003) Maintenance of heterochromatin by RNA interference of tandem repeats. *Nat. Genet. 35*: 213–214.

Martienssen, R. (2010a) Molecular biology. Small RNA makes its move. *Science 328*: 834–835.

Martienssen, R.A. (2010b) Heterochromatin, small RNA and post-fertilization dysgenesis in allopolyploid and interploid hybrids of *Arabidopsis*. *New Phytol. 186*: 46–53.

Martienssen, R., Barkan, A., Taylor, W.C., and Freeling, M. (1990) Somatically heritable switches in the DNA modification of *Mu* transposable elements monitored with a suppressible mutant in maize. *Genes Dev. 4*: 331–343.

Martienssen, R., and Baron, A. (1994) Coordinate suppression of mutations caused by Robertson's mutator transposons in maize. *Genetics 136*: 1157–1170.

Mathieu, O., and Bender, J. (2004) RNA-directed DNA methylation. *J. Cell Sci. 117*: 4881–4888.

Mathieu, O., Reinders, J., Caikovski, M., Smathajitt, C., and Paszkowski, J. (2007) Transgenerational stability of the *Arabidopsis* epigenome is coordinated by CG methylation. *Cell 130*: 851–862.

Matzke, M.A., and Birchler, J.A. (2005) RNAi-mediated pathways in the nucleus. *Nature Rev. Genet. 6*: 24–35.

Matzke, M., Kanno, T., Daxinger, L., Huettel, B., and Matzke, A.J. (2009) RNA-mediated chromatin-based silencing in plants. *Curr. Opin. Cell Biol. 21*: 367–376.

McClintock, B. (1956) Controlling elements and the gene. *Cold Spring Harb. Symp. Quant. Biol. 21*: 197–216.

Mi, S., Cai, T., Hu, Y., Chen, Y., Hodges, E., Ni, F., et al. (2008) Sorting of small RNAs into *Arabidopsis* argonaute complexes is directed by the 5′ terminal nucleotide. *Cell 133*: 116–127.

Mirouze, M., Reinders, J., Bucher, E., Nishimura, T., Schneeberger, K., Ossowski, S., Cao, J., Weigel, D., Paszkowski, J., and Mathieu, O. (2009) Selective epigenetic control of retrotransposition in *Arabidopsis*. *Nature 461*: 427–430.

Mittelsten Scheid, O., Afsar, K., and Paszkowski, J. (1998) Release of epigenetic gene silencing by trans-acting mutations in *Arabidopsis*. *Proc. Natl. Acad. Sci. USA 95*: 632–637.

Miura, A., Yonebayashi, S., Watanabe, K., Toyama, T., Shimada, H., and Kakutani, T. (2001) Mobilization of transposons by a mutation abolishing full DNA methylation in *Arabidopsis*. *Nature 411*: 212–214.

Morales-Ruiz, T., Ortega-Galisteo, A.P., Ponferrada-Marin, M.I., Martinez-Macias, M.I., Ariza, R.R., and Roldan-Arjona, T. (2006) *DEMETER* and *REPRESSOR OF SILENCING 1* encode 5-methylcytosine DNA glycosylases. *Proc. Natl. Acad. Sci. USA 103*: 6853–6858.

Morin, R.D., Aksay, G., Dolgosheina, E., Ebhardt, H.A., Magrini, V., Mardis, E.R., Sahinalp, S.C., and Unrau, P.J. (2008) Comparative analysis of the small RNA transcriptomes of *Pinus contorta* and *Oryza sativa*. *Genome Res. 18*: 571–584.

Mosher, R.A., Melnyk, C.W., Kelly, K.A., Dunn, R.M., Studholme, D.J., and Baulcombe, D.C. (2009) Uniparental expression of PolIV-dependent siRNAs in developing endosperm of *Arabidopsis*. *Nature 460*: 283–286.

Nobuta, K., Lu, C., Shrivastava, R., Pillay, M., De Paoli, E., Accerbi, M., et al. (2008) Distinct size distribution of endogeneous siRNAs in maize: Evidence from deep sequencing in the mop1-1 mutant. *Proc. Natl. Acad. Sci. USA 105*: 14958–14963.

Nonomura, K., Morohoshi, A., Nakano, M., Eiguchi, M., Miyao, A., Hirochika, H., and Kurata, N. (2007) A germ cell specific gene of the ARGONAUTE family is essential for the progression of premeiotic mitosis and meiosis during sporogenesis in rice. *Plant Cell 19*: 2583–2594.

Okada, T., Endo, M., Singh, M.B., and Bhalla, P.L. (2005) Analysis of the histone H3 gene family in *Arabidopsis* and identification of the male-gamete-specific variant *AtMGH3*. *Plant J. 44*: 557–568.

Olmedo-Monfil, V., Duran-Figueroa, N., Arteaga-Vazquez, M., Demesa-Arevalo, E., Autran, D., Grimanelli, D., Slotkin, R.K., Martienssen, R.A., and Vielle-Calzada, J.-P. (2010) Control of female gamete formation by a small RNA pathway in *Arabidopsis*. *Nature 464*: 628–632.

Onodera, Y., Haag, J.R., Ream, T., Nunes, P.C., Pontes, O., and Pikaard, C.S. (2005) Plant nuclear RNA polymerase IV mediates siRNA and DNA methylation-dependent heterochromatin formation. *Cell 120*: 613–622.

Ooi, S.K., Qiu, C., Bernstein, E., Li, K., Jia, D., Yang, Z., et al. (2007) DNMT3L connects unmethylated lysine 4 of histone H3 to de novo methylation of DNA. *Nature 448*: 714–717.

Ostertag, E.M., DeBerardinis, R.J., Goodier, J.L., Zhang, Y., Yang, N., Gerton, G.L., and Kazazian, H.H., Jr. (2002) A mouse model of human L1 retrotransposition. *Nat. Genet. 32*: 655–660.

Pasyukova, E., Nuzhdin, S., Li, W., and Flavell, A.J. (1997) Germ line transposition of the *copia* retrotransposon in *Drosophila melanogaster* is restricted to males by tissue-specific control of copia RNA levels. *Mol. Gen. Genet. 255*: 115–124.

Peaston, A.E., Evsikov, A.V., Graber, J.H., de Vries, W.N., Holbrook, A.E., Solter, D., and Knowles, B.B. (2004) Retrotransposons regulate host genes in mouse oocytes and preimplantation embryos. *Dev. Cell 7*: 597–606.

Pikaard, C.S., Haag, J.R., Ream, T., and Wierzbicki, A.T. (2008) Roles of RNA polymerase IV in gene silencing. *Trends Plant Sci. 13*: 390–397.

Piriyapongsa, J., and Jordan, I.K. (2008) Dual coding of siRNAs and miRNAs by plant transposable elements. *RNA 14*: 814–821.

Pontes, O., Costa-Nunes, P., Vithayathil, P., and Pikaard, C.S. (2009) RNA polymerase V functions in *Arabidopsis* interphase heterochromatin organization independently of the 24-nt siRNA-directed DNA methylation pathway. *Mol. Plant 2*: 700–710.

Popp, C., Dean, W., Feng, S., Cokus, S.J., Andrews, S., Pellegrini, M., Jacobsen, S.E., and Reik, W. (2010) Genome-wide erasure of DNA methylation in mouse primordial germ cells is affected by AID deficiency. *Nature 463*: 1101–1105.

Qi, Y., He, X., Wang, X.-J., Kohany, O., Jurka, J., and Hannon, G.J. (2006) Distinct catalytic and non-catalytic roles of ARGONAUTE4 in RNA-directed DNA methylation. *Nature 443*: 1008–1012.

Rabinowicz, P.D., Citek, R., Budiman, M.A., Nunberg, A., Bedell, J.A., Lakey, N., O'Shaughnessy, A.L., Nascimento, L.U., McCombie, W.R., and Martienssen, R.A. (2005) Differential methylation of genes and repeats in land plants. *Genome Res. 15*: 1431–1440.

Rai, K., Huggins, I.J., James, S.R., Karpf, A.R., Jones, D.A., and Cairns, B.R. (2008) DNA demethylation in zebrafish involves the coupling of a deaminase, a glycosylase, and gadd45. *Cell 135*: 1201–1212.

Rando, O.J., and Verstrepen, K.J. (2007) Timescales of genetic and epigenetic inheritance. *Cell 128*: 655–668.

Rangwala, S.H., and Richards, E.J. (2004) The value-added genome: building and maintaining genomic cytosine methylation landscapes. *Curr. Opin. Genet. Dev. 14*: 686–691.

Ream, T.S., Haag, J.R., Wierzbicki, A.T., Nicora, C.D., Norbeck, A.D., Zhu, J.-K., Hagen, G., Guilfoyle, T.J., Paaa-Toli, L., and Pikaard, C.S. (2009) Subunit compositions of the RNA-silencing enzymes Pol IV and Pol V reveal their origins as specialized forms of RNA polymerase II. *Mol. Cell 33*: 192–203.

Ronemus, M., Vaughn, M.W., and Martienssen, R.A. (2006) MicroRNA-targeted and small interfering RNA-mediated mRNA degradation is regulated by argonaute, dicer, and RNA-dependent RNA polymerase in *Arabidopsis*. *Plant Cell 18*: 1559–1574.

Saze, H., and Kakutani, T. (2007) Heritable epigenetic mutation of a transposon-flanked *Arabidopsis* gene due to lack of the chromatin-remodeling factor DDM1. *EMBO J. 26*: 3641–3652.

Saze, H., and Kakutani, T. (2011) Differentiation of epigenetic modifications between transposons and genes. *Curr. Opin. Plant Biol. 14*: 81–87.

Saze, H., Mittelsten Scheid, O., and Paszkowski, J. (2003) Maintenance of CpG methylation is essential for epigenetic inheritance during plant gametogenesis. *Nature Genet. 34*: 65–69.

Saze, H., Shiraishi, A., Miura, A., and Kakutani, T. (2008) Control of genic DNA methylation by a jmjC domain-containing protein in *Arabidopsis thaliana*. *Science 319*: 462–465.

Schalch, T., Job, G., Shanker, S., Partridge, J.F., and Joshua-Tor, L. (2011) The Chp1-Tas3 core is a multifunctional platform critical for gene silencing by RITS. *Nature Struct. Mol. Biol. 18*: 1351–1357.

Schmitz, R.J., Schultz, M.D., Lewsey, M.G., O'Malley, R.C., Urich, M.A., Libiger, O., Schork, N.J., and Ecker, J.R. (2011) Transgenerational epigenetic instability is a source of novel methylation variants. *Science 334*: 369–373.

Schoft, V.K., Chumak, N., Mosiolek, M., Slusarz, L., Komnenovic, V., Brownfield, L., Twell, D., Kakutani, T., and Tamaru, H. (2009) Induction of RNA-directed DNA methylation upon decondensation of constitutive heterochromatin. *EMBO Rep. 10*: 1015–1021.

Schwab, R., Maizel, A., Ruiz-Ferrer, V., Garcia, D., Bayer, M., Crespi, M., Voinnet, O., and Martienssen, R.A. (2009) Endogenous TasiRNAs mediate non-cell autonomous effects on gene regulation in *Arabidopsis thaliana*. *PLoS One 4*: e5980.

Scott, R.J., and Spielman, M. (2006) Genomic imprinting in plants and mammals: how life history constrains convergence. *Cytogenet. Genome Res. 113*: 53–67.

Settles, A.M., Baron, A., Barkan, A., and Martienssen, R.A. (2001) Duplication and suppression of chloroplast protein translocation genes in maize. *Genetics 157*: 349–360.

Sidorenko, L., Dorweiler, J.E., Cigan, A.M., Arteaga-Vazquez, M., Vyas, M., Kermicle, J., Jurcin, D., Brzeski, J., Cai, Y., and Chandler, V.L. (2009) A dominant mutation in mediator of paramutation2, one of three second-largest subunits of a plant-specific RNA polymerase, disrupts multiple siRNA silencing processes. *PLoS Genet. 5*: e1000725.

Singer, T., Yordan, C., and Martienssen, R.A. (2001) Robertson's *Mutator* transposons in *A. thaliana* are regulated by the chromatin-remodeling gene *Decrease in DNA Methylation* (*DDM1*). *Genes Dev. 15*: 591–602.

Singh, M., Goel, S., Meeley, R.B., Dantec, C., Parrinello, H., Michaud, C., Leblanc, O., and Grimanelli, D. (2011) Production of viable gametes without meiosis in maize deficient for an ARGONAUTE protein. *Plant Cell 23*: 443–458.

Slotkin, R.K., Freeling, M., and Lisch, D. (2005) Heritable transposon silencing initiated by a naturally occurring transposon inverted duplication. *Nat. Genet. 37*: 641–644.

Slotkin, R.K., and Martienssen, R. (2007) Transposable elements and the epigenetic regulation of the genome. *Nat. Rev. Genet. 8*: 272–285.

Slotkin, R.K., Vaughn, M., Borges, F., Tanurdzic, M., Becker, J.D., Feijó, J.A., and Martienssen, R.A. (2009) Epigenetic reprogramming and small RNA silencing of transposable elements in pollen. *Cell 136*: 461–472.

Smith, L.M., Pontes, O., Searle, I., Yelina, N., Yousafzai, F.K., Herr, A.J., Pikaard, C.S., and Baulcombe, D.C. (2007) An SNF2 protein associated with nuclear RNA silencing and the spread of a silencing signal between cells in *Arabidopsis*. *Plant Cell 19*: 1507–1521.

Soppe, W.J., Jacobsen, S.E., Alonso-Blanco, C., Jackson, J.P., Kakutani, T., Koornneef, M., and Peeters, A.J. (2000) The late flowering phenotype of *fwa* mutants is caused by gain-of-function epigenetic alleles of a homeodomain gene. *Mol. Cell 6*: 791–802.

Stam, M., Belele, C., Dorweiler, J.E., and Chandler, V.L. (2002a) Differential chromatin structure within a tandem array 100 kb upstream of the maize *b1* locus is associated with paramutation. *Genes Dev. 16*: 1906–1918.

Stam, M., Belele, C., Ramakrishna, W., Dorweiler, J.E., Bennetzen, J.L., and Chandler, V.L. (2002b) The regulatory regions required for *B'* paramutation and expression are located far upstream of the maize *b1* transcribed sequences. *Genetics 162*: 917–930.

Stokes, T.L., Kunkel, B.N., and Richards, E.J. (2002) Epigenetic variation in *Arabidopsis* disease resistance. *Genes Dev. 16*: 171–182.

Stokes, T.L., and Richards, E.J. (2002) Induced instability of two *Arabidopsis* constitutive pathogen-response alleles. *Proc. Natl. Acad. Sci. USA 99*: 7792–7796.

Stonaker, J.L., Lim, J.P., Erhard, K.F., Jr., and Hollick, J.B. (2009) Diversity of Pol IV function is defined by mutations at the maize *rmr7* locus. *PLoS Genet. 5*: e1000706.

Tahiliani, M., Koh, K.P., Shen, Y., Pastor, W.A., Bandukwala, H., Brudno, Y., et al. (2009) Conversion of 5-methylcytosine to 5-hydroxymethylcytosine in mammalian DNA by MLL partner TET1. *Science 324*: 930–935.

Tanurdzic, M., Vaughn, M.W., Jiang, H., Lee, T.-J., Slotkin, R.K., Sosinski, B., Thompson, W.F., Doerge, R.W., and Martienssen, R.A. (2008) Epigenomic consequences of immortalized plant cell suspension culture. *PLoS Biol. 6*: e302.

Tariq, M., Saze, H., Probst, A.V., Lichota, J., Habu, Y., and Paszkowski, J. (2003) Erasure of CpG methylation in *Arabidopsis* alters patterns of histone H3 methylation in heterochromatin. *Proc. Natl. Acad. Sci. USA 100*: 8823–8827.

Teixeira, F.K., Heredia, F., Sarazin, A., Roudier, F., Boccara, M., Ciaudo, C., et al. (2009) A role for RNAi in the selective correction of DNA methylation defects. *Science 323*: 1600–1604.

Tran, R.K., Henikoff, J.G., Zilberman, D., Ditt, R.F., Jacobsen, S.E., and Henikoff, S. (2005) DNA methylation profiling identifies CG methylation clusters in *Arabidopsis* genes. *Curr. Biol. 15*: 154–159.

Tsukahara, S., Kobayashi, A., Kawabe, A., Mathieu, O., Miura, A., and Kakutani, T. (2009) Bursts of retrotransposition reproduced in *Arabidopsis*. *Nature 461*: 423–426.

Turck, F., Roudier, F.O., Farrona, S., Martin-Magniette, M.-L., Guillaume, E., Buisine, N., Gagnot, S.V., Martienssen, R.A., Coupland, G., and Colot, V. (2007) *Arabidopsis* TFL2/LHP1 specifically associates with genes marked by trimethylation of histone H3 lysine 27. *PLoS Genet. 3*: e86.

Vaughn, M.W., and Martienssen, R.A. (2005) Finding the right template: RNA Pol IV, a plant-specific RNA polymerase. *Mol. Cell 17*: 754–756.

Vaughn, M.W., Tanurdzic, M., Lippman, Z., Jiang, H., Carrasquillo, R., Rabinowicz, P.D., et al. (2007) Epigenetic natural variation in *Arabidopsis thaliana*. *PLoS Biol. 5*: e174.

Verdel, A., Jia, S.T., Gerber, S., Sugiyama, T., Gygi, S., Grewal, S.I.S., and Moazed, D. (2004) RNAi-mediated targeting of heterochromatin by the RITS complex. *Science 303*: 672–676.

Verdel, A., and Moazed, D. (2005) RNAi-directed assembly of heterochromatin in fission yeast. *FEBS Lett. 579*: 5872–5878.

Voinnet, O. (2005) Non-cell autonomous RNA silencing. *FEBS Lett. 579*: 5858–5871.

Vongs, A., Kakutani, T., Martienssen, R.A., and Richards, E.J. (1993) *Arabidopsis thaliana* DNA methylation mutants. *Science 260*: 1926–1928.

Watanabe, T., Totoki, Y., Toyoda, A., Kaneda, M., Kuramochi-Miyagawa, S., Obata, Y., et al. (2008) Endogenous siRNAs from naturally formed dsRNAs regulate transcripts in mouse oocytes. *Nature 453*: 539–543.

Wierzbicki, A.T., Haag, J.R., and Pikaard, C.S. (2008) Noncoding transcription by RNA polymerase Pol IVb/Pol V mediates transcriptional silencing of overlapping and adjacent genes. *Cell 135*: 635–648.

Woo, H.R., Dittmer, T.A., and Richards, E.J. (2008) Three SRA-domain methylcytosine-binding proteins cooperate to maintain global CpG methylation and epigenetic silencing in *Arabidopsis*. *PLoS Genet. 4*: e1000156.

Woo, H.R., Pontes, O., Pikaard, C.S., and Richards, E.J. (2007) VIM1, a methylcytosine-binding protein required for centromeric heterochromatinization. *Genes Dev. 21*: 267–277.

Wossidlo, M., Arand, J., Sebastiano, V., Lepikhov, K., Boiani, M., Reinhardt, R., Scholer, H., and Walter, J. (2010) Dynamic link of DNA demethylation, DNA strand breaks and repair in mouse zygotes. *EMBO J. 29*: 1877–1888.

Wu, L., Zhou, H., Zhang, Q., Zhang, J., Ni, F., Liu, C., and Qi, Y. (2010) DNA methylation mediated by a microRNA pathway. *Mol. Cell 38*: 465–475.

Yin, H., Zhang, X., Liu, J., Wang, Y., He, J., Yang, T., Hong, X., Yang, Q., and Gong, Z. (2009) Epigenetic regulation, somatic homologous recombination, and abscisic acid signaling are influenced by DNA polymerase epsilon mutation in *Arabidopsis*. *Plant Cell 21*: 386–402.

Zaratiegui, M., Castel, S.E., Irvine, D.V., Kloc, A., Ren, J., Li, F., et al. (2011) RNAi promotes heterochromatic silencing through replication-coupled release of RNA Pol II. *Nature 479*: 135–138.

Zaratiegui, M., Irvine, D.V., and Martienssen, R.A. (2007) Noncoding RNAs and gene silencing. *Cell 128*: 763–776.

Zemach, A., McDaniel, I.E., Silva, P., and Zilberman, D. (2010) Genome-wide evolutionary analysis of eukaryotic DNA methylation. *Science 328*: 916–919.

Zhang, K., Fischer, T., Porter, R.L., Dhakshnamoorthy, J., Zofall, M., Zhou, M., Veenstra, T., and Grewal, S.I. (2011) Clr4/Suv39 and RNA quality control factors cooperate to trigger RNAi and suppress antisense RNA. *Science 331*: 1624–1627.

Zhang, X., Bernatavichute, Y.V., Cokus, S., Pellegrini, M., and Jacobsen, S.E. (2009) Genome-wide analysis of mono-, di- and trimethylation of histone H3 lysine 4 in *Arabidopsis thaliana*. *Genome Biol. 10*: R62.

Zhang, X., Clarenz, O., Cokus, S., Bernatavichute, Y.V., Pellegrini, M., Goodrich, J., and Jacobsen, S.E. (2007a) Whole-genome analysis of histone H3 lysine 27 trimethylation in *Arabidopsis*. *PLoS Biol. 5*: e129.

Zhang, X., Richards, E.J., and Borevitz, J.O. (2007b) Genetic and epigenetic dissection of *cis* regulatory variation. *Curr. Opin. Plant. Biol. 10*: 142–148.

Zhang, X., Shiu, S., Cal, A., and Borevitz, J.O. (2008) Global analysis of genetic, epigenetic and transcriptional polymorphisms in *Arabidopsis thaliana* using whole genome tiling arrays. *PLoS Genet. 4*: e1000032.

Zhang, X., Yazaki, J., Sundaresan, A., Cokus, S., Chan, S.W., Chen, H., et al. (2006) Genome-wide high-resolution mapping and functional analysis of DNA methylation in *Arabidopsis*. *Cell 126*: 1189–1201.

Zheng, B., Wang, Z., Li, S., Yu, B., Liu, J.Y., and Chen, X. (2009) Intergenic transcription by RNA polymerase II coordinates Pol IV and Pol V in siRNA-directed transcriptional gene silencing in *Arabidopsis*. *Genes Dev. 23*: 2850–2860.

Zheng, X., Pontes, O., Zhu, J., Miki, D., Zhang, F., Li, W.X., Iida, K., Kapoor, A., Pikaard, C.S., and Zhu, J.K. (2008) ROS3 is an RNA-binding protein required for DNA demethylation in *Arabidopsis*. *Nature 455*: 1259–1262.

Zheng, X., Zhu, J., Kapoor, A., and Zhu, J.K. (2007) Role of *Arabidopsis* AGO6 in siRNA accumulation, DNA methylation and transcriptional gene silencing. *EMBO J. 26*: 1691–1701.

Zheng, Z., Xing, Y., He, X.J., Li, W., Hu, Y., Yadav, S.K., Oh, J., and Zhu, J.K. (2010) An SGS3-like protein functions in RNA-directed DNA methylation and transcriptional gene silencing in *Arabidopsis*. *Plant J. 62*: 92–99.

Zilberman, D., Gehring, M., Tran, R.K., Ballinger, T., and Henikoff, S. (2007) Genome-wide analysis of *Arabidopsis thaliana* DNA methylation uncovers an interdependence between methylation and transcription. *Nat. Genet. 39*: 61–69.

Zilberman, D., and Henikoff, S. (2004) Silencing of transposons in plant genomes: kick them when they're down. *Genome Biol. 5*: 249.

6 Transposons in Plant Gene Regulation

Damon R. Lisch

Introduction

A conspicuous feature of all plant genomes is the presence of vast numbers of transposable elements (TEs). Barbara McClintock named them "controlling elements" because they have the capacity to regulate expression of genes (McClintock, 1956a, 1956b). Given that very little was known about the nature of genes at the time, it was this feature of TEs even more than their propensity to move that most interested McClintock. Indeed, the observation that a TE can regulate expression of more than one gene coordinately elegantly demonstrated for the first time that gene regulation can be separated from gene function and independently altered to affect expression of multiple genes. However, although we now know that TEs can influence gene regulation in many ways, it is also clear that TEs are not the primary means by which genes are regulated. Indeed, most individual TEs are unlikely to have any specific function other than their own propagation. Viewed from this perspective, first put forward in 1980 (Doolittle and Sapienza, 1980; Orgel and Crick, 1980), TEs were labeled as "junk DNA" or "parasites," surviving only because they have an intrinsic replication advantage over other genetic material. They are not "for" anything with respect to their hosts, any more than are viruses; they exist because they can. This perspective has important implications for how we view evolution. Selection at the level of the individual organism may be the ultimate arbiter of the success of a given lineage, but not all changes within genomes are a consequence of that selection. This notion is in line with the neutral theory of evolution (Kimura, 1968; Nei et al., 2011), but further posits that competition among genes can drive changes not only in the genes themselves, but in the overall structure of a genome. Whatever else they are, genomes are an environment in which a menagerie of genetic elements compete for advantage (Brookfield, 2005). In many organisms, but especially in plants, a majority of DNA appears to be the product of that competition.

The observation that even closely related plant species can exhibit dramatically different numbers of TEs supports the idea that any single element in any given genome is essentially parasitic (Piegu et al., 2006; Vitte and Bennetzen, 2006). Further, the extraordinary lengths taken by all eukaryotes to prevent TE amplification via a variety of epigenetic silencing pathways underscores the potential damage represented by TE activity, and suggests that TEs can be seen as an intrinsic pathogen from which the host must be constantly defended (Slotkin and Martienssen, 2007; Lisch, 2009). Perhaps in a sign of the times, protecting the "integrity of the genome" has become a favored phrase, and there is evidence that unchecked TE activity can have profoundly negative consequences (Kidwell et al., 1977; Walbot, 1991; Belancio et al., 2010).

Plant Transposons and Genome Dynamics in Evolution, First Edition. Edited by Nina V. Fedoroff.

However, it is important to emphasize that TEs are far from being junk DNA. Junk is something without any function at all, but TEs, at least those that are still around and still recognized as TEs, have been selected for replication advantage over hundreds of millions of years in the face of intense pressure in the form of host-encoded silencing systems, as well as selection against the negative consequences of TE activity. As a result, these DNA sequences are constantly evolving, blooming into massive families that can take up significant portions of a given genome, parasitizing each other, evolving new capabilities, shifting to new hosts, or dying out. The extraordinary diversity of TEs, even within a single genome, is a testament to the capacity of these genetic elements to evolve in response to (or in spite of) complex combinations of selection at the level of the gene and the organism (Venner et al., 2009).

To the extent that they are selfish (and not all of them are), it is only through amplification that TEs continue to survive. In order to do so, they must carry information, the opposite of junk. If they are autonomous, they must carry at least minimal promoter elements in order to express mRNA that encodes proteins such as transposase or reverse transcriptase. In addition, there is ample evidence that autonomous elements can be exquisitely sensitive to a variety of other signals, including those involving cell cycle, developmental timing, and both biotic and abiotic stresses (Donlin et al., 1995; Wessler, 1996; Raizada et al., 2001; Grandbastien et al., 2005b; Naito et al., 2009). If they are Class II, cut-and paste-elements, both autonomous and nonautonomous elements must include transposase binding sites, which are often composed of terminal and subterminal repeats (Craig, 2002). The proteins encoded by the elements must have the capability to perform all of the steps required for transposition. For Class II elements, this includes both the excision and integration reactions. For Class I elements, this includes a reverse transcriptase and an integrase of some kind. For at least some types of elements, such as the chromoviruses, there are additional capabilities, such as the capacity to recognize and bind to specific forms of histones (Gao et al., 2008).

Collectively, then, TEs represent a vast pool of constantly evolving information. Most of this information is employed by the TEs for a replication advantage. However, there are tens of thousands of rapidly evolving TEs in most plant genomes, providing the host with new opportunities to co-opt TE-encoded functions. Further, as a consequence of their activity, TEs are prone to produce a wide variety of rearrangements, deletions, and insertions, many of which can turn out to be selectively beneficial for the host. Thus, selection at the level of the gene has resulted in an extraordinary capacity for genomes to evolve. Rather than just being a pathogen that must be endured, then, TEs also represent an opportunity to evolve, a vast reservoir of potentially useful information. This has led to a more balanced view of TEs as DNA sequences that can occupy any point on a spectrum that ranges from primarily parasitic to entirely beneficial (Kidwell and Lisch, 2001).

This chapter will present experimental evidence demonstrating the potential of TEs to contribute to plant evolution, as well as comparative and historical evidence that suggests that they have, in fact, contributed to plant evolution. This is particularly important if one is to seek a balanced perspective; the simple fact that geneticists can find numerous examples for a potential role of TEs in evolution does not mean that TEs have in fact been essential players. The examples presented in this chapter are intriguing, but we do not yet know how representative they are of the total set of mutations that, collectively, leads to organismal evolution. With that in mind, let us consider the evidence.

New Regulatory Functions

Changes in the regulation of gene expression are a key component of evolutionary change (Prud'homme et al., 2007; Brakefield, 2011). From McClintock's earliest experiments, it was clear that TEs can alter regulation of gene expression, and there is a long history of observations detailing the ways in which they can do so (King, 1985; Britten, 1996; Kidwell and Lisch, 2001; Dooner and Weil, 2007; Schumann et al., 2010). The details of McClintock's observations have been described extensively elsewhere (Fedoroff, 1989a; Fedoroff, 1989b), but the broad implications of those observations deserve to be emphasized here. In the absence of any of the molecular tools that we take for granted today, McClintock demonstrated that TEs can cause a variety of qualitative and quantitative changes in gene expression. Further, her analysis made very clear that once a TE inserts into or near a gene, that gene is subject to many additional changes.

Class II DNA transposons are often found inserted into single-copy genic sequences, and are often associated with 5′UTRs and promoters (Mao et al., 2000; Feschotte et al., 2002; Jiang et al., 2011). In part, this is due to the effects of selection; most exon insertions are deleterious and would be expected to be rapidly removed from populations, particularly in self-fertilizing species such as *Arabidopsis* and rice. However, analysis of de novo insertion profiles demonstrates that many, probably most, Class II elements, as well as at least some Class I elements, insert preferentially into or near genes, and often into regulatory regions (Miyao et al., 2003; Naito et al., 2006; Hsing et al., 2007; Liu et al., 2009). Given that TEs often contain their own promoter and enhancer elements, it is clear that they are at least potentially competent to reprogram gene expression.

One way that TEs can reprogram gene expression is to bring new forms of tissue specificity. In maize, anthocyanin pigment gene expression requires, among other factors, one of two basic helix-loop-helix *myc* genes (*r1* or *b1*) and one of two *myb* genes (*c1* or *pl*) (Dooner et al., 1991). The duplicate copies of these genes have become subfunctionalized, such that expression of this pathway generally requires *r1* and *c1* in the seed, and *b1* and *pl* in the plant. However, two variants of *b1*, *B-Peru* and *B-Bolivia*, are altered such that they express in the seed instead of in the body of the plant, presumably a trait that had been selected for regions where these alleles arose. This variation is due to the insertion of a TE that carries sequences necessary and sufficient for seed expression (Selinger et al., 1998; Selinger and Chandler, 1999). Given the sequence near-identity among these alleles and a third allele—*B-Gua31*—which does not have this insertion and does not express in the seed, it is likely that these insertions are quite recent. All of the allelic variation associated with dramatic changes in patterns of gene expression of *b1* has been due either to insertions or RFLPs in sequences upstream of the gene rather than point mutations, supporting the idea that TE activity is a major source of variation at this locus (Figure 6.1).

One common feature of TEs is their responsiveness to biotic and abiotic stress (Wessler, 1996). TE expression is induced by salt (Naito et al., 2009), wounding (Mhiri et al., 1997b), cold (Ivashuta et al., 2002; Naito et al., 2009), as well as infection by bacteria (Grandbastien et al., 2005a) and viruses (Buchmann et al., 2009). The reasons for this are poorly understood. One possibility is that stress of various kinds results in a decrease in the ability of plants to maintain TEs in an epigenetically silenced state, rendering them "immunocompromised." Supporting this idea, 10 of 27 mutations that resulted in TE activation in *C. elegans* were involved in mitochondrial function, suggesting that TE control may be an energetically expensive process that might be neglected during times of stress (Vastenhouw et al., 2003). In addition, as part of their infectious cycle, many viruses actually down-regulate the production of small RNAs important for regulating both viruses and TEs (Buchmann et al., 2009). However, there is also evidence that TEs can be directly responsive to at least some stresses. The Tnt1 retroelement of tobacco, for instance, carries specific promoter elements that respond to wounding and abiotic stress (Mhiri et al., 1997a). Similarly, the *TLC1.1* element of *Solanum chilense* contains sequences that confer responsiveness to ethylene and a number of stress elicitor molecules, including salicylic acid, abscisic acid, methyl jasmonate, hydrogen peroxide, and 2,4-D (Salazar et al., 2007). The conclusion from these and other studies is that TEs can be

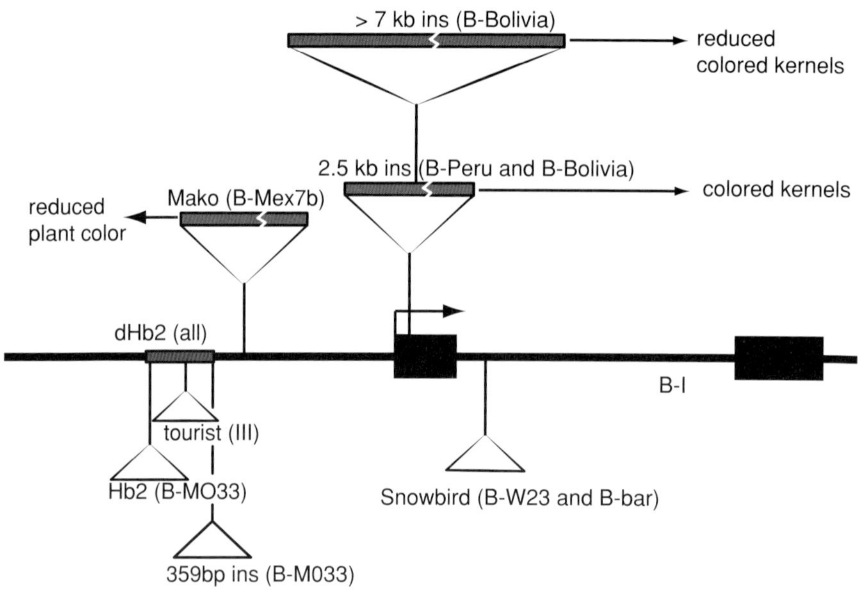

Figure 6.1 Insertions associated with alterations in expression at the *b1* locus in maize. Coding sequences are indicated by black boxes; TE insertions of the indicated types by triangles. Specific alleles of the *b1* locus are as indicated, as are the phenotypic effects of the insertions. Adapted from Selinger and Chandler, 1999.

exquisitely sensitive to stress. Recent work shows that this can have profound effects on expression of genes that are near them.

Rice is host to a wide variety of TEs, including some that are currently active, both in laboratory settings and in the field. MITE elements are capable of rapid and massive amplification in many species. Unlike some other currently active elements, such as *Mu* elements in maize (Candela and Hake, 2008), however, MITE insertions are not especially mutagenic. Indeed, they seem to be particularly facile at minimizing their impact by avoiding exonic sequences, so null mutations caused by these elements are relatively rare (Naito et al., 2006; Naito et al., 2009). However, they can have an effect on neighboring gene expression. Naito and colleagues undertook a large-scale examination of recent and de novo insertions of MITE elements in a line of rice exhibiting high levels of activity. They identified more than 1,500 recent insertions of the mPing MITE, more than 90% of which were into single-copy, genic sequences, but less than 5% of which were in exons. The vast majority of insertions had no effect on normal levels of gene expression. However, a number of the genes with MITE insertions less than 100 bp from the transcriptional start site were induced by cold or salt or both. Alleles of the genes in closely related accessions that lacked the insertions did not exhibit induction, showing that the TE insertions had conferred stress inducibility on these genes. Given the massive and ongoing expansion of MITES in these rice accessions (Naito et al., 2006), it is reasonable to assume that thousands of genes in plants in the field are becoming stress inducible as a consequence of MITE activity. This process may have facilitated adaptation in these accessions, many of which are from geographically isolated populations that may have been subject to particular selection regimes (Naito et al., 2006).

When considering stress, in addition to specific examples like mPing, it is worth considering global effects. Given the large numbers of TEs in many plant genomes and the fact that many of them are located near coding sequences, the global effects of stress on gene expression may be far more extensive and far less predictable than one might expect. Certainly, severe effects would be expected to be winnowed out through selection, and there are doubtless many mechanisms to sequester TE and gene expression (Lippman et al., 2004; Penterman et al., 2007; Saze and Kakutani, 2011). However, a genome that is experiencing rapid TE amplification may not have the luxury, or the capacity, to ameliorate these effects.

This last possibility raises an important point concerning the role of TEs in evolution of host gene expression. TEs are an essentially destabilizing factor; when active, they can cause massive increases in mutation rates and, in some cases, sterility (Kidwell et al., 1977; Robertson, 1978). The average TE that is activated by stress is far more likely to cause a deleterious change than a selectively beneficial change. Thus, the capacity for TEs to cause beneficial mutations is unlikely to have been selected directly, and indeed, those beneficial mutations come with potentially steep costs. In fact a stressed population of plants with large numbers of recently amplified TEs could be at a severe disadvantage when competing with a population that lacks those TEs. We cannot simply assume that the increase in variability associated with TE activity necessarily outweighs the costs that are also a consequence of that activity, or that enhancement of TE activity necessarily represents a response by the host in order to produce new variants (McClintock, 1984; Jorgensen, 2004).

TE-Induced Down-Regulation

In addition to up-regulation of genes, TEs can also down-regulate expression. Most LTR retroelements are methylated and transcriptionally inactive in adult plant tissues (Lister et al., 2008; Gehring et al., 2009; Zemach et al., 2010). However, when retroelements become unmethylated they can express transcripts from each of their Long Terminal Repeats (LTRs). This can result in expression of transcripts originating in the retrotransposon and extending into flanking sequences (Kashkush et al., 2003; Saze and Kakutani, 2007). *Dasheng* is a recently amplified retroelement that is present but polymorphic in a number of species of rice (Jiang et al., 2002; Ammiraju et al., 2007). Most of these elements are methylated and transcriptionally inactive, but some of them are unmethylated and expressed, and this expression is associated with down-regulation of adjacent rice genes (Kashkush and Khasdan, 2007). Remarkably, this hypomethylation can be tissue specific, such that these genes are now regulated in a tissue-specific manner due to differences in TE methylation. Observed differences between *Oryza sativa japonica* and *O. sativa indica* can be attributed to differences in the presence of the *Dasheng* element adjacent to a given gene or to differences in methylation of the element.

These results suggest that elements near genes can be largely invisible under some circumstances, but can profoundly affect those genes under other circumstances. When comparing *japonica* and *indica*, it is also worth pointing out that an element at the same position in these two subspecies implies that this element has remained at that position for the several hundred thousand years since the genes in these subspecies diverged (Molina et al., 2011). If these elements had been deleterious, they would not have been expected to remain at their positions for that long. This suggests that despite their effects on flanking genes, these insertions are either neutral or perhaps even positive. In addition, two subspecies can have the same element at the same position, but the pattern of LTR methylation and expression of the flanking gene can differ. This suggests that the element insertion may have endowed the host with a certain flexibility with respect to epigenetic regulation of that gene. Given that stress can alter expression of LTR retroelements, it would be fascinating to see whether or not stress affects these *Dasheng*-adjacent genes, as well. Reexamination of recently obtained global methylation and expression data for *japonica* and *indica* with these results in mind should also be quite informative (He et al., 2010).

In addition to the kind of dramatic effects described, there is also circumstantial evidence that the presence of silenced TEs near genes can have more subtle and global effects on gene expression. Excellent curation of TE distribution, methylation, and gene expression in *Arabidopsis* has made it possible to correlate these factors. The results of a recent analysis are striking. Proximity to methylated TEs is strongly and negatively correlated with levels of gene expression: the closer a gene is to a methylated TE, the lower its level of expression (Hollister and Gaut, 2009; Hollister et al., 2011). In contrast, there is no such correlation when comparing unmethylated TEs and gene expression. Further, unmethylated TEs are far more likely to be near genes than are methylated TEs, and methylated TEs near genes (but not far from them) are much more likely to show evidence of purifying selection. Ancient (>5 mya) methylated TEs are much more distant from genes than are younger methylated TEs, but the same is not true for unmethylated TEs. Finally, although orthologs of genes in *A. thaliana* and *A. lyrata* that lack nearby silenced TEs express at roughly equivalent levels, those with silenced TEs express at significantly lower levels than do orthologs in the other species that lack those insertions (Hu et al., 2011). In sum, on average, having methylated TEs near a gene reduces its expression, and over time such methylated TEs are lost.

Overall, the authors of these studies suggest that methylation of TEs represents a tradeoff between the benefits of reduced TE-induced mutations and the cost of inducing local islands of heterochromatin. This global cost may be an unavoidable consequence of TE amplification. As with all TE-induced changes, local down-regulation of genes may in some instances have been adaptive (Fujimoto et al., 2008), but it is likely that most of them must simply be endured. By analogy, staying in bed with a fever when we get sick does not represent an important new ability to remain horizontal and hot; it is simply what we need to do to fight off the virus.

Deletions and Rearrangements

Once a TE inserts into a particular position, it may have no immediate effect on nearby genes. However, the presence of the TE can add the potential for subsequent changes that can have profound effects on gene expression. This can be due to recombination between LTRs in retrotransposons, aberrant transposition events, or small changes at the original site of insertion due to imprecise repair of the break caused by an excision.

Some of the best evidence for the ongoing effects of TE insertions comes from maize, particularly from analysis of the genes that regulate expression of anthocyanins. For various sociological and religious reasons, Native Americans selected for a number of mutant phenotypes in these genes. Polymorphism at the *r1* locus, for example, has given rise to a wide variety of color patterns. Although a comparison with sorghum and rice suggests that this locus was ancestrally single copy, some modern alleles are quite complex (Swigonova et al., 2005). The reference *R-r* allele, for instance, is composed of four tandem duplicates of the *r1* coding sequence in the following order: one complete copy with full homology to the rice and sorghum homologues, one truncated duplicate carrying only the 5′ end of the coding sequences juxtaposed directly with a fragment of a *Doppia* transposon, and two final truncated versions of the gene that have a short portion of the 5′ end of the gene missing (Walker et al., 1995) (Figure 6.2). These inverted versions of the gene are separated by a full-length copy of the *Doppia* element. At this locus the ancestral version of the gene is responsible for expression in the plant (the *P* component of the locus). Together, the last three duplicate copies are responsible for seed expression (the *S* component), which is dependent on

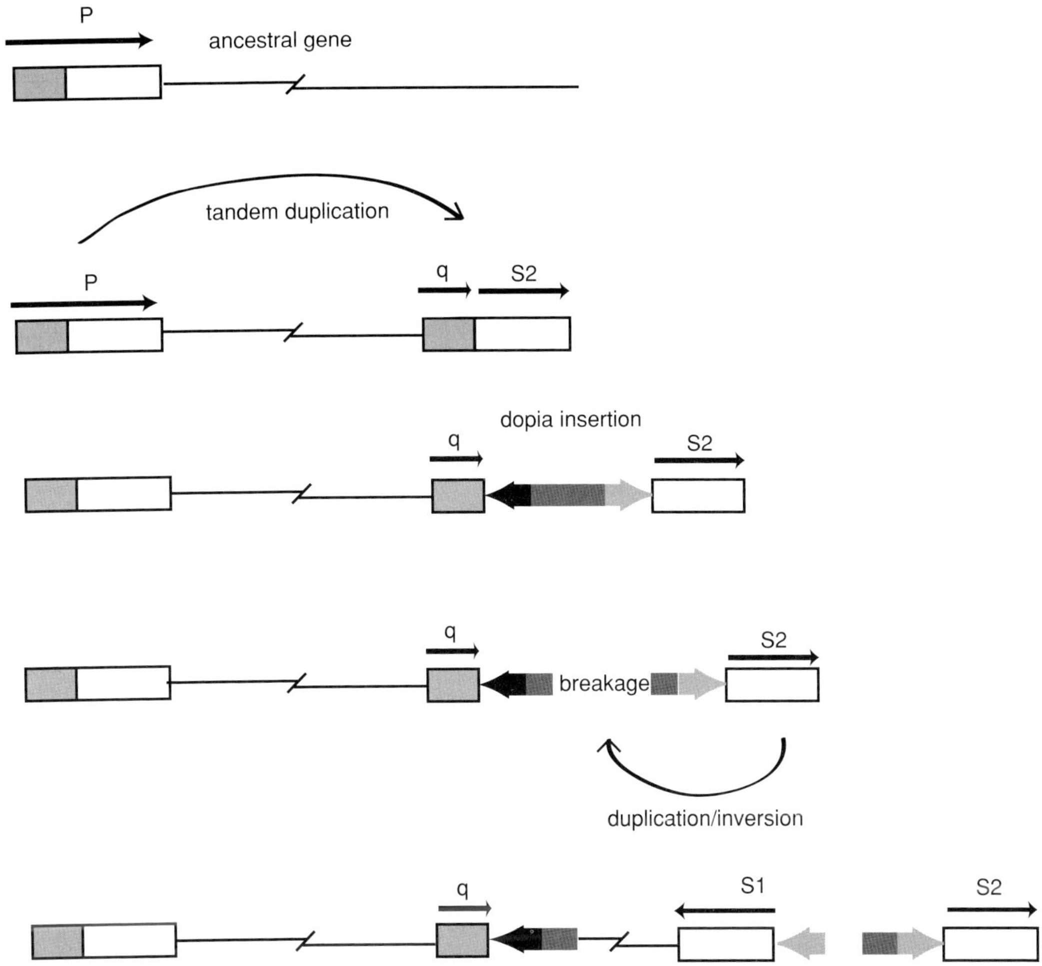

Figure 6.2 Rearrangements associated with changes at the *r1* locus in maize. Rectangles are coding sequences. Shaded boxes and arrows are *Doppia* TE sequences. Adapted from Walker et al., 1995.

the 5′ end of the second, truncated version of the gene in juxtaposition with the *Doppia* fragment. Although a direct involvement of the *Doppia* element in this rearrangement has not been proved in this case, given the propensity for similar Class II elements to induce rearrangements of this kind, it is likely that insertion of this element led to subsequent instability that in turn led to new patterns of tissue specificity (Zhang et al., 2011; Yu et al., 2011). What is particularly interesting in this case is that a single TE may have been involved in first producing functional redundancy and then reprogramming gene expression of one of the copies.

A classic example of rewiring of gene expression comes from *Antirrhinum majus*, the common snapdragon. A *Tam3* element inserted into the promoter of the *nivea* gene resulted in a tissue-specific reduction of pigment in the flower. This loss of pigment was due to displacement of portions of the normal promoter of the *nivea* gene. An aberrant transposition of this *Tam3* element resulted in two linked copies of the element with an inversion between them that caused an increase in expression but retention of tissue specificity. Subsequent excision of the element closest to the *nivea* coding region resulted in a new pattern of expression that is responsive to *cycloidearadialis*, a gene that is required for proper floral polarity (Lister et al., 1993). The author's interpretation of these data is that the initial insertion separated functional component parts of the promoter, and that the subsequent rearrangement and excision events introduced a novel regulatory region to the *nivea* gene.

In some cases, TE activity subsequent to insertion can cause even more dramatic changes at the site of insertion, but leave no trace of the original element behind. The *adh1* gene in maize is required for survival in anoxic conditions (Schwartz, 1969). Plants carrying a mutant allele of this gene with a *Mu* element insertion in the TATA box were screened for revertants. One revertant allele was striking in that its expression differed both quantitatively and qualitatively from both the progenitor and the mutant alleles, resulting in a unique pattern of organ specificity (Kloeckener-Gruissem and Freeling, 1995). The reversion was due to a rather complex rearrangement of the promoter region of the *adh1* gene, which involved deletion, inversion, duplication, and insertion of new sequences, along with the loss of the original *Mu* insertion. The authors of this study suggested that all of these changes resulting from the introduction of a double-stranded gap following the excision of the element, and dubbed the process "promoter scrambling." What is striking about this example is how dramatic the changes were, and the fact that there was no trace of the element following the creation of this new mutant allele. If this allele had ended up being selectively advantageous, there would be little evidence left implicating a TE in creating it, other than a very oddly scrambled promoter. If this kind of event is a common source of important new novelty, we may never know how important TEs are in creating them. It may be informative, however, to carefully examine homologous genes in closely related species for evidence of this kind of change, particularly if it is associated with the loss of a TE in one of the two species.

Suppressible Alleles

One way that TEs can alter genes is by making expression of those genes dependent on the presence of the gene products of autonomous elements. McClintock's observations on the interactions between autonomous *Spm* elements and nonautonomous *dSpm* elements inserted into genes remains to this day some of the most richly nuanced descriptions of the phenomenology of suppressible alleles. These are instances in which the mutant phenotype of a TE insertion depends on the availability of proteins derived from an active element. In one case, a *dSpm* element in the *a1* gene caused a partial loss of pigmentation, likely due to reduced gene expression. In the presence of an active *Spm* element, expression of the *a1* gene is completely suppressed except in somatic sectors in which the *dSpm* element has excised from the *a1* gene, resulting in normal levels of expression of that gene (Masson et al., 1987). Other insertion alleles exhibit the opposite effect in that gene expression at a low level is observed only in the presence of an *Spm* element and also completely restored in revertant sectors. McClintock found that these two functions (suppressor and mutator) were separable, and that both *dSpm* and *Spm* elements were subject to both irreversible (state) and reversible (phase) changes (reviewed by Fedoroff, 1989a; and Federoff, Chapter 5 in this volume).

More recent observations have made it clear that suppressibility is a common feature of TE-induced mutations (Figure 6.3). At a molecular level, the best understood examples of this are from the *Mutator* system in maize. In each case, a nonautonomous element is inserted into a gene, and the phenotype of that insertion depends on the presence of the autonomous element (*MuDR*). A wide variety of suppressible alleles have been identified (Figure 6.3). In nearly all cases, gene expression is blocked due to the presence of the transposase. In several examples, the insertion of a

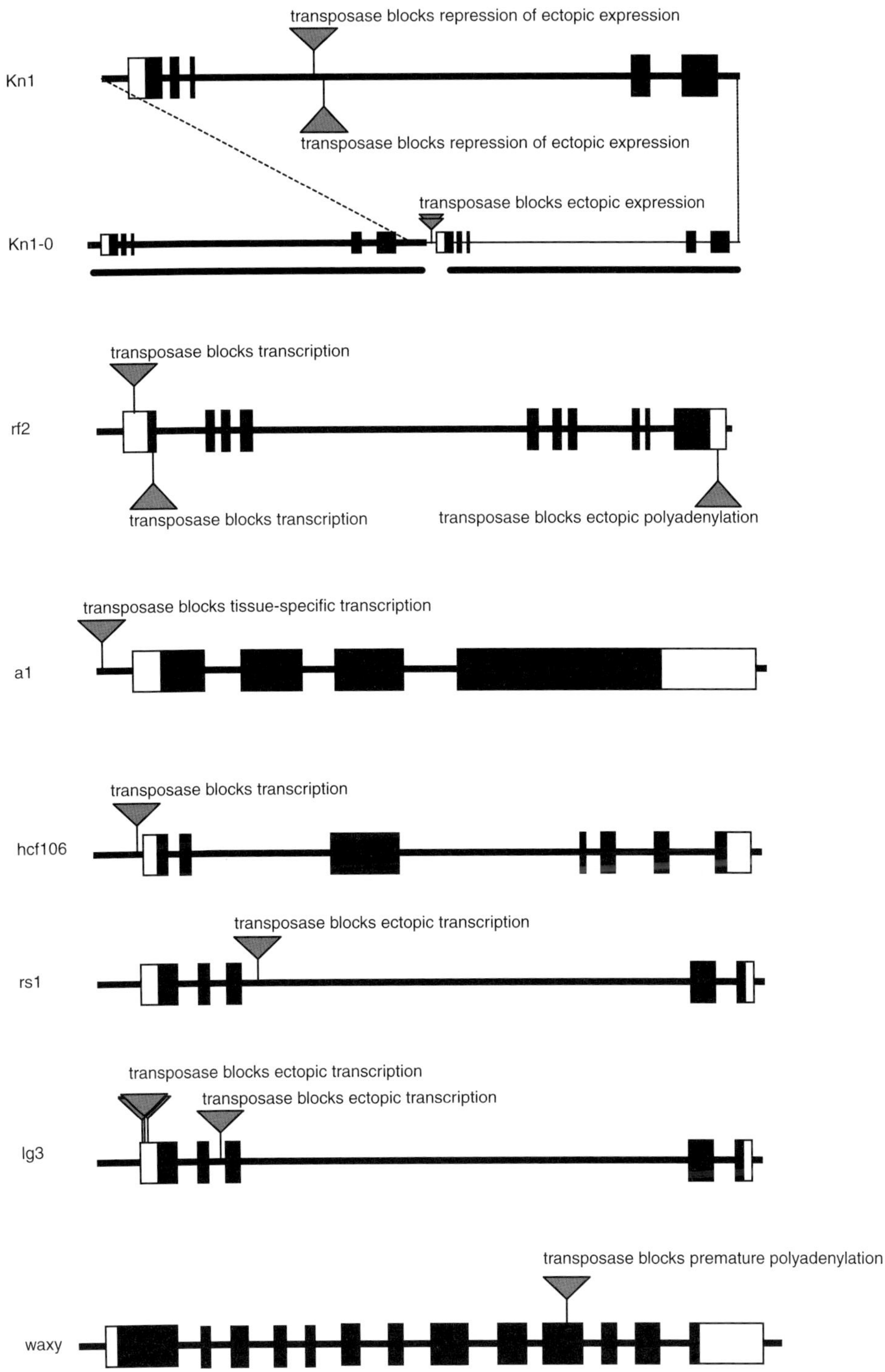

Figure 6.3 Suppressible alleles in maize. Grey triangles represent TE insertions. Black boxes represent coding sequences, and white boxes represent untranslated regions. The effects of the insertions are as indicated.

nonautonomous element causes a mutant phenotype only if the transposase is present. This is the case for an allele of *high chlorophyll floresence 106*, in which a *Mu1* insertion in the promoter of the *hcf106* confers a mutant phenotype only in the presence of *Mu* activity (Barkan and Martienssen, 1991). In its absence, transcription is initiated from both downstream of the insertion and from within the *Mu1* element's terminal inverted repeat. Similarly, the *a1-mum2* allele of the *a1* gene in maize shows a mutant phenotype only in the presence of *MuDR* activity (Chomet et al., 1991; Pooma et al., 2002). Interestingly, this allele also shows tissue specificity. By contrast to the plant, which shows robust *a1* expression in the absence of the transposase, the aleurone does not show expression at all.

Suppression can also affect polyadenylation. Analysis of a supressible allele of *rf2* reveals that *Mu* insertions can cause premature polyadenylation within the transposon. However, because this new polyadenylation site is just 30 bp downstream from the stop codon in the 3′UTR, the result is a functional transcript and a nonmutant phenotype. In the presence of active *MuDR* elements, however, a nonfunctional, aberrant transcript is produced, resulting in a mutant phenotype (Cui et al., 2003).

Mu insertions can also cause ectopic expression. Insertions into the third intron of *knotted1* cause ectopic expression of that gene in leaves (Greene et al., 1994). Two of these mutations are suppressible, such that ectopic expression only occurs when the transposase is present. Greene et al.'s interpretation of these observations is that the insertion on its own has no effect, but binding of the transposase to the inserted *Mu* element results in interference with a negative regulator in the intron.

If gene expression is ectopic to begin with, the presence of a new insertion, in combination with transposase, can suppress the mutant phenotype. This is the case for insertions into dominant mutant alleles of *roughsheath1*, *liguless3*, and *knotted1*. Screens for reversion of the mutants that exhibited ecotopic expression of these homeobox genes led to the discovery of a number of new *Mu*-suppressible phenotypes (Lowe et al., 1992; Girard and Freeling, 1999, 2000). In each case, ectopic expression, but not normal expression, was blocked by the presence of the transposase.

Suppressible alleles in maize clearly illustrate that gene function can be tied directly to the presence or absence of a transposase. In many cases, gene expression is only inhibited when the autonomous element is active, but the converse can also be true. This has led to the hypothesis that transposases may in some cases be co-opted into gene regulatory pathways. According to this scenario, down-regulation of a given gene at a particular time or place becomes selectively advantageous. Given that autonomous transposable elements can be activated by a wide variety of environmental conditions and can exhibit tissue specificity, one could imagine cases in which the presence of the transposase becomes useful to the host. Take, for instance, a gene that produces a surface-binding motif recognized by a pathogen. Suppression of this gene after infection could then be beneficial. Given that TEs are known to be responsive to pathogens (Wessler, 1996), infection could result in production of the transposase, which could in turn result in conditional down-regulation of a given insertion allele of that gene. Over time, only those aspects of the system that are beneficial (the binding site for the transposase and at least those domains of the protein required for suppression) would be expected to be retained. Of course, this represents only one of many scenarios for co-optation of suppression pathways, but the presence of vast numbers of TEs within a few hundred base pairs of the start of transcription of large numbers of genes in many plant species certainly suggests that the raw material for suppression co-optation is available.

However, the net effect of suppression may be a limitation on TE activity due to an increased cost of expressing the transposase. TEs in close proximity to genes generally appear to have minimal effects on gene expression (Lippman et al., 2004). However, this conclusion has been reached by examining plants in which there is little evidence of TE activity. Work on suppressible mutations in maize suggests that many otherwise benign TE insertions may become deleterious once the appropriate transposase becomes available. This in the end may limit the expansion of TE families by increasing selection against transposase expression. It would be interesting to examine the effects of transposase on expression of genes that contain otherwise silent insertions that are competent to bind the cognate transposase. MITE elements in rice would be a good place to start because they are present in large numbers, are often closely associated with genes, and because rice lines have been identified that carry or lack active autonomous elements (Jiang et al., 2003; Naito et al., 2009). A similar phenomenon may underlie "*Mu*-syndrome," a characteristic suite of morphological features associated with high levels of *Mutator* activity and dramatic changes in gene expression profiles (Walbot et al., 1988; Skibbe et al., 2009).

TEs and Plant Domestication

Evidence that TEs activity has resulted in positively selected changes in gene expression are largely anecdotal, but often striking, particularly among domesticated plant species. Domestication represents an excellent opportunity to examine the roles of TEs in evolution because selection was relatively intense and the time frame, tens of thousands rather than millions of years, makes it possible to catch TEs in the act, while sequences that are still clearly TEs are recognizable. Much older TE-induced changes are far more difficult to identify because selection is likely to retain only sequences that continue to be important for function.

The simplest and perhaps the most common form of TE-induced mutation involves loss of function alleles, which are often associated with plant domestication (Doebley et al., 2006). One such example takes us back to the birth of genetic analysis, Mendel's wrinkled peas. This trait had been cultivated long before Mendel, presumably because the wrinkled peas taste sweeter (Bhattacharyya et al., 1993). As it turns out, the wrinkled pea phenotype is due to the insertion of a short derivative of a *h*AT element in the *rr* locus, resulting in a null allele of the SBEI gene, which is involved in starch branching (Bhattacharyya et al., 1990).

Knock-out mutations are also a way to get grains with a stickier texture. In species such as rice, sorghum, and millet, variants with lower amylase content have been repeatedly and independently selected due to local preferences for that texture (sticky rice, for instance). The most common mutations associated with this trait are lesions in the *waxy* gene, which encodes granule-bound starch synthase 1 (GBSS 1) (Sano, 1984). A comprehensive analysis of 871 landraces of foxtail millet revealed a striking correlation between TE insertional activity and the appearance and reversion of *waxy* mutations associated with sticky varieties over a wide geographical range (Kawase et al., 2005). Kawase and coworkers found that all *waxy* and low-amylase types of foxtail millet arose as a consequence of TE insertions, and that these phenotypes arose independently several times (Figure 6.4). This is perhaps the best example of the capacity for TEs to rapidly alter gene function in a domesticated crop, and suggests that—at least for this gene in this species—TE insertion is the primary means by which null mutants are generated. The involvement of several types of elements suggests that this was not due to targeting of this gene by a particular element. It is worth noting, however, that *waxy* mutations in other domesticated species, including rice and barley, appear to have occurred only once or a few times and did not involve TE insertions (Domon et al., 2002; Olsen and Purugganan, 2002; Yamanaka et al., 2004). This may be due to the availability of preexisting polymorphisms in these species, a lower level of TE activity, or different selective regimes. This raises an important point. Based on the data from foxtail millet, TEs are certainly competent to

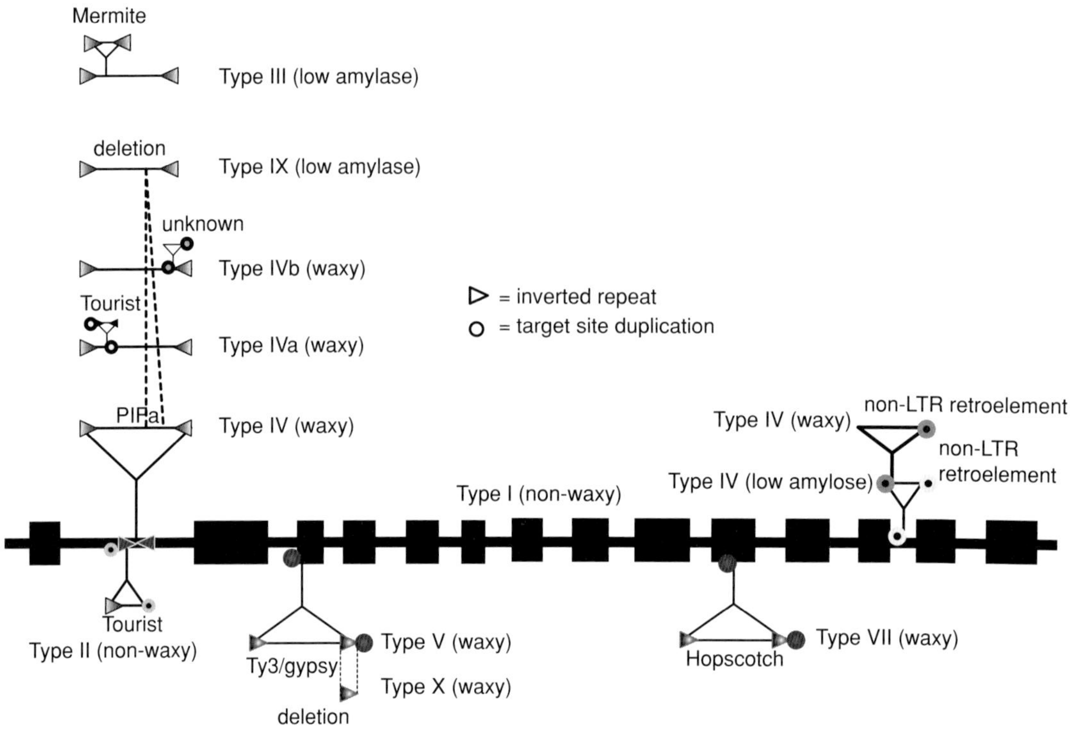

Figure 6.4 TE insertions associated with expression variation at the *waxy* locus in foxtail millet. Black boxes represent exons. TE insertion points are as indicated. TE inverted repeats are represented by triangles; target site duplications by circles. The type of TE is indicated next to the insertion. Adapted from Kawasi et al., 2005.

induce a high frequency of mutations in response to selection, but whether or not they actually do may depend on a wide variety of factors. We have no reason to assume a priori that they have done so in all or even most species.

Unlike many other cultivated species, grapes are largely vegetatively propagated, with new variation often being due to newly appearing somatic mutations that can be propagated clonally (Pelsy, 2010). Therefore, grapes represent an excellent opportunity to examine the frequency with which TEs can contribute to the ongoing production of phenotypic diversity as a consequence of somatic events. Most white grape varietals appear to be the result of a single insertion of a retrotransposon, *Gret1*, into the *VvmybA1* gene, a *myb* transcription factor required for production of anthocyanins (Kobayashi et al., 2004; This et al., 2007). *Gret1* is a typical LTR retroelement, with long terminal repeats flanking coding sequences for gag-pol and reverse transcriptase. LTR retrotransposons are prone to deletion via illegitimate recombination between their LTRs (Bennetzen et al., 2005), and it appears that this propensity has been exploited by wine growers. Two red-skinned cultivars, Ruby Okuyama and Flame Muscat, were independently derived from the white-skinned cultivars Italia and Muscat of Alexandria, respectively (Pelsy, 2010). In both cases, the reversion events were due to recombination between the *Gret1* LTRs, restoring gene function. Thus, not only did the initial insertion event cause the loss of gene expression, but the structure of the insertion insured that this variation was intrinsically unstable. Asian pink grape varietals also appear to be due to a TE insertion into the same gene, in this case of a short sequence (Shimazaki et al., 2011). Although the identity of this sequence is not known, it is likely to be a transposon or a derivative thereof. Interestingly, TEs, particularly MITE elements, appear to have been particularly active following grape domestication, and are the source of many extant polymorphisms between different varieties (Benjak et al., 2009).

Given that apples, like grapes, are often propagated as clonal varieties, it will be interesting to determine the causes of morphological variation in that species, as well. In fact, there is already evidence that at least one trait in some apple cultivars is due to transposon insertions. Rae Ime, Spencer Seedless, and Wellington Bloomless are cultivars that produce fruit parthogenically, without fertilization or seeds. These mutant phenotypes are due to retrotransposon insertions into intron 4 (for Rae Ime) and intron 6 (for Spencer Seedless and Wellington Bloomless) of the apple homolog of the *Arabidopsis pistillata* (*pi*) gene that encodes a MADS box transcription factor (Yao et al., 2001). In each case, the insertion results in the loss of functional transcript.

One species whose domestication has been the subject of intense scrutiny is maize. Modern maize was derived from its progenitor species, *teosinte*, roughly 5,000 years ago (Zarrillo et al., 2008). Analysis of the differences between these species using reciprocal backcrossed introgression lines has been used to identify a surprisingly limited number of key domestication loci, including, among others, *teosinte branched1* (*tb1*) and *teosinte glume architecture* (*tga1*) (Doebley et al., 2006). At *tb1*, two regions far upstream of the coding sequence are responsible for morphological differences between maize and teosinte. The first region, located 64–69 kb upstream, controls maize basal branching. The second, 58–64 kb upstream, controls the ear phenotype (Clark et al., 2006). Insertions of a *Hopscotch* retroelement in the proximal region is tightly associated with an increase in expression of the *tb1* gene in maize and with the decreased branching of this species relative to its progenitor (Studer et al., 2011). The results of transient transformation assays revealed that the *Hopscotch* element can act as an enhancer of *tb1*. A second insertion, this time of a MITE element in the distal region, is tightly associated with the ear phenotype in maize and is hypothesized to interrupt a repressor function. Unlike the insertions into the *waxy* gene in millet, the *Hopscotch* insertion confers a neomorphic semidominant phenotype, illustrating the capacity of TEs to do more than simply disrupt gene function. It should be noted, however, that the *Hopscotch* insertion predated domestication, suggesting that humans selected for a preexisting polymorphic trait in *teosinte*. These data show that TE insertions quite distant from coding sequences can have profound effects on gene expression. It appears that TEs were first discovered in a species whose evolution was directly shaped by their activity, an observation that Barbara McClintock would no doubt find profoundly gratifying.

Interestingly, this is not the only example of insertions into distant regulatory regions in maize. The *vgt1* locus is the source of a significant portion of variation in flowering time in this species (Salvi et al., 2002). *Vgt1* is a conserved noncoding sequence approximately 70 kb upstream of an *Ap2*-like transcription factor (*ZmRap2.7*) that is known to be involved in control of flowering time (Salvi et al., 2007). A polymorphic MITE insertion into *vgt1*, which interrupts the conserved sequences, appears to be the cause of the flowering time variation at this locus.

The insertions into *vgt1* and *tb1* suggests that efforts to identify variation associated with trait polymorphism should focus not just on coding sequences, but should also focus on associated conserved noncoding sequences. Recent research has identified thousands of such sequences in a variety of plants (Freeling and Subramaniam, 2009). It would be interesting to determine the number of polymorphic insertions into these regions in various accessions of plants and to correlate those insertions with known loci associated with morphological variation.

As the example from grapes suggests, TEs have been involved not only in the process of domestication, but in subsequent trait selection, as well. Three vernalization genes have been shown to be responsible for differences between winter and spring varieties of wheat and barley: VRN1, VRN2, and VRN3 (Trevaskis et al., 2007). Dominant forms of the VRN1 gene are responsible for the spring growth habit, while dominant VRN2 alleles are responsible for the winter growth habit, which is believed to be the ancestral state (Yan et al., 2004). Analysis of many *Triticum* species and domesticated variants has demonstrated that the spring growth habit arose multiple times as a consequence of multiple alterations of VRN1, including deletions and insertions of sequences in the promoter and first intron of this gene (Cockram et al., 2007). At least one of these variants was caused by insertion of a TE, which supports the potential involvement of these elements in the process of domestication. However, most of the changes involved loss of negative control regions within the VRN1 gene as a consequence of deletions at short direct repeats, probably due to illegitimate recombination. This suggests that illegitimate recombination, rather than TE activity, is the primary cause of allelic variation in this gene. It is interesting to note, however, that in 3 of the 20 variants, the deletion was adjacent to a TE, suggesting that in at least some of these cases TE activity was involved in the deletion, perhaps due to the induction of a double-stranded break.

A recent report on pigmentation in blood oranges illustrates how TEs can provide both stress responsiveness and tissue specificity to a gene, creating a new selectable trait. The deep red color of the blood orange is governed by the activity of an R2R3 Myb transcription factor. A survey of blood oranges revealed that the enhanced tissue-specific expression of this gene is associated with the presence of an LTR retrotransposon in its promoter (Butelli et al., 2012). In three related lineages, this is due to a single ancestral insertion, with variation in expression levels associated with rearrangements subsequent to the TE insertion. In these variants, transcription is initiated within the LTR and proceeds into the R2R3 Myb gene. Remarkably, a fourth, independently derived blood orange lineage exhibits a similar pattern of expression due to the insertion of an unrelated retrotransposon whose LTR activates expression of the same Myb gene. Both the tissue specificity and the cold inducibility of the transcription factor are attributable to the regulatory properties of the inserted retrotransposon sequences. This example beautifully illustrates the capacity for TEs to simultaneously introduce multiple promoter elements that would be difficult to acquire any other way.

As we have seen, TEs can move genes into new contexts via aberrant transposition. Since patterns of gene expression can depend on local context, this movement can be associated with changes in expression. The involvement of this process in domestication can be observed in the development of oblong variety of tomatoes, such as Roma. The *SUN* locus is responsible for a significant proportion of the differences in fruit shape between such varieties of *Solanum lycopersicum* and its round-fruited wild relative, *S. pimpinellifolium* (van der Knaap et al., 2004). The key gene at the *SUN* locus is *IQD12* (Xiao et al., 2008). Both the domesticated and wild tomato species have a copy of *IQD12* on chromosome 10, but oblong varieties of S. *lycopsersicum* have an addition copy on chromosome 7, and it is this copy that accounts for the difference in fruit shape (Xiao et al., 2008). Analysis of the version on chromosome 7 revealed that the transposed copy of *IQD12* is flanked by a retrotransposon, *Rider*, along with several other genes. Sequence analysis of this transposed DNA sequence strongly suggests that it is the result of an aberrant retrotransposition of the *Rider* element (Figure 6.5). The retrotransposition resulted in integration of the *IQD12* gene into a new location. Because the aberrant transcript was antisense relative to the normal *IQD12* gene, the normal introns were not spliced during the transposition process, essentially preserving an exact copy of the origin gene. Insertion was into a gene (*DEFL*) that is expressed at high levels in developing fruit, and

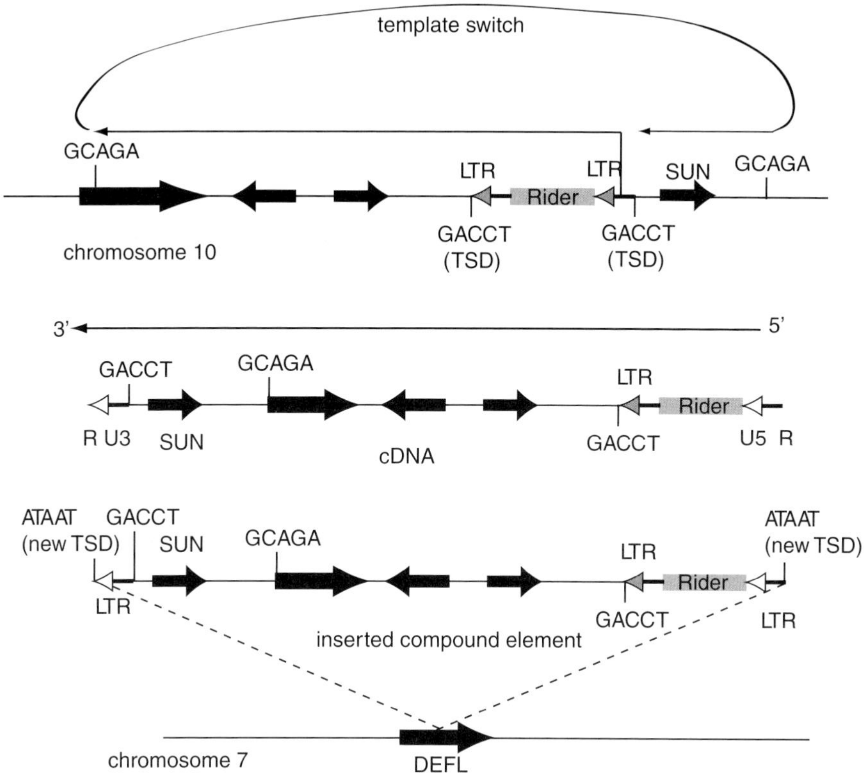

Figure 6.5 Retrotransposition of *SUN* in tomato. Genes are represented by black-arrowed rectangles. The retrotransposon *Rider* is a grey box flanked by long terminal repeats (LTR = grey-filled arrows), flanked in turn by a five–base-pair target-site duplication (TSD). Transcription is hypothesized to have begun within the left LTR of *Rider*, and then proceeded past the second LTR, where it would normally stop, into adjacent genes. Template switching at "GCAGA" resulted in a transcript that included sequences immediately 5′ to the *Rider* element. Reverse transcription resulted in a cDNA containing the *Rider* element along with several adjacent genes flanked by a regenerated pair of identical LTRs (white-filled arrows, designated R U3 and U5 R). This cDNA was then integrated into the DEFL locus on chromosome 7, resulting in the production of a new five–base-pair TSD. Adapted from Xiao et al., 2008.

expression of the inserted *IQD12* gene mimics that of the ancestral form of the *DEFL* gene. Thus, in domesticated tomatoes, the duplicate copy of *LQD12* came under cis regulation of *DEFL* due to an insertion that co-opted a new promoter element.

All of these examples suggest that TEs have certainly been a factor during domestication of a number of crop species, and the number of such examples will increase. However, we should keep in mind that many of the mutations known to be involved in domestication were caused by other changes, such as point mutations and small deletions. Indeed, TEs have been implicated in only a minority of them (Doebley et al., 2006). It remains unclear how crucial TEs were for the process, as they have been just one of many sources of the mutations involved in domestication. This is particularly true for simple null mutants, which can be easily produced via point mutations or small deletions.

The Dynamic Genome

Transposable elements are hardly insulated from the rest of the genome; there is a constant exchange of information from TEs to their hosts, and back again. TEs can capture gene sequences, and in turn any part of TEs, from their regulatory regions to their enzymatic functions, can be co-opted for use by the host (Dooner and Weil, Chapter 9 in this volume). Whether or not we see them as distinct from the rest of the genome depends on the extent to which the presence of any given TE sequence provides a selective advantage to the organism as a whole. Rather than being assignable to either selfish or altruistic categories, any given DNA sequence occupies a position on a continuum between these extremes. In fact, many plant genes mimic aspects of TE behavior and evolution. These are genes that can be present in multiple copies, readily move from place to place in the genome, and evolve rapidly. The best studied is provided by the *R*-genes (McDowell and Simon, 2006), but there are other, similar classes of genes, including F-box genes (Xu et al., 2009; Hua et al., 2011) and the defensins (Silverstein et al., 2005). A recent analysis of these genes in *Arabidopsis* revealed that they are rarely found at the same position in related *Brassicacae*, and that the differences in position are consistent with movement of these genes into otherwise syntenic portions of the genome; that is, they had transposed (Freeling et al., 2008). Interestingly, these genes also accumulated in regions of the genome heavily populated by TEs, such as pericentromeric heterochromatin, and many of them are pseudogenes (Hua et al., 2011). The mechanism of this movement is unknown, although analysis of recent replicative transposition of some of these genes in plants suggests that it may be mediated by ectopic recombination between short, direct repeats (Woodhouse et al., 2010). Similar gene transposition events in *Drosophila* appear to be mediated by TE sequences flanking the genes (Yang et al., 2008). In addition to their propensity to move and to rapidly increase in copy number, R-genes exhibit additional similarities to TEs. Like TEs, their expression can be modulated by DNA methylation and chromatin modification, and like TEs, stress in the form of infection can result in a loss of methylation and increased instability (Grandbastien et al., 2005b; Boyko et al., 2007; Durrant et al., 2007).

These observations demonstrate that some gene families are clearly mobile and can increase their copy number in a manner reminiscent of TEs. To the extent that these genes carry structural features that predispose them to transpose and amplify their copy number, it is possible to think of them as "selfish." However, these gene families are clearly associated with intense selection at the level of the host. R-genes and defensins are key to the ability for plants to cope with infectious agents, and the presence of a population of these genes, each accumulating point mutations and each with its own chromosomal environment, ensures enough variation to cope with a variety of pathogens (Friedman and Baker, 2007). This may represent an instance in which selfishness—as judged by the propensity to increase copy number even in the absence of external selection—and altruism—as judged by their importance in protecting the plant from disease—blend seamlessly. The behavior of gene families such as R-genes reveals that although gene amplification can be deleterious, it also contributes importantly to both the sophistication and flexibility of higher eukaryotes. The increased complexity that is a characteristic feature of higher organisms has undoubtedly been facilitated by the growth and diversification of many gene families (Lynch and Conery, 2003). From this point of view, the vast expansion of TEs could actually result from a propensity to accommodate, or even facilitate, gene duplication in the first place (Fedoroff, 2000).

From the very beginning, geneticists have noted with excitement the potential for TEs to foster evolutionary change (Starlinger, 1977; Berg et al., 1984; McClintock, 1984). They are an endogenous, inducible mutagen, capable of altering gene expression in remarkably subtle ways or reorganizing and transforming the structure of entire genomes. It is hard not to get enthusiastic about their potential. I can remember quite clearly the first time I learned about the existence of TEs as an undergraduate. A professor teaching a lecture on the immune system mentioned, only in passing, that some genes could move from place to place in the genome and in doing so mutate other genes. I was floored. Genes that mutate other genes? That has to matter for evolution. I doubt that there are many of us who study TEs that have not had a similar epiphany early in our careers. That has to matter. Over the years, an avalanche of data has been produced suggesting that it does indeed matter, and many reviewers, including myself, have enthusiastically promoted the idea that TEs are important players in evolutionary change.

This chapter contains just some of the examples of the many, many ways TEs have been implicated in plant gene evolution, and the list will certainly get longer in the future (Feschotte and Pritham, 2007). At this point, it is certainly conceivable that nearly any beneficial mutation in any organism could have been caused by a TE. This is particularly true of mutations that would be difficult to evolve in any other way, such as those that result in rapid changes in inducibility and tissue specificity (Kidwell and Lisch, 2001). The questions that remain are much more difficult: were TEs necessary for the mode and tempo of evolution as it actually occurred? Did TE activity provide a functionality, a degree of evolvability, that would not have been available if they did not exist (Pigliucci, 2008; Desai and Fisher, 2011)? Would the absence of TEs be a limitation on the capacity of genomes to evolve? Of course, we cannot rerun the experiment; it is impossible to find a form of life that does not contain them, and even genomes with relatively few TEs now may have had many TEs in the past. Further, associative data, such as that between speciation, adaptation, and TE copy number increases, do not prove a causal connection between TEs and the capacity to speciate (Ungerer et al., 1998; Ungerer et al., 2006).

The big questions may not be answerable. Yet in the age of high-throughput sequencing and comparative genomics, we can begin to assess the relative contribution of TEs to evolutionary novelty. Whole genome sequencing, sophisticated phenotypic analysis, and exceedingly dense recombinant inbred and backcross lines should make it possible to genetically isolate vast numbers of allelic and genic variants that contribute to morphological differences within and between plant species. This will make it possible to determine the relative contribution of TEs to morphological change precisely and comprehensively (Bentsink et al., 2010).

Plant lineages have the additional advantage of having experienced a series of relatively recent whole genome duplications (Jiao et al., 2011). Each duplication produces a paired set of genes, one of which is initially redundant. Over time, one or the other of each pair is lost, or the two genes become diversified in some beneficial way (Yogeeswaran et al., 2005). This process represents an excellent opportunity to examine the relative role of TEs in diversification of gene function. Since the whole genome duplications occurred at various times—from the tens of millions of years to the last few thousand, or even yesterday—and since we can induce them in the laboratory, it is possible to examine changes in duplicate genes at many time scales. The most recent events can be used to examine direct and immediate changes caused by TEs, which are often activated during polyploidization (Kashkush et al., 2002; Kraitshtein et al., 2010; Petit et al., 2010; Yaakov and Kashkush, 2011). Somewhat older duplications, such as those associated with domestication, can allow us to examine the degree to which new TE insertions have been selected by humans, and even older duplications can show us how often TE insertions have been fixed in species over long periods of time. A careful and systematic analysis of changes that have actually contributed to morphological evolution represents our best bet to find out exactly how important TEs have been over time.

Finally, it is worth considering the blurry line between host and parasite, plant and transposon. It can certainly be useful to think of TEs as autonomous parasites, but the association between TEs and the rest of the genome is far too intimate and long-standing to conveniently separate genomes into "us" and "them." If TEs must adapt to their environment in order to maximize their amplification while minimizing the damage they do, so must host genes adapt to their environment, which is often largely composed of TEs. There is also good evidence that large-scale structural components of the genome are profoundly influenced by TEs. Centromeres and pericentromeric regions (Wolfgruber et al., 2009), subtelomeres (Kuo et al., 2006), and vast regions between genes in plants such as maize and wheat are composed largely of mobile elements (Vitte and Bennetzen, 2006; Schnable et al., 2009). Therefore, it is likely that the functions of these regions and the genes within them have become dependent on their presence. Further, TEs exist along a spectrum, from primarily parasitic to entirely mutualistic, and there is every reason to believe that a given TE lineage can move along that spectrum in response to selection at the level of the gene and at the level of the organism. A given TE may act both to increase its copy number and to repair DNA breaks (Srikanta et al., 2009), or maintain telomeres (Levis et al., 1993; George et al., 2010), or stabilize centromeres (Du et al., 2010). Selfishness and mutualism are not mutually exclusive because simultaneous selection at multiple levels is possible, and perhaps inevitable. By contrast, genes such as R-genes may have an intrinsic propensity to replicate themselves, but this propensity has been harnessed to provide a rapidly changing pool of genes that provide resistance to pathogens.

It has been suggested that TEs represent an unavoidable cost of having a genome; the capacity to replicate one's DNA may inevitably lead to the possibility that some DNA will cheat by making extra copies of itself regardless of its adaptive benefit. This is the core of the selfish DNA hypothesis, and it should not be dismissed lightly. It is tempting to assign functionality to all biological phenomena, an impulse that has lead many scientists to suggest that TEs wouldn't exist if they were not beneficial in some way (Faulkner and Carninci, 2009). It is both counterintuitive and disconcerting to accept the possibility that selection at different levels could be operating at cross purposes. TEs are messy and disruptive, and ultimately perhaps even detrimental from the point of view of the host (Vinogradov, 2003; Rankin et al., 2010). I suggest that the messiness is precisely the point. TEs may be an essential force in evolution not despite their selfishness, but because of it. The intense competition, and cooperation, between hosts and TEs has produced novelty that would not otherwise exist. The critical point is that genomes can, and demonstrably do, rapidly evolve new functions even in the absence of external selective pressures. Thus, the imperfect internal dynamics of TE replication and transposition enhance evolvability, providing much of the variation on which selection acts in evolution. The activity of TEs has resulted in genomes that are intrinsically dynamic, poised for both subtle and dramatic change not necessarily because they need to be, but because they have to be. It is ironic that the part of the genome most often dismissed as peripheral, or even junk, may in fact lie at the heart of our capacity to evolve.

References

Ammiraju, J.S., Zuccolo, A., Yu, Y., Song, X., Piegu, B., Chevalier, F., Walling, J.G., Ma, J., Talag, J., Brar, D.S., et al. (2007) Evolutionary dynamics of an ancient retrotransposon family provides insights into evolution of genome size in the genus *Oryza*. *Plant J.* *52*: 342–351.

Barkan, A., and Martienssen, R.A. (1991) Inactivation of maize transposon *Mu* suppresses a mutant phenotype by activating an outward-reading promoter near the end of *Mu1*. *Proc. Natl. Acad. Sci. USA 88*: 3502–3506.

Belancio, V.P., Roy-Engel, A.M., and Deininger, P.L. (2010) All y'all need to know 'bout retroelements in cancer. *Semin. Cancer Biol. 20*: 200–210.

Benjak, A., Boue, S., Forneck, A., and Casacuberta, J.M. (2009) Recent amplification and impact of MITEs on the genome of grapevine (*Vitis vinifera* L.). *Genome Biol. Evol. 1*: 75–84.

Bennetzen, J.L., Ma, J., and Devos, K.M. (2005) Mechanisms of recent genome size variation in flowering plants. *Ann. Bot. 95*: 127–132.

Bentsink, L., Hanson, J., Hanhart, C.J., Blankestijn-de Vries, H., Coltrane, C., Keizer, P., El-Lithy, M., Alonso-Blanco, C., de Andres, M.T., Reymond, M., et al. (2010) Natural variation for seed dormancy in *Arabidopsis* is regulated by additive genetic and molecular pathways. *Proc. Natl. Acad. Sci. USA 107*: 4264–4269.

Berg, D.E., Berg, C.M., and Sasakawa, C. (1984) Bacterial transposon *Tn5*: evolutionary inferences. *Mol. Biol. Evol. 1*: 411–422.

Bhattacharyya, M., Martin, C., and Smith, A. (1993) The importance of starch biosynthesis in the wrinkled seed shape character of peas studied by Mendel. *Plant Mol. Biol. 22*: 525–531.

Bhattacharyya, M.K., Smith, A.M., Ellis, T.H., Hedley, C., and Martin, C. (1990) The wrinkled-seed character of pea described by Mendel is caused by a transposon-like insertion in a gene encoding starch-branching enzyme. *Cell 60*: 115–122.

Boyko, A., Kathiria, P., Zemp, F.J., Yao, Y., Pogribny, I., and Kovalchuk, I. (2007) Transgenerational changes in the genome stability and methylation in pathogen-infected plants: virus-induced plant genome instability. *Nucleic Acids Res. 35*: 1714–1725.

Brakefield, P.M. (2011) Evo-devo and accounting for Darwin's endless forms. *Philos. Trans. R. Soc. Lond. B Biol. Sci. 366*: 2069–2075.

Britten, R.J. (1996) DNA sequence insertion and evolutionary variation in gene regulation. *Proc. Natl. Acad. Sci. USA 93*: 9374–9377.

Brookfield, J.F. (2005) The ecology of the genome - mobile DNA elements and their hosts. *Nature Rev. Genet. 6*: 128–136.

Buchmann, R.C., Asad, S., Wolf, J.N., Mohannath, G., and Bisaro, D.M. (2009) Geminivirus AL2 and L2 proteins suppress transcriptional gene silencing and cause genome-wide reductions in cytosine methylation. *J. Virol. 83*: 5005–5013.

Butelli, E., Licciardello, C., Zhang, Y., Liu, J., Mackay, S., Bailey, P., Reforgiato-Recupero, G., and Martin, C. (2012) Retrotransposons control fruit-specific, cold-dependent accumulation of anthocyanins in blood oranges. *Plant Cell 24*: 1242–1255.

Candela, H., and Hake, S. (2008) The art and design of genetic screens: maize. *Nature Rev. Genet. 9*: 192–203.

Chomet, P., Lisch, D., Hardeman, K.J., Chandler, V.L., and Freeling, M. (1991) Identification of a regulatory transposon that controls the *mutator* transposable element system in maize. *Genetics 129*: 261–270.

Clark, R.M., Wagler, T.N., Quijada, P., and Doebley, J. (2006) A distant upstream enhancer at the maize domestication gene *tb1* has pleiotropic effects on plant and inflorescent architecture. *Nature Genet. 38*: 594–597.

Cockram, J., Mackay, I.J., and O'Sullivan, D.M. (2007) The role of double-stranded break repair in the creation of phenotypic diversity at cereal VRN1 loci. *Genetics 177*: 2535–2539.

Craig, N.L. (2002) Mobile DNA: an Introduction. In *Mobile DNA II*, N.L. Craig, R. Craigie, M. Gellert, and A.M. Lambowitz, eds. (Washington, DC: American Society for Microbiology Press), pp. 3–10.

Cui, X., Hsia, A.P., Liu, F., Ashlock, D.A., Wise, R.P., and Schnable, P.S. (2003) Alternative transcription initiation sites and polyadenylation sites are recruited during *Mu* suppression at the *rf2a* locus of maize. *Genetics 163*: 685–698.

Desai, M.M., and Fisher, D.S. (2011) The balance between mutators and nonmutators in asexual populations. *Genetics 188*(4) 997–1014.

Doebley, J.F., Gaut, B.S., and Smith, B.D. (2006) The molecular genetics of crop domestication. *Cell 127*: 1309–1321.

Domon, E., Fuijita, M., and Ishikawa, N. (2002) The insertion/deletion polymorphisms in the *waxy* gene of barley genetic resources from East Asia. *Theor. Appl. Genet. 104*: 132–138.

Donlin, M.J., Lisch, D., and Freeling, M. (1995) Tissue-specific accumulation of MURB, a protein encoded by *MuDR*, the autonomous regulator of the *mutator* transposable element family. *Plant Cell 7*: 1989–2000.

Doolittle, W.F., and Sapienza, C. (1980) Selfish genes, the phenotype paradigm and genome evolution. *Nature 284*: 601–603.

Dooner, H.K., Robbins, T.P., and Jorgensen, R.A. (1991) Genetic and developmental control of anthocyanin biosynthesis. *Annu. Rev. Genet. 25*: 173–199.

Dooner, H.K., and Weil, C.F. (2007) Give-and-take: interactions between DNA transposons and their host plant genomes. *Curr. Opin. Genet. Dev. 17*: 486–492.

Du, Y., Topp, C.N., and Dawe, R.K. (2010) DNA binding of centromere protein C (CENPC) is stabilized by single-stranded RNA. *PLoS Genet. 6*: e1000835.

Durrant, W.E., Wang, S., and Dong, X. (2007) *Arabidopsis* SNI1 and RAD51D regulate both gene transcription and DNA recombination during the defense response. *Proc. Natl. Acad. Sci. USA 104*: 4223–4227.

Faulkner, G.J., and Carninci, P. (2009) Altruistic functions for selfish DNA. *Cell Cycle 8*: 2895–2900.

Fedoroff, N. (1989a) Maize transposable elements. In *Mobile DNA*, M. Howe and D. Berg, eds. (Washington: American Society for Microbiology), pp. 375–411.

Fedoroff, N.V. (1989b) About maize transposable elements and development. *Cell 56*: 181–191.

Fedoroff, N. (2000) Transposons and genome evolution in plants. *Proc. Natl. Acad. Sci. USA 97*: 7002–7007.

Feschotte, C., and Pritham, E.J. (2007) DNA transposons and the evolution of eukaryotic genomes. *Annu. Rev. Genet. 41*: 331–368.

Feschotte, C., Jiang, N., and Wessler, S.R. (2002) Plant transposable elements: where genetics meets genomics. *Nature Rev. Gen. 3*: 329–341.

Freeling, M., and Subramaniam, S. (2009) Conserved noncoding sequences (CNSs) in higher plants. *Curr. Opin. Plant Biol. 12*: 126–132.

Freeling, M., Lyons, E., Pedersen, B., Alam, M., Ming, R., and Lisch, D. (2008) Many or most genes in *Arabidopsis* transposed after the origin of the order *Brassicales*. *Genome Res. 18*: 1924–1937.

Friedman, A.R., and Baker, B.J. (2007) The evolution of resistance genes in multi-protein plant resistance systems. *Curr. Opin. Genet. Dev. 17*: 493–499.

Fujimoto, R., Kinoshita, Y., Kawabe, A., Kinoshita, T., Takashima, K., Nordborg, M., Nasrallah, M.E., Shimizu, K.K., Kudoh, H., and Kakutani, T. (2008) Evolution and control of imprinted *FWA* genes in the genus *Arabidopsis*. *PLoS Genet. 4*: e1000048.

Gao, X., Hou, Y., Ebina, H., Levin, H.L., and Voytas, D.F. (2008) Chromodomains direct integration of retrotransposons to heterochromatin. *Genome Res. 18*: 359–369.

Gehring, M., Bubb, K.L., and Henikoff, S. (2009) Extensive demethylation of repetitive elements during seed development underlies gene imprinting. *Science 324*: 1447–1451.

George, J.A., Traverse, K.L., DeBaryshe, P.G., Kelley, K.J., and Pardue, M.L. (2010) Evolution of diverse mechanisms for protecting chromosome ends by *Drosophila TART* telomere retrotransposons. *Proc. Natl. Acad. Sci. USA 107*: 21052–21057.

Girard, L., and Freeling, M. (1999) Regulatory changes as a consequence of transposon insertion. *Dev. Genet. 25*: 291–296.

Girard, L., and Freeling, M. (2000) Mutator-suppressible alleles of rough *sheath1* and *liguleless3* in maize reveal multiple mechanisms for suppression. *Genetics 154*: 437–446.

Grandbastien, M., Audeon, C., Bonnivard, E., Casacuberta, J., Chalhoub, B., Costa, A., Le, Q., Melayah, D., Petit, M., Poncet, C., et al. (2005a) Stress activation and genomic impact of *Tnt1* retrotransposons in Solanaceae. *Cytogenet. Genome Res. 110*: 229–241.

Grandbastien, M.A., Audeon, C., Bonnivard, E., Casacuberta, J.M., Chalhoub, B., Costa, A.P., Le, Q.H., Melayah, D., Petit, M., Poncet, C., et al. (2005b) Stress activation and genomic impact of *Tnt1* retrotransposons in Solanaceae. *Cytogenet. Genome Res. 110*: 229–241.

Greene, B., Walko, R., and Hake, S. (1994) Mutator insertions in an intron of the maize *knotted1* gene result in dominant suppressible mutations. *Genetics 138*: 1275–1285.

He, G., Zhu, X., Elling, A.A., Chen, L., Wang, X., Guo, L., Liang, M., He, H., Zhang, H., Chen, F., et al. (2010) Global epigenetic and transcriptional trends among two rice subspecies and their reciprocal hybrids. *Plant Cell 22*: 17–33.

Hollister, J.D., and Gaut, B.S. (2009) Epigenetic silencing of transposable elements: a trade-off between reduced transposition and deleterious effects on neighboring gene expression. *Genome Res. 19*: 1419–1428.

Hollister, J.D., Smith, L.M., Guo, Y.L., Ott, F., Weigel, D., and Gaut, B.S. (2011) Transposable elements and small RNAs contribute to gene expression divergence between *Arabidopsis thaliana* and *Arabidopsis lyrata*. *Proc. Natl. Acad. Sci. USA 108*: 2322–2327.

Hsing, Y., Chern, C., Fan, M., Lu, P., Chen, K., Lo, S., Sun, P., Ho, S., Lee, K., Wang, Y., et al. (2007) A rice gene activation/knockout mutant resource for high throughput functional genomics. *Plant Mol. Biol. 63*: 351–364.

Hu, T.T., Pattyn, P., Bakker, E.G., Cao, J., Cheng, J.F., Clark, R.M., Fahlgren, N., Fawcett, J.A., Grimwood, J., Gundlach, H., et al. (2011) The *Arabidopsis lyrata* genome sequence and the basis of rapid genome size change. *Nature Genet. 43*: 476–481.

Hua, Z., Zou, C., Shiu, S.H., and Vierstra, R.D. (2011) Phylogenetic comparison of *F-Box* (*FBX*) gene superfamily within the plant kingdom reveals divergent evolutionary histories indicative of genomic drift. *PLoS One 6*: e16219.

Ivashuta, S., Naumkina, M., Gau, M., Uchiyama, K., Isobe, S., Mizukami, Y., and Shimamoto, Y. (2002) Genotype-dependent transcriptional activation of novel repetitive elements during cold acclimation of alfalfa (*Medicago sativa*). *Plant J. 31*: 615–627.

Jiang, N., Bao, Z., Zhang, X., Hirochika, H., Eddy, S.R., McCouch, S.R., and Wessler, S.R. (2003) An active DNA transposon family in rice. *Nature 421*: 163–167.

Jiang, N., Jordan, I.K., and Wessler, S.R. (2002) Dasheng and RIRE2. A nonautonomous long terminal repeat element and its putative autonomous partner in the rice genome. *Plant Physiol. 130*: 1697–1705.

Jiang, N., Ferguson, A.A., Slotkin, R.K., and Lisch, D. (2011) *Pack-Mutator*-like transposable elements (*Pack-MULEs*) induce directional modification of genes through biased insertion and DNA acquisition. *Proc. Natl. Acad. Sci. USA 108*: 1537–1542.

Jiao, Y., Wickett, N.J., Ayyampalayam, S., Chanderbali, A.S., Landherr, L., Ralph, P.E., Tomsho, L.P., Hu, Y., Liang, H., Soltis, P.S., et al. (2011) Ancestral polyploidy in seed plants and angiosperms. *Nature 473*: 97–100.

Jorgensen, R.A. (2004) Restructuring the genome in response to adaptive challenge: McClintock's bold conjecture revisited. *Cold Spring Harb. Symp. Quant. Biol. 69*: 349–354.

Kashkush, K., Feldman, M., and Levy, A.A. (2002) Gene loss, silencing and activation in a newly synthesized wheat allotetraploid. *Genetics 160*: 1651–1659.

Kashkush, K., Feldman, M., and Levy, A.A. (2003) Transcriptional activation of retrotransposons alters the expression of adjacent genes in wheat. *Nat. Genet. 33*: 102–106.

Kashkush, K., and Khasdan, V. (2007) Large-scale survey of cytosine methylation of retrotransposons and the impact of readout transcription from long terminal repeats on expression of adjacent rice genes. *Genetics 177*: 1975–1985.

Kawase, M., Fukunaga, K., and Kato, K. (2005) Diverse origins of waxy foxtail millet crops in East and Southeast Asia mediated by multiple transposable element insertions. *Mol. Genet. Genomics 274*: 131–140.

Kidwell, M.G., Kidwell, J.F., and Sved, J.A. (1977) Hybrid dysgenesis in *Drosophila melanogaster*: A syndrome of aberrant traits including mutation, sterility and male recombination. *Genetics 86*: 813–833.

Kidwell, M.G., and Lisch, D.R. (2001) Perspective: transposable elements, parasitic DNA, and genome evolution. *Evolution 55*: 1–24.

Kimura, M. (1968) Evolutionary rate at the molecular level. *Nature 217*: 624–626.

King, C.C. (1985) A model for transposon-based eucaryote regulatory evolution. *J. Theor. Biol. 114*: 447–462.

Kloeckener-Gruissem, B., and Freeling, M. (1995) Transposon-induced promoter scrambling: a mechanism for the evolution of new alleles. *Proc. Natl. Acad. Sci. USA 92*: 1836–1840.

Kobayashi, S., Goto-Yamamoto, N., and Hirochika, H. (2004) Retrotransposon-induced mutations in grape skin color. *Science 304*: 982.

Kraitshtein, Z., Yaakov, B., Khasdan, V., and Kashkush, K. (2010) Genetic and epigenetic dynamics of a retrotransposon after allopolyploidization of wheat. *Genetics 186*: 801–812.

Kuo, H.F., Olsen, K.M., and Richards, E.J. (2006) Natural variation in a subtelomeric region of *Arabidopsis*: Implications for the genomic dynamics of a chromosome end. *Genetics 173*: 401–417.

Levis, R.W., Ganesan, R., Houtchens, K., Tolar, L.A., and Sheen, F.M. (1993) Transposons in place of telomeric repeats at a *Drosophila* telomere. *Cell 75*: 1083–1093.

Lippman, Z., Gendrel, A., Black, M., Vaughn, M., Dedhia, N., McCombie, W., Lavine, K., Mittal, V., May, B., Kasschau, K., et al. (2004) Role of transposable elements in heterochromatin and epigenetic control. *Nature 430*: 471–476.

Lisch, D. (2009) Epigenetic regulation of transposable elements in plants. *Annu. Rev. Plant Biol. 60*: 43–66.

Lister, C., Jackson, D., and Martin, C. (1993) Transposon-induced inversion in *Antirrhinum* modifies nivea gene expression to give a novel flower color pattern under the control of cycloidearadialis. *Plant Cell 5*: 1541–1553.

Lister, R., O'Malley, R.C., Tonti-Filippini, J., Gregory, B.D., Berry, C.C., Millar, A.H., and Ecker, J.R. (2008) Highly integrated single-base resolution maps of the epigenome in *Arabidopsis*. *Cell 133*: 523–536.

Liu, S., Yeh, C.T., Ji, T., Ying, K., Wu, H., Tang, H.M., Fu, Y., Nettleton, D., and Schnable, P.S. (2009) *Mu* transposon insertion sites and meiotic recombination events co-localize with epigenetic marks for open chromatin across the maize genome. *PLoS Genet. 5*: e1000733.

Lowe, B., Mathern, J., and Hake, S. (1992) Active *Mutator* elements suppress the knotted phenotype and increase recombination at the *Kn1-O* tandem duplication. *Genetics 132*: 813–822.

Lynch, M., and Conery, J.S. (2003) The origins of genome complexity. *Science 302*: 1401–1404.

Mao, L., Wood, T.C., Yu, Y., Budiman, M.A., Tomkins, J., Woo, S., Sasinowski, M., Presting, G., Frisch, D., Goff, S., et al. (2000) Rice transposable elements: a survey of 73,000 sequence-tagged-connectors. *Genome Res. 10*: 982–990.

Masson, P., Surosky, R., Kingsbury, J.A., and Fedoroff, N.V. (1987) Genetic and molecular analysis of the *Spm*-dependent *a-m2* alleles of the maize *a* locus. *Genetics 117*: 117–137.

McClintock, B. (1956a) Intranuclear systems controlling gene action and mutation. *Brookhaven Symp. Biol. 8*: 58–74.

McClintock, B. (1956b) Controling elements and the gene. *Brookhaven Symp. Biol. 21*: 197–216.

McClintock, B. (1984) The significance of responses of the genome to challenge. *Science 226*: 792–801.

McDowell, J.M., and Simon, S.A. (2006) Recent insights into *R* gene evolution. *Mol. Plant. Pathol. 7*: 437–448.

Mhiri, C., Morel, J.B., Vernhettes, S., Casacuberta, J.M., Lucas, H., and Grandbastien, M.A. (1997a) The promoter of the tobacco *Tnt1* retrotransposon is induced by wounding and by abiotic stress. *Plant Mol. Biol. 33*: 257–266.

Mhiri, C., Morel, J., Vernhettes, S., Casacuberta, J., Lucas, H., and Grandbastien, M. (1997b) The promoter of the tobacco *Tnt1* retrotransposon is induced by wounding and by abiotic stress. *Plant Mol. Biol. 33*: 257–266.

Miyao, A., Tanaka, K., Murata, K., Sawaki, H., Takeda, S., Abe, K., Shinozuka, Y., Onosato, K., and Hirochika, H. (2003) Target site specificity of the *Tos17* retrotransposon shows a preference for insertion within genes and against insertion in retrotransposon-rich regions of the genome. *Plant Cell 15*: 1771–1780.

Molina, J., Sikora, M., Garud, N., Flowers, J.M., Rubinstein, S., Reynolds, A., Huang, P., Jackson, S., Schaal, B.A., Bustamante, C.D., et al. (2011) Molecular evidence for a single evolutionary origin of domesticated rice. *Proc. Natl. Acad. Sci. USA 108*: 8351–8356.

Naito, K., Cho, E., Yang, G., Campbell, M.A., Yano, K., Okumoto, Y., Tanisaka, T., and Wessler, S.R. (2006) Dramatic amplification of a rice transposable element during recent domestication. *Proc. Natl. Acad. Sci. USA 103*: 17620–17625.

Naito, K., Zhang, F., Tsukiyama, T., Saito, H., Hancock, C.N., Richardson, A.O., Okumoto, Y., Tanisaka, T., and Wessler, S.R. (2009) Unexpected consequences of a sudden and massive transposon amplification on rice gene expression. *Nature 461*: 1130–1134.

Nei, M., Suzuki, Y., and Nozawa, M. (2011) The neutral theory of molecular evolution in the genomic era. *Annu. Rev. Genomics Hum. Genet. 11*: 265–289.

Olsen, K.M., and Purugganan, M.D. (2002) Molecular evidence on the origin and evolution of glutinous rice. *Genetics 162*: 941–950.

Orgel, L.E., and Crick, F.H. (1980) Selfish DNA: the ultimate parasite. *Nature 284*: 604–607.

Pelsy, F. (2010) Molecular and cellular mechanisms of diversity within grapevine varieties. *Heredity 104*: 331–340.

Penterman, J., Uzawa, R., & Fischer, R.L. (2007) Genetic interactions between DNA demethylation and methylation in *Arabidopsis*. *Plant Physiol. 145*: 1549–1557.

Petit, M., Guidat, C., Daniel, J., Denis, E., Montoriol, E., Bui, Q.T., Lim, K.Y., Kovarik, A., Leitch, A.R., Grandbastien, M.A., et al. (2010) Mobilization of retrotransposons in synthetic allotetraploid tobacco. *New Phytol. 186*: 135–147.

Piegu, B., Guyot, R., Picault, N., Roulin, A., Saniyal, A., Kim, H., Collura, K., Brar, D.S., Jackson, S., Wing, R.A., et al. (2006) Doubling genome size without polyploidization: dynamics of retrotransposition-driven genomic expansions in *Oryza australiensis*, a wild relative of rice. *Genome Res. 16*: 1262–1269.

Pigliucci, M. (2008) Is evolvability evolvable? *Nature Rev. Genet. 9*: 75–82.

Pooma, W., Gersos, C., and Grotewold, E. (2002) Transposon insertions in the promoter of the *Zea mays a1* gene differentially affect transcription by the Myb factors P and C1. *Genetics 161*: 793–801.

Prud'homme, B., Gompel, N., and Carroll, S.B. (2007) Emerging principles of regulatory evolution. *Proc. Natl. Acad. Sci. USA 104* Suppl 1, 8605–8612.

Raizada, M., Benito, M.-I., and Walbot, V. (2001) The *MuDR* transposon terminal inverted repeat contains a complex plant promoter directing distinct somatic and germinal programs. *Plant J. 25*: 79–91.

Rankin, D.J., Bichsel, M., and Wagner, A. (2010) Mobile DNA can drive lineage extinction in prokaryotic populations. *J. Evol. Biol. 23*: 2422–2431.

Robertson, D. (1978) Characterization of a mutator system in maize. *Mutat. Res. 51*: 21–28.

Salazar, M., Gonzalez, E., Casaretto, J.A., Casacuberta, J.M., and Ruiz-Lara, S. (2007) The promoter of the *TLC1.1* retrotransposon from *Solanum chilense* is activated by multiple stress-related signaling molecules. *Plant Cell Rep. 26*: 1861–1868.

Salvi, S., Sponza, G., Morgante, M., Tomes, D., Niu, X., Fengler, K.A., Meeley, R., Ananiev, E.V., Svitashev, S., Bruggemann, E., et al. (2007) Conserved noncoding genomic sequences associated with a flowering-time quantitative trait locus in maize. *Proc. Natl. Acad. Sci. USA 104*: 11376–11381.

Salvi, S., Tuberosa, R., Chiapparino, E., Maccaferri, M., Veillet, S., van Beuningen, L., Isaac, P., Edwards, K., and Phillips, R.L. (2002) Toward positional cloning of *Vgt1*, a QTL controlling the transition from the vegetative to the reproductive phase in maize. *Plant Mol. Biol. 48*: 601–613.

Sano, Y. (1984) Differential regulation of *waxy* gene expression in rice endosperm. *Theor. Appl. Genet. 68*: 467–473.

Saze, H., and Kakutani, T. (2007) Heritable epigenetic mutation of a transposon-flanked *Arabidopsis* gene due to lack of the chromatin-remodeling factor DDM1. *EMBO J. 26*: 3641–3652.

Saze, H., and Kakutani, T. (2011) Differentiation of epigenetic modifications between transposons and genes. *Curr. Opin. Plant Biol. 14*: 81–87.

Schnable, P.S., Ware, D., Fulton, R.S., Stein, J.C., Wei, F., Pasternak, S., Liang, C., Zhang, J., Fulton, L., Graves, T.A., et al. (2009) The B73 maize genome: complexity, diversity, and dynamics. *Science 326*: 1112–1115.

Schumann, G.G., Gogvadze, E.V., Osanai-Futahashi, M., Kuroki, A., Munk, C., Fujiwara, H., Ivics, Z., and Buzdin, A.A. (2010) Unique functions of repetitive transcriptomes. *Int. Rev. Cell Mol. Biol. 285*: 115–188.

Schwartz, D. (1969) An example of gene fixation from selective advantage in suboptimal conditions. *Am. Nat. 103*: 479–481.

Selinger, D.A., and Chandler, V.L. (1999) Major recent and independent changes in levels and patterns of expression have occurred at the *b* gene, a regulatory locus in maize. *Proc. Natl. Acad. Sci. USA 96*: 15007–15012.

Selinger, D.A., Lisch, D., and Chandler, V.L. (1998) The maize regulatory gene *B-Peru* contains a DNA rearrangement that specifies tissue-specific expression through both positive and negative promoter elements. *Genetics 149*: 1125–1138.

Shimazaki, M., Fujita, K., Kobayashi, H., and Suzuki, S. (2011) Pink-colored grape berry is the result of short insertion in intron of color regulatory gene. *PLoS One 6*: e21308.

Silverstein, K.A., Graham, M.A., Paape, T.D., and VandenBosch, K.A. (2005) Genome organization of more than 300 defensin-like genes in *Arabidopsis*. *Plant Physiol. 138*: 600–610.

Skibbe, D.S., Fernandes, J.F., Medzihradszky, K.F., Burlingame, A.L., and Walbot, V. (2009) Mutator transposon activity reprograms the transcriptomes and proteomes of developing maize anthers. *Plant J. 59*: 622–633.

Slotkin, R.K., and Martienssen, R. (2007) Transposable elements and the epigenetic regulation of the genome. *Nature Rev. Genet. 8*: 272–285.

Srikanta, D., Sen, S.K., Huang, C.T., Conlin, E.M., Rhodes, R.M., and Batzer, M.A. (2009) An alternative pathway for Alu retrotransposition suggests a role in DNA double-strand break repair. *Genomics 93*: 205–212.

Starlinger, P. (1977) DNA rearrangements in procaryotes. *Annu. Rev. Genet. 11*: 103–126.

Studer, S., Qiong Z., Ross-Ibarra, J., and Doebley, J. (2011) A transposon insertion in the maize gene, *tb1* functions as an enhancer and was a causative variant for change in plant architecture during maize domestication. *Nat. Genet. 43*: 1160–1163.

Swigonova, Z., Bennetzen, J.L., and Messing, J. (2005) Structure and evolution of the *r/b* chromosomal regions in rice, maize and sorghum. *Genetics 169*: 891–906.

This, P., Lacombe, T., Cadle-Davidson, M., and Owens, C.L. (2007) Wine grape (*Vitis vinifera* L.) color associates with allelic variation in the domestication gene *VvmybA1*. *Theor. Appl. Genet. 114*: 723–730.

Trevaskis, B., Hemming, M.N., Dennis, E.S., and Peacock, W.J. (2007) The molecular basis of vernalization-induced flowering in cereals. *Trends Plant Sci. 12*: 352–357.

Ungerer, M., Baird, S., Pan, J., and Rieseberg, L. (1998) Rapid hybrid speciation in wild sunflowers. *Proc. Natl. Acad. Sci. USA 95*: 11757–11762.

Ungerer, M.C., Strakosh, S.C., and Zhen, Y. (2006) Genome expansion in three hybrid sunflower species is associated with retrotransposon proliferation. *Curr. Biol. 16*: R872–R873.

van der Knaap, E., Sanyal, A., Jackson, S.A., and Tanksley, S.D. (2004) High-resolution fine mapping and fluorescence in situ hybridization analysis of *sun*, a locus controlling tomato fruit shape, reveals a region of the tomato genome prone to DNA rearrangements. *Genetics 168*: 2127–2140.

Vastenhouw, N.L., Fischer, S.E.J., Robert, V.J.P., Thijssen, K.L., Fraser, A.G., Kamath, R.S., Ahringer, J., and Plasterk, R.H.A. (2003) A genome-wide screen identifies 27 genes involved in transposon silencing in *C. elegans*. *Curr. Biol. 13*: 1311–1316.

Venner, S., Feschotte, C., and Biemont, C. (2009) Dynamics of transposable elements: towards a community ecology of the genome. *Trends Genet. 25*: 317–323.

Vinogradov, A.E. (2003) Selfish DNA is maladaptive: evidence from the plant Red List. *Trends Genet. 19*: 609–614.

Vitte, C., and Bennetzen, J.L. (2006) Analysis of retrotransposon structural diversity uncovers properties and propensities in angiosperm genome evolution. *Proc. Natl. Acad. Sci. USA 103*: 17638–17643.

Walbot, V. (1991) The *mutator* transposable element family of maize. *Genet. Eng. 13*: 1–37.

Walbot, V., Britt, A., Luehrsen, K., McLaughlin, M., and Warren, C. (1988) Regulation of mutator activities in maize. *Basic Life Sci. 47*: 121–135.

Walker, E.L., Robbins, T.P., Bureau, T.E., Kermicle, J., and Dellaporta, S.L. (1995) Transposon-mediated chromosomal rearrangements and gene duplications in the formation of the maize *R-r* complex. *EMBO J. 14*: 2350–2363.

Wessler, S.R. (1996) Turned on by stress. Plant retrotransposons. *Curr. Biol. 6*: 959–961.

Wolfgruber, T.K., Sharma, A., Schneider, K.L., Albert, P.S., Koo, D.H., Shi, J., Gao, Z., Han, F., Lee, H., Xu, R., et al. (2009) Maize centromere structure and evolution: sequence analysis of centromeres 2 and 5 reveals dynamic Loci shaped primarily by retrotransposons. *PLoS Genet. 5*: e1000743.

Woodhouse, M.R., Pedersen, B., and Freeling, M. (2010) Transposed genes in *Arabidopsis* are often associated with flanking repeats. *PLoS Genet. 6*: e1000949.

Xiao, H., Jiang, N., Schaffner, E., Stockinger, E.J., and van der Knaap, E. (2008) A retrotransposon-mediated gene duplication underlies morphological variation of tomato fruit. *Science 319*: 1527–1530.

Xu, G., Ma, H., Nei, M., and Kong, H. (2009) Evolution of F-box genes in plants: different modes of sequence divergence and their relationships with functional diversification. *Proc. Natl. Acad. Sci. USA 106*: 835–840.

Yaakov, B., and Kashkush, K. (2011) Massive alterations of the methylation patterns around DNA transposons in the first four generations of a newly formed wheat allohexaploid. *Genome 54*: 42–49.

Yamanaka, S., Nakamura, I., Watanabe, K.N., and Sato, Y. (2004) Identification of SNPs in the *waxy* gene among glutinous rice cultivars and their evolutionary significance during the domestication process of rice. *Theor. Appl. Genet. 108*: 1200–1204.

Yan, L., Loukoianov, A., Blechl, A., Tranquilli, G., Ramakrishna, W., SanMiguel, P., Bennetzen, J.L., Echenique, V., and Dubcovsky, J. (2004) The wheat *VRN2* gene is a flowering repressor down-regulated by vernalization. *Science 303*: 1640–1644.

Yang, S., Arguello, J.R., Li, X., Ding, Y., Zhou, Q., Chen, Y., Zhang, Y., Zhao, R., Brunet, F., Peng, L., et al. (2008) Repetitive element-mediated recombination as a mechanism for new gene origination in *Drosophila*. *PLoS Genet. 4*: e3.

Yao, J., Dong, Y., and Morris, B.A. (2001) Parthenocarpic apple fruit production conferred by transposon insertion mutations in a MADS-box transcription factor. *Proc. Natl. Acad. Sci. USA 98*: 1306–1311.

Yogeeswaran, K., Frary, A., York, T.L., Amenta, A., Lesser, A.H., Nasrallah, J.B., Tanksley, S.D., and Nasrallah, M.E. (2005) Comparative genome analyses of *Arabidopsis* spp.: Inferring chromosomal rearrangement events in the evolutionary history of A. thaliana. *Genome Res. 15*: 505–515.

Yu, C., Zhang, J., and Peterson, T. (2011) Genome rearrangements in maize induced by alternative transposition of reversed *Ac/Ds* termini. *Genetics 188*: 59–67.

Zarrillo, S., Pearsall, D.M., Raymond, J.S., Tisdale, M.A., and Quon, D.J. (2008) Directly dated starch residues document early formative maize (*Zea mays* L.) in tropical Ecuador. *Proc. Natl. Acad. Sci. USA 105*: 5006–5011.

Zemach, A., Kim, M.Y., Silva, P., Rodrigues, J.A., Dotson, B., Brooks, M.D., and Zilberman, D. (2010) Local DNA hypomethylation activates genes in rice endosperm. *Proc. Natl. Acad. Sci. USA 107*: 18729–18734.

Zhang, J., Yu, C., Krishnaswamy, L., and Peterson, T. (2011) Transposable elements as catalysts for chromosome rearrangements. *Methods Mol. Biol. 701*: 315–326.

7 Imprinted Gene Expression and the Contribution of Transposable Elements

Mary A. Gehring

Transposable elements (TEs) are sources of genetic and epigenetic variation that can be acted upon positively or negatively by evolutionary selection. TEs were discovered in part because their excision in endosperm tissue of developing maize kernels produced an easily scorable phenotype. This chapter focuses on the role of transposable elements in gene imprinting, an epigenetic phenomenon that takes place during endosperm development. Imprinting describes allelic differences in gene expression that depend on whether an allele is inherited from the male or female parent (Fig. 7.1). The process is considered epigenetic because alleles that reside in the same nucleus are functionally distinct at the transcriptional level, even though their DNA sequence might be identical.

Endosperm is a seed tissue that provides nutrients to the developing embryo or germinating seedling. In addition to being a critical source of amino acids and sugars for plants and people, genetic and genomic studies point to the endosperm as a dynamic environment subject to large-scale epigenetic reprogramming, much of which is initiated before fertilization and directed at TEs. One consequence of reprogramming is imprinted gene expression, which often involves differential methylation of maternal and paternal alleles and allele-specific regulation by Polycomb group proteins.

TEs have profoundly shaped the genetic and epigenetic landscape of the genome. Evidence is increasingly emerging that TEs are integrated components of the endosperm transcriptome. In general, TEs can serve as promoters, fodder for DNA-binding proteins, or, because they are subject to DNA methylation and other epigenetic modifications, epigenetic modifiers of gene expression (Slotkin and Martienssen, 2007; Feschotte, 2008). There is abundant evidence that TE insertions can have negative consequences for the host organism, leading to their removal from genic space. However, insertion of a TE that causes an adaptative change in gene expression can be selected for, such that the TE becomes an integral aspect of gene regulation (Feschotte, 2008). The expression of a large fraction of imprinted genes may be directly tied to the epigenetic status of transposable elements that have been co-opted or domesticated to regulate gene expression. Genes may become imprinted when a TE inserts nearby and alters the gene's expression pattern. In both plants and animals, TEs might drive imprinted gene expression because they attract epigenetic attention from the genome, which, combined with tissue-specific expression of epigenetic modifiers, allows maternal and paternal alleles to be differentially marked. This chapter considers why and how genes are imprinted and the impact of transposable elements on the process, using detailed data from single-gene analyses and more recent whole-genome studies. Most of the molecular details come from *Arabidopsis thaliana*, with important early genetic work from *Zea mays*. A critical challenge for

Plant Transposons and Genome Dynamics in Evolution, First Edition. Edited by Nina V. Fedoroff.

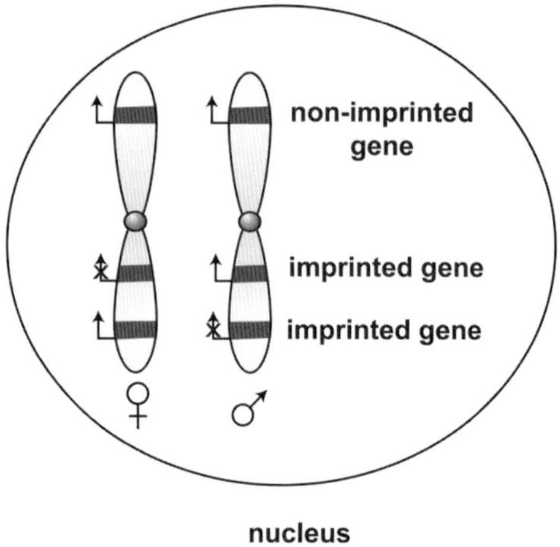

Figure 7.1 Imprinted genes are expressed from only one parental chromosome. Alleles that reside in the same nucleus can be expressed differently depending on which parent transmits them. Artwork by Tom DiCesare.

the future will be to understand how relevant the insights gleaned from the study of *Arabidopsis thaliana*, a highly inbred plant with a small genome and relatively few TEs, are for plants with very different genome structures.

Why are Genes Imprinted?

Genomic imprinting is a startling phenomenon from an evolutionary perspective. Why discard the advantages of diploidy in favor of monoallelic expression at certain loci? Monoallelic expression could place an organism in a genetically precarious situation if the single expressed allele it inherits from its parents is mutated. And, in fact, inheritance of mutant versions of some imprinted genes is lethal during seed development in plants (Grossniklaus et al., 1998; Kiyosue et al., 1999; Chaudhury et al., 1997). Yet in both flowering plants and mammals, around 1%–2% of genes (several hundred in absolute numbers) exhibit parentally biased expression at some stage of development. Several theories have been proposed to explain why imprinted gene expression is evolutionarily advantageous at certain loci. It is important to note that these theories do not address the molecular mechanism of parent-of-origin specific expression, only why selection favors imprinted expression over nonimprinted expression once it arises. Many of the theories rely on an understanding of the developmental processes that create offspring. Therefore, it is important to first consider the process of seed development and the developmental origin of endosperm.

The Developmental Origin of Endosperm

The development of endosperm is a unique innovation of flowering plants, which arose around 200 million years ago. Meiosis produces megaspores (female) and microspores (male) in developing flower buds. Megasporogenesis takes place inside an ovule, and only one of the four meiotic products, the megaspore, survives in most species. The haploid megaspore undergoes three rounds of mitosis to form an uncellularized female gametophyte containing eight nuclei, with four nuclei

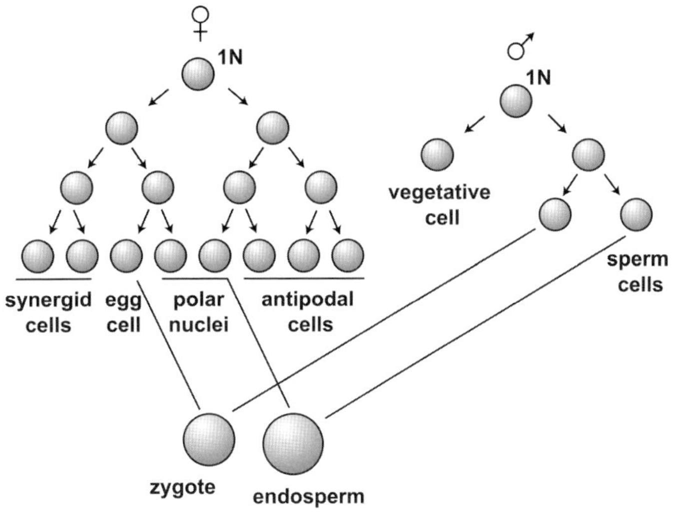

Figure 7.2 Nuclear division during gametophyte formation. Haploid megaspores (female) undergo three rounds of mitosis to generate all eight nuclei of the female gametophyte. Haploid microspores (male) undergo two rounds of mitosis. During fertilization, one haploid sperm fuses with the haploid egg, creating a diploid zygote. The other haploid sperm fuses with the two haploid polar nuclei, creating a triploid endosperm. Artwork by Tom DiCesare.

each spatially segregated at opposite ends of the gametophyte (Figures 7.2 and 7.3). After mitosis is complete, two nuclei (the polar nuclei) migrate from opposite poles of the female gametophyte toward its center. One of these nuclei is the immediate mitotic sister of the egg cell nucleus, while the other is the sister of an antipodal cell nucleus (Figure 7.2).

The female gametophyte cellularizes to form seven cells: the egg cell, two synergid cells, the diploid central cell, and three antipodal cells (Figure 7.3). The two polar nuclei of the central cell undergo karyogamy before or at the time of fertilization, depending on the species, creating a homodiploid nucleus. Several fascinating variations on the constitution of the female gametophyte exist among angiosperms—proliferation of antipodal cells, fusion of multiple nuclei in the central cell—but the 8-nuclei, 7-celled female gametophyte is the most common form (Maheshwari, 1950). Meanwhile, on the male side, the haploid microspore undergoes two rounds of mitosis to generate a pollen grain that contains two haploid sperm and a vegetative, or pollen tube, nucleus (Figures 7.2 and 7.3). During fertilization, one sperm fertilizes the egg cell, creating a diploid zygote, and the other sperm fertilizes the central cell, creating a triploid primary endosperm nucleus (Figures 7.2 and 7.3).

Except for the difference in ploidy, the resultant embryo and endosperm are genetically identical. As the process of seed formation begins, the primary endosperm nucleus divides rapidly. Several endosperm nuclei are present before the zygote undergoes its first division, and both the central cell and endosperm appear more transcriptionally active than the egg cell and zygote, respectively, at early stages (Grimanelli et al., 2005; Pillot et al., 2010). In the most common type of endosperm development, observed in the model plants *Arabidopsis thaliana* and *Zea mays*, nuclei divide synchronously for several rounds of mitosis without any cellularization, forming a syncytium (Lopes and Larkins, 1993; Brown et al., 1999; Berger, 2003; Sabelli and Larkins, 2009). Once cellularization proceeds, three distinct endosperm domains can be distinguished, the micropylar, peripheral, and chalazal endosperms. Cells within either the micropylar or peripheral domains divide

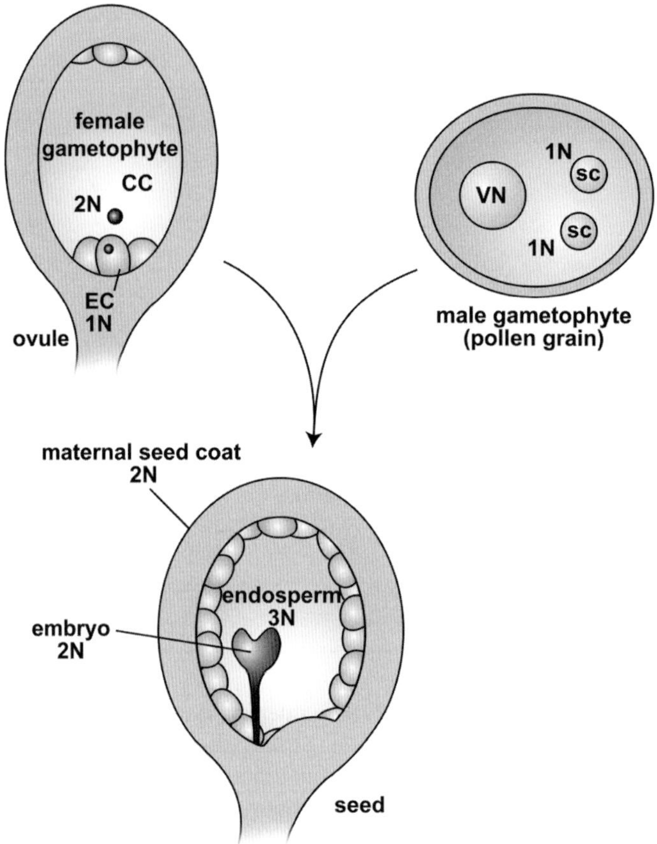

Figure 7.3 Double fertilization. Seeds consist of three distinct components, the diploid embryo and triploid endosperm, formed by fertilization, and the diploid seed coat, derived from maternal ovule tissues. EC, egg cell; CC, central cell; VN, vegetative nucleus; SC, sperm cells. Artwork by Tom DiCesare.

synchronously, but asynchronously with the other domain (Boisnard-Lorig et al., 2001). In some species, like maize, this is followed by massive endoreduplication of endosperm nuclei (Kowles and Phillips, 1985; Sabelli and Larkins, 2009). The different endosperm domains have distinct gene expression programs and functions (Le et al., 2010). The chalazal endosperm, which sits atop the endpoint of maternal vasculature, remains partially uncellularized, and in maize forms a specialized transfer cell layer with extensive cell wall ingrowths that likely facilitates nutrient uptake from maternal tissues. The *Arabidopsis* chalazal endosperm exhibits dense cytoplasm, extensive endoplasmic reticulum cisternae, and large nuclei of higher ploidy because of nuclear fusion and endoreduplication (Nguyen et al., 2000; Boisnard-Lorig et al., 2001; Baroux et al., 2004). Many imprinted genes exhibit prolonged expression in the chalazal endosperm compared to the other domains (Wolff et al., 2011). The ultimate fate of the endosperm depends on the lineage, although in all species it is a terminal tissue. In dicots, endosperm is ephemeral and reserves are ultimately absorbed and stored in the embryo's cotyledons, with perhaps only one or two layers of endosperm remaining at seed maturity. In cereals, endosperm is persistent and is used as an energy source during seedling germination.

Selection for Imprinted Expression

The most widely discussed theory to explain the adaptive advantage of imprinted expression is the kinship or parental conflict theory of imprinting (Haig and Westoby, 1989; Haig and Westoby, 1991; Haig, 2000; Haig, 2004). The theory proposes that imprinted expression has evolved "as a mechanism of transcriptional control at loci whose expression has fitness consequences for asymmetric kin," (Haig, 2000, p. 11) and is most often explained in terms of conflicting interests between maternal and paternal alleles in an offspring over resource allocation from the mother to the offspring. However, the theory does apply more broadly to any asymmetric kin, not just mother and offspring (Haig, 2004). An intriguing feature of imprinting that fits well with the tenets of the theory is that it has been observed in placental mammals and flowering plants. In both organisms, offspring are directly nourished by the maternal parent, in mammals via the placenta and in plants through the endosperm.

Imprinted expression may be the evolutionary stable strategy in the following way. Mothers may have offspring by multiple fathers, either simultaneously or in the future. The mother is equally related to all of her offspring and strives to partition resources equally among them. In contrast, it is in the father's interest for his offspring to receive more nutrients from the mother than do other maternal half-siblings. Paternally expressed imprinted genes will promote transmission of paternal alleles by extracting maximal resources from the mother. Maternally expressed imprinted genes counteract this effect by restricting nutrient flow so that all offspring receive adequate nutrients. The endosperm, like the placenta, is the site of nutrient transfer between generations and, under the theory, is expected to be subject to imprinting. One might predict that other plant tissues, like the embryonic suspensor, which transfers nutrients between the maternal parent and embryo during the early stages of seed development, might also be subject to imprinting, but this has not been experimentally explored. The function of several imprinted genes in both plants and mammals conform to the expectations of the kinship theory (Tycko and Morison, 2002; Grossniklaus et al., 1998; Kiyosue et al., 1999; Chaudhury et al., 1997). In *Arabidopsis thaliana*, the maternally expressed imprinted genes *MEA* and *FIS2* cause endosperm overgrowth when mutant alleles are inherited maternally, suggesting that these genes normally repress growth of the endosperm.

The triploid nature of the endosperm itself could be due to parental conflict (Haig and Westoby, 1989). Extant basal angiosperms have a 4-nuclei, 4-cell female gametophyte that produces a diploid endosperm with one maternal and one paternal genome (Williams and Friedman, 2002; Friedman, 2006). If a molecular mechanism for parent-of-origin specific gene existed, doubling the maternal genomic contribution to the endosperm might have given the maternal parent an advantage in the conflict with the paternal parent (Haig and Westoby, 1989). In this regard, it would be interesting to determine whether or not epigenetic modifiers known to be active during endosperm imprinting function in diploid endosperm, as well.

Other theories suggest imprinting is an outcome of coadaptation rather than conflict. The maternal-offspring coadaptation theory states that imprinting is expected for loci "whenever there is genetic variation at a locus affecting the maternal and offspring traits involved in the maternal-offspring interaction" (Wolf and Hager, 2006, p. 2240). Enhancing coadaptation between the maternal parent and developing offspring might increase offspring fitness. This explanation is favored by Costa et al., who recently demonstrated that a maize maternally expressed imprinted gene, *Meg1*, promotes nutrient allocation to offspring. *Meg1* function is necessary for endosperm transfer cell layer specification. Reducing *Meg1* transcript levels causes the production of small kernels with reduced levels of glucose and fructose, whereas increasing *Meg1* dosage leads to a large kernel phenotype (Costa et al., 2012).

Both the kinship theory and the coadaptation theory make certain predictions about the processes that imprinted genes should impact, the proportion of genes that should be maternally or paternally expressed, and the degree of imprinting at a given locus. Multiple theories would appear to fit several known imprinted genes. One could easily imagine a gene involved in transfer of nutrients from the maternal parent to the seed that might fit equally well under the kinship or maternal-offspring coadaptation theory. Detailed understanding of the evolutionary history and function of imprinted genes is necessary to understand why imprinted expression was selected for, but only a few studies have examined this in either plants or animals. The following quote from Brandvain (Brandvain et al., 2011) is informative to keep in mind: “It is straightforward to construct a plausible-sounding evolutionary argument about why a particular gene in a given species should be imprinted. However, plausibility does not constitute strong evidence for any particular model.” There is likely to be more than one reason imprinted gene expression is adaptive, and a unifying theory explaining all instances of imprinting seems biologically unlikely.

Finally, another idea, which has received comparatively little attention, posits that selection is not acting at the level of the transcriptional control of gene expression, but rather on chromatin structure differences between maternally and paternally inherited chromosomes (Pardo-Manuel de Villena et al., 2000; Paldi, 2003). The theory has mainly considered the mammalian imprinting data, in which imprinted genes lie in large chromosomal clusters (hundreds of kilobases to a megabase) with distinct chromatin structure and replication timing. The hypothesis is that sex-specific chromatin differences were selected for as a mechanism to distinguish homologous chromosomes from each other and from nonhomologous chromosomes, and facilitate chromosome pairing during meiosis (Pardo-Manuel de Villena et al., 2000). This specific explanation seems unlikely to be a factor in plants, as imprinting takes place in a tissue that never has a chance to participate in meiosis and contribute to germ cells. Moreover, plant imprinted genes do not appear to be in large chromosomal clusters, although a few mini clusters on the order of two to three genes have been identified (Wolff et al., 2011; Gehring et al., 2011; Luo et al., 2011). However, the idea that maternal and paternal specific epigenetic differences could have functions besides simply establishing or reflecting differential gene expression is an important one. Not a single plant imprinted gene has been shown to have detrimental phenotypic effects when expressed biallelically or overexpressed from one allele, albeit very few have been explicitly tested. Given that imprinted expression is part and parcel of a larger remodeling of maternal chromatin, to be discussed momentarily, the idea that differential allelic expression might not be the only functional aspect of imprinting or the only process under selection should not be dismissed.

Principles Derived from the First Imprinted Gene

The first imprinted gene discovered in any organism was the *R* locus of maize (Kermicle, 1970). *R* controls anthocyanin production in the aleurone, the outer layer of endosperm found in grasses. Observations in the beginning of the last century indicated a reciprocal aleurone phenotype depending on whether *RR* (pigmented aleurone phenotype) or *rr* (colorless aleurone phenotype, recessive to *R*) served as the male or the female parent in a genetic cross (Kempton, 1919). Pollinating *RR* females with *rr* males leads to deep-red, solidly colored kernels. In contrast, when *rr* females are pollinated by *RR* males, the aleurone of progeny kernels is mottled due to irregular anthocyanin distribution (Figure 7.4A).

A reasonable hypothesis to explain these results is that the reciprocal pigmentation phenotype is due to *R* gene dosage in the endosperm. Since the endosperm inherits two maternal and one

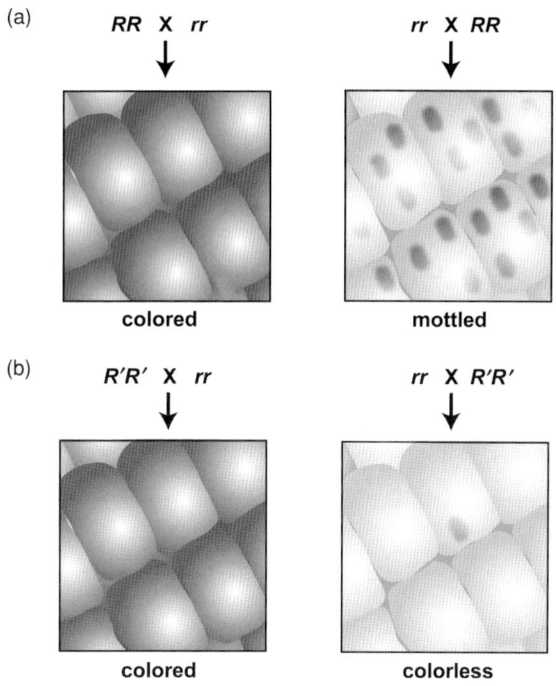

Figure 7.4 **The *R* locus is imprinted in maize.** A) Maize aleurone is differentially pigmented depending on whether *R* is transmitted through the maternal or paternal parent. B) The paramutant *R′* allele is subject to increased silencing compared to *R* when inherited through the paternal parent. Crosses are written with the maternal genotype on the left and the paternal genotype on the right. Artwork by Tom DiCesare.

paternal genome, the genotype of the aleurone in reciprocal crosses is either *RR/r* (two doses of *R*) or *rr/R* (one dose of *R*). Kermicle (Kermicle, 1970) tested this hypothesis by using B-A chromosome translocations to generate endosperm with varying numbers of *R* alleles inherited from the male or female parent. These experiments conclusively showed that *R* gene dosage was not responsible for the reciprocal aleurone phenotypes. Additional crosses ruled out other potential explanations such as a maternal sporophytic effect, cytoplasmic inheritance, *r* dosage, or dosage balance between *r* and *R* (Table 7.1). Instead, the data showed that the *R* alleles carried by the polar nuclei were functionally

Table 7.1 Summary of key results from Kermicle, 1970.

Aleurone Genotype	Aleurone Pigmentation	
Female/Male Origin	Mottled	Solid
rr/R	+	
rr/rR	+	
rr/RR	+	
RR/R		+
*RR/*r		+
RR/rr		+
(-)*R/r*		+

different from *R* alleles transmitted by the sperm. The phenomenon was termed paragenetic by Kermicle, which has since been supplanted by the term epigenetic.

Further experiments in another paper in the same issue of *Genetics* cleverly demonstrated that only the endosperm, and not the embryo or seedling, was affected (Brink et al., 1970).

Four key principles were explicitly derived from these elegant genetic experiments (Kermicle, 1970; Brink et al., 1970):

(1) Genes can be transmitted in different forms through the male and female gametophytes.
(2) The difference affects the primary endosperm nucleus but not the egg nucleus.
(3) Expression of the phenomenon is peculiar to the endosperm. It remains stable, or nearly so, during seed development.
(4) Reversibility of the difference is not necessary if the phenomenon is restricted to the endosperm.

Remarkably, more than 40 years later, these principles, as will be demonstrated in the subsequent pages, still frame current thinking about the logic and mechanism of gene imprinting as the molecular details underlying the genetics and epigenetics of these processes have emerged.

Gene Imprinting and Parent-of-Origin Effects on Seed Development

Although *R* gene imprinting was described in 1970, there was little progress on discovering additional specific plant imprinted genes until the late 1990s. However, it was clear that the endosperm was subject to parent-of-origin effects on seed development, most often involving seed size (Lin, 1982, 1984), and endosperm failure had long been pegged as a major reason for hybrid incompatibility (Brink and Cooper, 1947). Working with B-A chromosome translocations in maize, Lin showed that small seeds were produced when chromosome 10L was not inherited from the male, but inheriting extra copies of 10L from the female did not ameliorate the effect (Lin, 1982). This suggested that there were genes active on paternally inherited chromosome 10 that were not active on maternally inherited chromosome 10. These and other experiments showed that in maize and potatoes the ratio of maternal to paternal endosperm genomes must be 2:1 in order for viable seeds to be produced (Johnston et al., 1980; Lin, 1984). Other plants, such as *Arabidopsis thaliana*, are more tolerant of ploidy alterations, although not without parent-of-origin phenotypic effects. Maternal genomic excess in interploidy crosses leads to smaller seeds that undergo earlier endosperm cellularization, while paternal genomic excess causes the opposite phenotype (Scott et al., 1998). These differences could be due to the action of differentially expressed imprinted genes that function in endosperm growth, as predicted by the kinship theory. However, the lack of identification of the specific genes underlying the phenotype precludes any firm conclusions. Parent-of-origin effects might also be ascribed to differential contributions of dosage-sensitive regulators to the seed (Dilkes and Comai, 2004).

What Genes are Imprinted?

The first imprinted gene in *A. thaliana* was described in 1998–1999. In the ensuing decade, an additional 10 imprinted genes were identified in *A. thaliana* (Table 7.2), with a similar number known in maize. Many encode the kind of potential regulatory proteins that tend to excite developmental biologists. Most prominently these included members of the Polycomb group genes, which maintain repressed states of gene expression. When mutant alleles of these genes are inherited maternally, seeds abort due to endosperm overproliferation. This indicated a direct relationship between parent-of-origin effects on seed phenotype and gene imprinting, and fit nicely with the kinship theory of imprinting. Additionally, not only are *MEA* and *FIS2*—which form part of the repressive Polycomb group complex—imprinted, they in turn regulate silencing of other imprinted genes, including themselves (Köhler et al., 2005; Gehring et al., 2006; Jullien et al., 2006a). At least one copy of the maize *Mez* and *FIE* genes, as well as rice *FIE1*—all of which encode components of the PcG complex—are also endosperm imprinted in those species (Danilevskaya et al., 2003; Gutierrez-Marcos et al., 2006; Hermon et al., 2007; Haun et al., 2007). Imprinting of members of the same Polycomb group complex in diverse angiosperm species strongly suggests a functional role for imprinting at these loci, and that these genes are key regulators of the endosperm developmental program. In 2009, additional *Arabidopsis* imprinted genes were predicted based on epigenetic and expression criteria, with some validated (Gehring et al., 2009). These included a large number of transcription factors, many part of the same family of class IV leucine zipper homeodomain genes (Gehring et al., 2009), two of which have been identified as seed-specific transcription factors (Le et al., 2010) (Table 7.2). Overall, the identification of the first handful of imprinted genes suggested that imprinting affected known or potential regulatory proteins that were, or could potentially be, important for orchestrating endosperm development.

The pace of imprinted gene discovery has recently increased rapidly. The advent of high through-put mRNA sequencing technologies has made large-scale identification of imprinted genes and further understanding of the regulation of imprinting possible. Three studies in *Arabidopsis* (Hsieh et al., 2011; Wolff et al., 2011; Gehring et al., 2011) and one in rice (Luo et al., 2011) have recently identified between 50 and 250 imprinted genes in endosperm. By performing reciprocal crosses between strains with single nucleotide polymorphisms, transcripts can be identified as derived from the maternally or paternally inherited allele. The true number of imprinted genes in each species is likely two or three times what has been described because not all genes can be analyzed due to lack of sequence polymorphisms.

While allele-specific mRNA sequencing (mRNA-seq) represents a powerful method to identify imprinted genes, it is important to realize the limitations and challenges associated with these data. Like the RT-PCR and RT-qPCR approaches that have come before it to assess imprinted gene expression at individual loci, mRNA-seq only reports on steady-state transcript levels. Therefore,

Table 7.2 *A. thaliana* imprinted genes discovered prior to the application of high-throughput mRNA sequencing.

Gene	Type	Discovered through	Gene Function
MEA	MEG	Genetic screen for seed phenotype.	Polycomb group gene
FIS2	MEG	Genetic screen for seed phenotype.	Polycomb group gene
PHE1	PEG	Expression deregulated in *fis* mutants.	MADS-box transcription factor
FWA/HDG6	MEG	Candidate gene based on reproductive expression.	homeodomain transcription factor
MPC	MEG	Expression pattern like *FWA* in interploidy crosses and *mea* mutants.	poly-A binding protein
FH5	MEG	Candidate gene based on endosperm expression.	actin-organizing formin
MYB3R2	MEG	Gene less methylated in endosperm than embryo.	MYB domain transcription factor
HDG3	PEG	Gene less methylated in endosperm than embryo.	homeodomain transcription factor
HDG8	MEG	Gene less methylated in endosperm than embryo.	homeodomain transcription factor
HDG9	MEG	Gene less methylated in endosperm than embryo.	homeodomain transcription factor
AT5G62110	PEG	Gene less methylated in endosperm than embryo.	homeodomain-like

MEG = maternally expressed imprinted gene; PEG = paternally expressed imprinted gene.

although genes might be called imprinted, it is not truly known, only assumed, that maternal and paternal alleles are differentially transcribed. A larger issue is what one calls imprinted when enormous sequencing depth is possible. Each gene will represent a continuum of maternal:paternal expression ratios, making a sharp distinction between imprinted and nonimprinted gene expression difficult. Is an imprinted gene one that is entirely monoallelic? Or is statistically significant parentally biased expression also considered gene imprinting? As sequencing depth increases, the incidence of monoallelic expression should decrease, as detecting extremely rare transcripts from the other parental allele becomes more likely. More importantly, is there a difference in the biological significance of monoallelic versus biased expression? According to the kinship theory of imprinting, the evolutionary stable strategy for an imprinted gene is complete monoallelic expression. If this is not observed, shall a gene be rejected as not imprinted, or shall we conclude that the evolutionary pressures acting at the locus do not involve conflict among asymmetric kin? It is also impossible to be entirely certain that sequenced transcripts are not already present in gametes before fertilization, although the stages at which imprinting has been assayed, at least four days after fertilization, make this seem unlikely except for the hardiest transcripts. All of these challenges remain to be resolved by the field, and it is anticipated that the continual application of new technologies (for example, whole-genome nuclear run-on assays or profiling of transcripts associated with polysomes) will help address these issues. The four recent studies have taken slightly different approaches to each of these issues, using various intuitive and statistical criteria to define a gene as exhibiting imprinted expression.

Given these considerations, what broad themes have emerged from these data? All global gene expression studies confirm that imprinting is largely endosperm specific. One imprinted locus has been identified in maize embryos (Jahnke and Scholten, 2009) and one in rice (Luo et al., 2011). Both genes are transiently active from the maternal allele in the embryo but persistently imprinted in the endosperm. No embryo-imprinted genes have been confirmed in *Arabidopsis*, at least at the later stages of seed development that have been examined (Hsieh et al., 2011; Gehring et al., 2011). Thus, the principles derived from the *R* gene still hold even when thousands of genes are considered. The majority of imprinted genes are maternally biased rather than paternally biased, perhaps suggesting greater control by the maternal parent over seed development. Both maternally and paternally expressed imprinted genes are expressed more highly in endosperm than embryo, tend to have lower levels of expression in vegetative tissues, and are often highly expressed in chalazal endosperm. Imprinted genes are also often expressed in pollen, although this is likely to be in the vegetative nucleus and not the sperm cells (Gehring et al., 2009; Hsieh et al., 2011; Wolff et al., 2011; Gehring et al., 2011; Luo et al., 2011). The *Arabidopsis* experiments confirm that genes involved in nucleic acid binding (e.g., transcription factors and chromatin-related proteins) are indeed enriched among imprinted genes, particularly those that are paternally biased (Wolff et al., 2011; Gehring et al., 2011). Intriguingly, there are a number of epigenetic regulators subject to imprinting, including proteins involved in maintenance DNA methylation, RNA-directed DNA methylation, and histone methylation. Thus, as suggested by the study of imprinted Polycomb group genes, complex circuits may exist within the endosperm to regulate the regulators of imprinting. Of course, most of the imprinted loci appear far more pedestrian: oxidoreductases, glycosyl hydrolases, pectin methylesterases, ubiquitin-dependent proteases, and lipid-binding proteins, to name a few.

The rice study examined all regions of the genome for imprinting, not only known transcripts. Interestingly, about a third of the candidate imprinted loci represent intergenic regions (Luo et al., 2011). While these could simply be unannotated genes, they might also represent noncoding RNAs, which have important functions in mammalian imprinting but have not been extensively characterized in plants (Luo et al., 2011). There are also many new imprinted genes that have been well studied in other developmental contexts, like ethylene, auxin, and jasmonic acid hormone pathway genes (e.g., *EIN2*). There has been little indication for a role for these genes during endosperm development, although it is known that auxin and ethylene contribute to the initiation of programmed cell death in maize endosperm. Determining whether or not the newly discovered imprinted genes are functionally relevant during endosperm development will be important for understanding why genes are imprinted.

Epigenome Dynamics during Seed Development

Imprinted gene expression takes place in the context of an endosperm that is epigenetically distinct from its sister, the embryo, and other sporophytic tissues (Johnson and Bender, 2009; Feng et al., 2010b). The endosperm is characterized by altered chromatin structure and genome-wide hypomethylation that specifically affects euchromatic transposable elements and repeated sequences.

Epigenetic Landscape in Vegetative Tissues

A great deal is known about the basic properties of the epigenome in sporophytic tissues, which will be briefly described here. These principles have primarily been derived from experiments utilizing vegetative tissues consisting of many different cell types, like seedlings or flower buds. Most research has been performed in *Arabidopsis thaliana*, with comparable data more recently obtained from rice and maize. Many epigenetic modifications are directed at TEs and other repeated sequences.

In *Arabidopsis* interphase nuclei, chromosomes are organized into gene-rich euchromatic loops that emanate from a heterochromatic chromocenter comprised of centromeric and pericentromeric repeats, including rDNA loci and transposable elements (Fransz et al., 2002). Zooming in to the molecular level, heterochromatin is characterized by high levels of repressive histone modifications and 5-methylcytosine, the best-studied epigenetic mark (Roudier et al., 2011). Often, although not exclusively, 5-methylcytosine is associated with transcriptionally inactive regions of the genome, and is concentrated at chromocenters, but is also distributed throughout the chromosome arms. In plants, 5-methylcytosine is most frequently found in the context of CpG dinucleotides, but also at CHG and CHH positions (H being any base except C) (Cokus et al., 2008; Lister et al., 2008).

A key property of DNA methylation is its heritability. After DNA replication, maintenance DNA methyltransferases and accessory proteins recognize hemimethylated CpG sites and methylate the new strand of DNA. The primary maintenance methyltransferase in plants is *MET1*. CHG methylation is maintained by the chromomethylase CMT3, which also recognizes a modified histone, linking DNA methylation and histone methylation states (Law and Jacobsen, 2010). Perpetuation of asymmetric DNA methylation (CHH) requires an elaborate RNA-directed DNA methylation system (Law and Jacobsen, 2010; Gehring and Henikoff, 2008). Through mechanisms that are still not precisely biochemically or molecularly defined, de novo DNA methyltransferases are targeted to specific sequences by 24-nucleotide small RNAs to deposit CG and non-CG methylation at homologous sequences. The small RNAs are generated from aberrant RNAs via specialized plant RNA polymerases, DICERs, RNA-DEPENDENT RNA POLYMERASEs, and ARGONAUTE enzymes (Matzke et al., 2009). Transposable elements and other nonunique sequences serve as sources of small RNAs and are the most frequent targets of DNA methylation in the genome (Lisch, 2009; Teixeira and Colot, 2010). TEs within heterochromatin are densely methylated at both CG and non-CG sites, match corresponding small RNAs, and become transcriptionally active when genome-wide DNA methylation levels are severely reduced due to mutation of the maintenance methyltransferase (Zilberman et al., 2007; Zhang et al., 2006). Thus, one function of DNA methylation is to repress TE transcription. Despite this, individual TEs that lose methylation can remain transcriptionally silent (Woodhouse et al., 2006) and about a quarter of *Arabidopsis* TEs lack DNA methylation, primarily due to depletion in CpG sites (Ahmed et al., 2011). Loss of multiple methyltransferases, or mutations in a chromatin remodeler that also impacts DNA methylation patterns, can lead to new TE insertions (Miura et al., 2001; Kato et al., 2003). Methylated TEs associated with small RNAs are also distributed, though somewhat less frequently, in gene-rich euchromatic regions. Seventeen percent of euchromatic TEs overlap annotated genes. TEs accumulate close to the ends of genes, but unmethylated TEs are more likely than methylated TEs to be close to the 5′ ends (Ahmed et al., 2011).

Genes are targets of DNA methylation independent of TEs. Genic methylation is primarily in the CpG context, and there is no correlation with the presence of small RNAs (Cokus et al., 2008; Lister et al., 2008). Within genes, DNA methylation is distributed away from the 5′ end (Figure 7.5A). This may be due to the repressive effects of DNA methylation when it is found around the transcriptional start site of genes. Genes associated with promoter methylation are expressed at low levels in *Arabidopsis* (Zhang et al., 2006; Zilberman et al., 2007). Methylation patterns in both genes and TEs in the flowering plants rice and poplar are very similar to *Arabidopsis* (Zemach et al., 2010a; Feng et al., 2010a).

The methylation patterns observed in the genome are the product of both DNA methyltransferase and DNA demethylase activity (Gehring and Henikoff, 2008). *Arabidopsis* contains four plant-specific DNA glycosylases that remove 5-methylcytosine from DNA: *ROS1, DME, DML2,* and *DML3*. *DME* is primarily expressed during reproductive development and is an essential gene, while the other three are expressed ubiquitously and lack any obvious morphological phenotype when mutated (Choi et al., 2002; Gong et al, 2002; Penterman et al., 2007a). DNA glycosylases recognize damaged or mispaired bases in DNA and remove them by base excision repair. The substrate of these particular enzymes is 5-methylcytosine, which they excise from the DNA, leaving an abasic site. Other enzymes then complete DNA repair by replacing the base with unmodified cytosine and

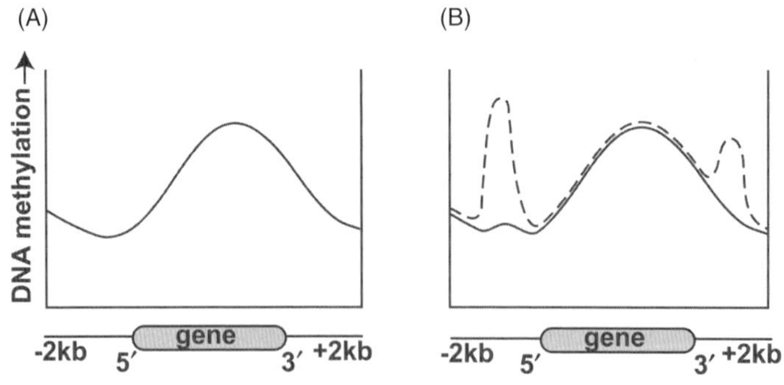

Figure 7.5 DNA methylation is distributed away from the 5′ and 3′ ends of genes. A) Schematic of average DNA methylation levels within and flanking all genes in vegetative tissues. B) Schematic of average DNA methylation levels within and flanking all genes associated with embryo-endosperm differentially methylated regions in embryo (dashed line) and endosperm (solid line). Artwork by Tom DiCesare.

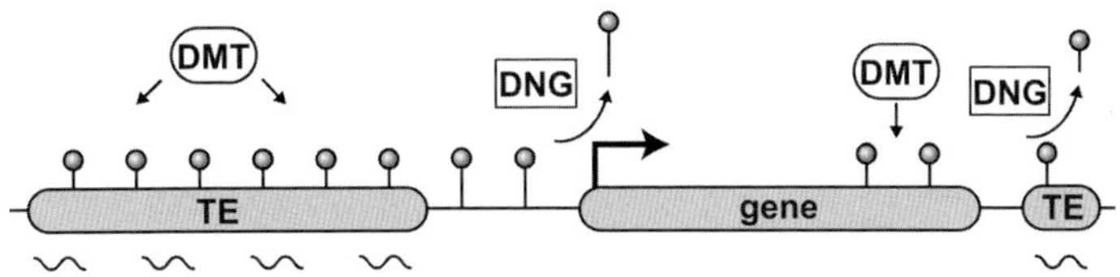

Figure 7.6 DNA glycosylases protect genes from DNA methylation directed at transposable elements. DNA methyltransferases (DMT) methylate transposable elements (TEs) targeted by small RNAs. The 5-methylcytosine DNA glycosylases (DNG) remove methylation that encroaches on genes. DNA methyltransferases also methylate the coding sequence of genes, but this is not targeted by DNA glycosylases. Lollipops, DNA methylation; squiggly lines, small RNAs. Artwork by Tom DiCesare.

sealing the nick in the DNA backbone. It is presently unknown how the glycosylases recognize 5-methylcytosine or how they are targeted to certain sequences. Mapping DNA methylation in *ros1 dml2 dml3* mutants showed that hundred of sites are subject to DNA demethylation in whole plants or flower buds (Penterman et al., 2007a; Lister et al., 2008). Interestingly, demethylation is often targeted towards the 5′ and 3′ ends of genes (Penterman et al., 2007a). Moreover, these regions correspond to sequences enriched in small RNAs and are targeted by RNA-directed DNA methylation (i.e., CG and non-CG methylation) in the absence of the demethylases (Penterman et al., 2007b).

A key concept to emerge from the study of the methylation and demethylation systems in *Arabidopsis* is that they appear to balance the need to keep transposable elements silent with the necessity of maintaining adequate gene expression levels. One function of the demethylases is to protect genes from methylation-induced silencing directed at TEs (Figure 7.6). Methylated TE density is negatively correlated with gene expression in *Arabidopsis thaliana*, and orthologous genes between closely related *Arabidopsis* species show lower levels of expression when one gene is associated with a methylated TE located within 500 base pairs of the gene (Hollister and Gaut, 2009; Hollister et al., 2011). A vigorous TE defense system may require a form of self-protection, which is provided by the demethylases (Figure 7.6).

Cytological Observations of Chromatin in Seeds

Cytological observations indicate that chromatin in endosperm is organized very differently compared to in interphase leaf nuclei. Endosperm nuclear volume is significantly greater than nuclear volume in diploid tissues, even when the increased DNA content is corrected for, suggesting that endosperm chromatin is less compacted compared to other cell types (Baroux et al., 2007). Moreover, the usual distribution of heterochromatin and euchromatin within the nucleus is altered. In addition to the chromocenters, endosperm chromatin contains many smaller heterochromatic foci interspersed among euchromatin, especially at early stages of endosperm development. This chromatin seems to be preferentially, although not exclusively, formed by the maternal fraction of the genome (Baroux et al., 2007). Interestingly, the interspersed heterochromatin is not formed by redistribution of sequences that normally contribute to chromocenters; rather, new regions of the genome are being recruited into heterochromatin (Baroux et al., 2007). Presently, the sequences underlying gross cytological differences in chromatin structure are unknown.

The differences in chromatin structure between endosperm nuclei and somatic nuclei are presaged by differences that arise in the female gametophyte. The repressive histone mark H3K9me2 (histone

H3 lysine 9 dimethylation) is diffuse in the central cell rather than being concentrated in discrete chromocenters as is usual. This pattern is maintained during the first three divisions of the endosperm (Pillot et al., 2010). Interestingly, observations of paternally inherited chromosomes with labeled histone variants indicate that the paternal chromatin remains segregated from maternal chromatin during this time (Ingouff et al., 2007). Physical separation of parental chromosomes may be important for establishing or maintaining allele-specific expression.

Epigenomic Profiling in Seeds

What might be the molecular underpinnings of observed differences in endosperm chromatin structure? Recent efforts to profile DNA methylation, histone marks, and small RNAs at high resolution in endosperm tissue have begun to address this question. The results are directly relevant for understanding TE dynamics and the mechanism of imprinted expression.

The endosperm genome is hypomethylated compared to the embryo genome or the genome of somatic tissues. Maize endosperm contains 13% less 5-methylcytosine than embryo or leaf tissue (Lauria et al., 2004). Recent studies have used whole-genome profiling methods to precisely determine embryo and endosperm methylation patterns in *Arabidopsis thaliana* and rice (Gehring et al., 2009; Hsieh et al., 2009; Zemach et al., 2010b). Specific features of the endosperm methylation profile are broadly similar to methylation patterns in vegetative tissues: higher methylation in transposable elements than in genes and low methylation around the transcriptional start site of genes. While embryo and endosperm methylation patterns are highly correlated, the endosperm genome is overall hypomethylated compared to the embryo genome, especially at sites of CG methylation (Gehring et al., 2009; Hsieh et al., 2009; Zemach et al., 2010b). Endosperm DNA is hypomethylated at discrete sites compared to the embryo or leaf tissue (Lauria et al., 2004; Kinoshita et al., 2004; Gehring et al., 2006; Gehring et al., 2009). Loss of DNA methylation, especially at CpG sites, occurs primarily on the maternal genome (Hsieh et al., 2009; Gehring et al., 2009), consistent with observed cytological observations of altered maternal chromatin structure in endosperm (Baroux et al., 2007). Genome-wide mapping has allowed regions with less methylation in the endosperm than embryo (differentially methylated regions or DMRs) to be defined (Gehring et al., 2009). Most DMRs have matching siRNAs in floral tissues (Lister et al., 2008), indicating they are normally targeted by RNA-directed DNA methylation at other stages of development (Gehring et al., 2009). DMRs are enriched for transposable elements, especially short transposable elements distributed in euchromatin (Gehring et al., 2009; Hsieh et al., 2009), and more than half of the largest differentially methylated regions occur within 2 kilobases 5′ or 3′ of genes (Gehring et al., 2009). In particular, Helitron DNA transposons are slightly enriched in hypomethylated regions compared to other types of TEs (Gehring et al., 2009). Helitrons are relatively young TEs and are more likely to be found in genic space compared to other TEs (Hollister and Gaut, 2007; Hollister and Gaut, 2009).

Although hypomethylation occurs primarily outside of the coding sequence of genes, more than a thousand genes are associated with DMRs either within the gene or 1 kilobase 5′ or 3′ (Gehring et al., 2009). Thus, the genes that are demethylated in the endosperm represent a class of genes associated with unusual (or not average) methylation patterns in the embryo and other sporophytic tissue, where the genes accumulate methylation near their 5′ and 3′ ends (Figure 7.5B). In the endosperm, these genes lose methylation so that their methylation profile is more like the average methylation profile of all genes in vegetative tissues (Figure 7.5A). It is unlikely that the methylation is normally directly inhibitory to transcription, as it accumulates about 700 base pairs upstream of the transcriptional start site, and most of the genes associated with DMRs are expressed in other tissues. However, genes with endosperm-preferred expression (Day et al., 2008) are less methylated at 5′ sequences in the endosperm than embryo (Gehring et al., 2009). Genes preferentially expressed in rice endosperm are also hypomethylated in that tissue (Zemach et al., 2010b).

The histone mark H3K27me3 (histone H3 lysine 27 trimethylation), which is deposited by the Polycomb complex, has also been profiled in endosperm (Weinhofer et al., 2010). Most of the loci modified by H3K27me3 in endosperm are also modified in vegetative tissues. However, there are specific TEs that accumulate H3K27me3 in the endosperm. Interestingly, these TEs have less DNA methylation compared to all TEs in the endosperm. Loss of DNA methylation in endosperm might allow targeting by the repressive Polycomb complex (Weinhofer et al., 2010).

What is the genetic and molecular basis for endosperm DNA hypomethylation? It is a product of both active DNA demethylation and loss of maintenance methylation. These processes are initiated before fertilization in the female gametophyte. Expression of the maintenance methyltransferase *MET1* is reduced in the mature female gametophyte for at least one round of DNA replication, which would result in hemimethylated DNA (Jullien et al., 2008). The central cell is also the site of *DME* expression, a 5-methylcytosine DNA glycosylase that is essential for seed development (Choi et al., 2002). Seeds that inherit a mutant *dme* allele through the maternal parent abort after several days of abnormal embryo and endosperm development (Choi et al., 2002). However, *DME* is not imprinted, but is expressed specifically in the polar nuclei and central cell during female gametophyte development with little expression detected in seeds after fertilization (Choi et al., 2002). In vitro, *DME* excises 5-methylcytosine from DNA in all sequence contexts. This evidence, along with in vivo methylation data, has shown that *DME* functions as a DNA demethylase (Gehring et al., 2006). Expression of *DME* in the central cell is one factor that results in hypomethylation of maternal endosperm after fertilization, and is necessary for the expression of some imprinted genes (Choi et al., 2002; Kinoshita et al., 2004; Jullien et al., 2006b; Hsieh et al., 2011). Thus, at least one aspect of the "different forms" transmitted by the male and female parent, as described by Kermicle (1970), is differential DNA methylation.

Profiling DNA methylation in *dme* mutant endosperm indicated that endosperm hypomethylation 5′ and 3′ of genes (Figure 7.5B) is partially dependent on *DME* (Gehring et al., 2009; Hsieh et al., 2009). Yet, paradoxically, *dme* endosperm has lower levels of genome-wide DNA methylation than wild-type endosperm, despite regions of local CpG DNA hypermethylation. This phenotype is due to overall reduced levels of non-CG methylation, the type targeted by RNA-directed DNA methylation that is primarily found at heterochromatic TEs (Hsieh et al., 2009; Gehring et al., 2009). These results suggest that the efficacy of RNA-directed DNA methylation (RdDM) is compromised in *dme* endosperm. Interestingly, 24-nucleotide siRNAs are highly abundant in seeds, especially in the endosperm/seed coat fraction of the genome (Mosher et al., 2009). Active DNA demethylation might be needed, probably indirectly, to maintain high levels of RdDM in endosperm.

Many perplexing questions remain about the endosperm epigenome. How does the maternal genome remain hypomethylated compared to the paternal genome, when the presence of CHH methylation on paternal DNA indicates that RdDM must be occurring? This might be connected to the observed cytological changes in endosperm chromatin structure. Recent experiments in *Arabidopsis* have shown that DNA methylation is preferentially targeted to nucleosomal DNA (Chodavarapu et al., 2010). If maternal chromatin undergoes a fundamental change due to active DNA demethylation by base excision repair (which must involve the removal of nucleosomes), then DNA methyltransferase might not be properly recruited to maternal DNA despite an abundance of small RNAs.

Mechanisms of Gene Imprinting and the Relation to TEs

Many of the genes associated with differential endosperm methylation are imprinted (Gehring et al., 2009; Gehring et al., 2011). However, the emerging picture is that DNA demethylation is not directed at imprinted genes, but at transposable element sequences. The kinship and coadaptation theories of imprinting do not explain why imprinted expression arises at a locus, only why it is selected for. In some cases, as will become clear, imprinted expression may have little functional importance but arise and be lost frequently due to TE insertion. The idea that the origin of imprinting might be related to the host response to transposable elements was first described in 1993 (Barlow, 1993), and has been further expounded upon (McDonald et al., 2005). A host organism defends against TEs and other repetitive sequences by a complex repressive system of DNA methylation and chromatin modifications. The TE is then under pressure to escape repression (hiding near genes that the host needs to express might be one way to do so), and the host and TE experience a constant intragenomic conflict. The molecular mechanisms that drive imprinting thus likely evolved from preexisting mechanisms and machinery designed to deal with foreign DNA. Imprinted expression might have arisen fortuitously as a byproduct of intragenomic conflict between TEs and the host (McDonald et al., 2005), and then been selected for if the change in expression was adaptive.

Imprinting is thought to be a major force in early mammalian development and, increasingly, behavior (Bartolomei and Ferguson-Smith, 2011). Although imprinting in plants and animals evolved independently (Feil and Berger, 2007), many similar mechanisms are involved: differential DNA methylation of maternal and paternal alleles, regulation by Polycomb group proteins, and a host of other epigenetic mechanisms not yet described for individual plant imprinted genes, including regulation by long noncoding RNAs and microRNAs. Studies of the evolution of imprinting within the mammalian lineage, both at the locus specific and genome-wide level, also suggest that imprinting evolved as a byproduct of genome defense, although definitively determining causality remains a major challenge (Renfree et al., 2009). Among mammals, imprinting appears to be restricted to the therians—placental mammals and marsupials—but absent from the egg-laying monotremes (prototherians) such as platypus. Comparison of the sequence composition and methylation of imprinted domains among eutherians, metatherians, and prototherians has allowed some determinations to be made about the acquisition of gene imprinting. For example, *PEG10* is an imprinted gene essential for placental formation in mice and is part of a cluster of several imprinted genes. *PEG10* is derived from a retransposon. *PEG10* is also present in tamar wallaby and opposum (marsupials) but not in platypus, indicating insertion into the genome after the split between therians and prototherians (Suzuki et al., 2007). Unlike in mice, the adjacent genes, which are also present in platypus, are not imprinted in tammar, and differential methylation is confined to *PEG10.* This suggests that insertion of the retrotransposon drove the acquisition of imprinting for several genes in the region (Suzuki et al., 2007).

Sequencing of the platypus genome has facilitated comparative genomics of orthologous imprinted regions across mammals and between other vertebrates. While no significant difference is observed in terms of the proportion of repetitive sequences in syntenic eutherian imprinted regions and those in platypus, platypus has lower density of two particular classes of TEs in these regions, LTRs and DNA transposons, coincident with the acquisition of imprinting (Pask et al., 2009). This appears to be part of a general repeat expansion in the mammalian genome and not specific to imprinted regions (Pask et al., 2009).

The molecular mechanism of gene imprinting has been relatively well-characterized for three genes in *Arabidopsis*: *FWA, MEA, and PHE1*. From these studies key themes emerged that were subsequently confirmed by genome-wide studies: 1) maternal and paternal alleles of imprinted genes are differentially methylated in the endosperm; 2) differential methylation occurs on repetitive sequences; 3) demethylation is initiated before fertilization; and 4) imprinting is maintained after fertilization by additional chromatin-based mechanisms. For each of these genes, the site of epigenetic alteration is at repetitive sequences, some clearly derived from transposable elements normally targeted by RNA-directed DNA methylation. We first examine these three genes in detail, and then expand more broadly to recent genomic studies on imprinted expression and the state of the epigenome.

FWA

The *Arabidopsis FWA* locus represents one of the simplest mechanisms of imprinted expression and has been subject to intense study on the relationship between DNA methylation, a TE, and transcriptional silencing. *FWA* encodes a homeodomain transcription factor that is maternally expressed and paternally silent in the endosperm (Kinoshita et al., 2004). The gene has an unusually structured 5′ end, consisting of a set of short and long direct repeats that overlap the first two untranslated exons (Figure 7.7). The repeated region is similar to a SINE retrotransposon and produces small RNAs (Lippman et al., 2004; Chan et al., 2006). During vegetative growth and development, the repeats are methylated in both CG and non-CG contexts and *FWA* is not expressed. The *fwa* mutants, identified as having a dominant late-flowering phenotype, represent epialleles in which the repeats have lost methylation, and *FWA* is ectopically expressed (Soppe et al., 2000). Methylation of the SINE-related repeats is under the control of *MET1*, the maintenance DNA methyltransferse, and *met1* mutants bear a late-flowering phenotype due to ectopic expression of *FWA* (Kankel et al., 2003).

In wild-type plants, *FWA* is expressed during reproductive development (Soppe et al., 2000), specifically in the central cell before fertilization and after fertilization in the endosperm exclusively from the maternal allele (Kinoshita et al., 2004). The function of *FWA* in the endosperm is unknown, and mutants have been reported to lack a seed phenotype (Kinoshita et al., 2004). *FWA* expression is dependent on the 5-methylcytosine DNA glycosylase *DME*, and the 5′ region of *FWA* is less methylated in the endosperm than in the embryo. *FWA* loses imprinting and becomes biallelically expressed in endosperm when the pollen parent harbors a mutation in *MET1*. Interestingly, a

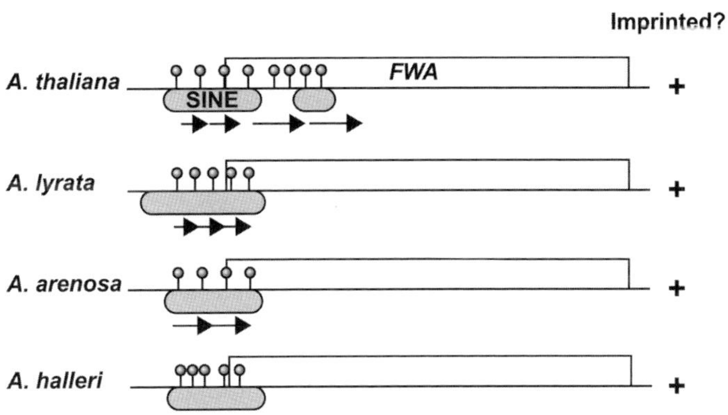

Figure 7.7 Gene structure and imprinting of *FWA* in the *Arabidopsis* genus. *FWA* is imprinted and associated with a methylated SINE sequence in several species within the *Arabidopsis* genus. Lollipops, DNA methylation; arrows, direct repeats. Artwork by Tom DiCesare.

hypomethylated pollen parent leads to paternal allele expression of *FWA* in the embryo, creating an imprinted situation in that tissue. Therefore, the response of *FWA* to methylation is the same in vegetative and in reproductive tissues: loss of methylation at 5′ repeats, whether due to lack of maintenance methylation by *MET1* or active DNA demethylation by *DME* (or both), leads to *FWA* expression. Thus, multiple lines of evidence indicate that *FWA* expression is controlled by methylation of the SINE-related direct repeats. But is it the tandem repeat structure that is important or the SINE sequence itself? Fujimoto et al. (2008) were able to address this question by examining *FWA* vegetative expression, methylation, and imprinting in three other species in the *Arabidopsis* genus, *A. lyrata, A. halleri*, and *A. arenosa*. The SINE-like sequence was found in the 5′ region of *FWA* in all species, but each had a unique constitution of tandem repeats within the element except for *A. halleri*, which lacked tandem direct repeats entirely. The SINE region was methylated in all species, regardless of whether tandem repeats were present, and *FWA* was imprinted (Figure 7.7). This indicates that the tandem repeat structure per se is not required for epigenetic *FWA* silencing, but points to the SINE transposable element sequence itself as the target of methylation and demethylation.

MEDEA

MEDEA (*MEA*) was the first imprinted gene described in *Arabidopsis thaliana* (Kinoshita et al., 1999; Vielle-Calzada et al., 1999). It is maternally expressed and paternally silent in the endosperm, although paternal allele silencing is not stably maintained at late stages of seed development in some combinations of intraspecific crosses (Kinoshita et al., 1999). Like *FWA*, the *MEA* gene is flanked by TEs and repeated sequences. A Helitron fragment, ATREP10, lies about 500 base pairs 5′ of the transcriptional start site, while the 3′ end of the gene contains a series of 117 bp tandem direct repeats that vary in number between strains (Martienssen et al., 2004). These sequences are methylated in vegetative tissues, although the 5′ Helitron is only sparsely methylated in CG contexts or, in some strains, not at all (Cao and Jacobsen, 2002; Xiao et al., 2003; Gehring et al., 2006). The 3′ tandem direct repeats are targeted by RNA-directed DNA methylation with corresponding small RNAs. *MET1* is required to maintain methylation of both regions during vegetative development (Xiao et al., 2003), as are components of the RNA-directed DNA methylation pathway at the 3′ end (Cao and Jacobsen, 2002). Unlike for *FWA*, *MEA* expression can be detected at low levels during vegetative development, and rarely obtained *mea/mea* plants exhibit subtle morphological phenotypes, suggesting additional functions outside of reproduction (Gehring, 2005). Moreover, *MEA* expression is not upregulated in vegetative tissues of *met1* mutants (Jullien et al., 2006a). These observations suggest that methylation of the 5′ and the 3′ repetitive regions is not sufficient for *MEA* silencing nor is the loss of methylation at these regions sufficient for expression. Indeed, *MEA* is also targeted by PcG complexes during vegetative development (Baroux et al., 2006). Yet, expression of the maternal *MEA* allele in the central cell and endosperm depends on the DNA glycosylase *DME* (Figure 7.8) (Choi et al., 2002; Xiao et al., 2003). In endosperm, the expressed maternal *MEA* allele is hypomethylated on repetitive sequences immediately 5′ and 3′ of the gene compared to the silent paternal allele and compared to both maternal and paternal alleles in the embryo (Gehring et al., 2006). Changes in chromatin structure due to the process of base excision repair, rather than the removal of DNA methylation per se, may be important for permitting the expression of the maternal allele. Unlike *FWA*, the paternal *MEA* allele remains silent in the endosperm even if inherited from *met1* pollen or pollen from RNA-directed DNA methylation mutants (Gehring et al., 2006, Jullien et al., 2006a), again indicating that loss of DNA methylation is not sufficient for gene

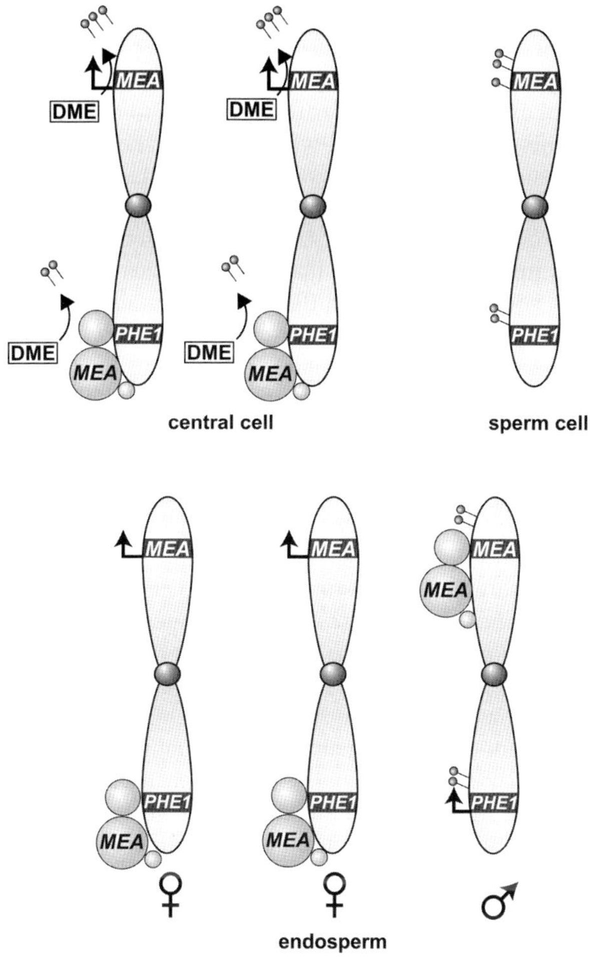

Figure 7.8 Mechanism of *MEA* and *PHE1* imprinting. Before fertilization, DME removes 5-methylcytosine (lollipops) from the maternal alleles of *MEA* and *PHE1*. *MEA* is transcribed. Polycomb group complexes (circles) that include MEA assemble at *PHE1* and repress maternal expression. Both *MEA* and *PHE1* remain methylated in sperm. In the endosperm, maternally produced MEA targets paternal *MEA* to maintain silencing. *PHE1* is paternally expressed. Artwork by Tom DiCesare.

expression. Rather, maternally inherited mutations in the Polycomb group genes *FIE* and *MEA* lead to derepression of the paternal allele. Correspondingly, *MEA* is marked by H3K27 methylation in wild-type siliques, specifically on the paternally inherited allele (Gehring et al., 2006). Thus, activation of the maternal *MEA* allele by DME in the central cell before fertilization is necessary to maintain silencing of the paternal *MEA* allele in the endosperm after fertilization (Figure 7.8).

PHERES1

PHERES1 (PHE1), a MADS-box domain protein, was the first imprinted plant gene discovered that is paternally expressed and maternally silent in the endosperm. Although *MEA* and *PHE1* are oppositely imprinted genes they share many similarities—both require PcG proteins and DNA demethylation for proper imprinted expression (Figure 7.8). *PHE1* was initially identified as a gene upregulated in siliques obtained from *MEA/mea* and *FIE/fie* plants, and down-regulation of *PHE1* by antisense targeting partially rescues the *mea* seed abortion phenotype (Köhler et al., 2003). Unlike *MEA* and *FWA*, *PHE1* is not expressed before fertilization. One to two days after pollination, *PHE1* is expressed in both the embryo and endosperm, with expression becoming restricted to the chalazal endosperm at the globular stage of embryogenesis (Köhler et al., 2003). Endosperm expression is predominantly from the paternal allele (Köhler et al., 2005). The 5′ region of the *PHE1* is directly targeted by the FIE-MEA PcG complex. In *MEA/mea*, mutant seeds expression from the maternal *PHE1* allele is substantially up-regulated, while the paternal allele remains unaffected (Köhler et al., 2005). The PcG complex is necessary, but not sufficient, to maintain *PHE1* imprinting (Makarevich et al., 2008).

In contrast to *FWA*, DNA methylation is required for expression of the paternal *PHE1* allele. When *met1* mutants are crossed as males to wild-type females, paternal *PHE1* expression is reduced and maternal *PHE1* expression increases (Makarevich et al., 2008). The *PHE1* gene does not contain any DNA methylation in the coding sequence or promoter, but a methylated remnant of a Helitron transposable element, ATREP11, lies about 2 kilobases 3′ of the gene and is less methylated in the endosperm than in the embryo, presumably due to loss of methylation on the maternally inherited allele (Makarevich et al., 2008; Gehring et al., 2009). The TE sequence also contains 3 short (54 bp) tandem direct repeats. Insertions that disrupt the sequence lead to activation of the maternal *PHE1* allele and repression of the paternal *PHE1* alelle; transgenes that lack the Helitron or contain a Helitron that is not properly methylated are not correctly imprinted (Makarevich et al., 2008; Villar et al., 2009). Thus, the 3′ TE region represents both a genetic and epigenetic requirement for proper imprinting of *PHE1*. Thus, similarly to *MEA*, both demethylation of a TE-related sequence and PcG targeting are critical for establishing and maintaining imprinted expression. In both cases, unlike *FWA*, DNA methylation probably does not directly impact transcription initiation, because the DMRs are not near the transcriptional start site. Secondary mechanisms involving the relationship between DNA methylation and higher order chromatin structure or noncoding RNA transcription might be important.

TEs and Allele-Specific Imprinting

Genes that are imprinted in an allele-specific manner (only some alleles within a species are imprinted) provide a natural test for the genetic and epigenetic requirements for imprinting. The involvement of TEs in imprinting suggests that allele-specific imprinting could be quite common because TE composition and epigenetic modification will vary between intraspecific strains as lineages diverge. Whole genome studies show that several genes are imprinted in an allele-specific manner in *A. thaliana* (Hsieh et al., 2011; Wolff et al., 2011; Gehring et al., 2011). Between two *A. thaliana* strains, only about one-third of the genes associated with the top embryo-endosperm differentially methylated regions are in common (Gehring et al., 2009).

Further work on *R* pointed to importance of TEs and DNA methylation in imprinting, although much of the primary data remains unpublished (Kermicle, 1978; Alleman and Doctor, 2000). As shown in Figure 7.4, inheritance of *R* alleles through the male leads to a mottled, rather than completely colorless, aleurone pigmentation phenotype. *R* is therefore not completely silenced when inherited paternally because anthocyanin pigment is produced in a subset of cells, indicating the gene has become active. *R* is subject to allele-specific imprinting. Some alleles of the *R* locus participate in paramutation, a heritable alteration in gene expression states due to allelic interactions (Chandler, 2010). Paramutant forms of *R*, *R′*, are more strongly paternally silenced in the aleurone (Alleman and Doctor, 2000). In crosses with *rr* males, *R′R′* females produce fully colored kernels. In the reciprocal cross the seeds are almost colorless, with much less mottling than when *R* is paternally inherited (Figure 7.4B). What is the difference between *R* and *R′*? The *R* locus is complex, consisting of one to five tightly linked homologous genes. Paramutable alleles of *R* are composed of multiple seed pigmenting genes arranged in a head-to-tail orientation with a centrally located promoter. This promoter contains small repeats derived from a *Doppia* transposable element. Paramutant forms of *R* are more methylated at the *Doppia* TE (Walker, 1998). Therefore, decreased paternal expression of *R′* is likely due to increased methylation of the TE, leading to stronger transcriptional silencing. Maternally inherited *R* is less methylated than paternally inherited *R* (Alleman and Doctor, 2000). The phenotype when *R* or *R′* is inherited maternally (colored kernels) would not be affected because both alleles are likely subject to demethylation before fertilization and thus expressed in the endosperm after fertilization.

Insights from Whole Genome Studies

Now that genome-wide allele-specific expression and seed methylation data are available, the lessons of *FWA*, *MEA*, and *PHE1* can be assessed for their general relevance. Are all genes with parent-of-origin specific gene expression associated with differentially methylated regions derived from repetitive elements? The answer is clearly no, but for a fraction of imprinted *Arabidopsis* genes, particularly those that are paternally expressed (PEGs), this is indeed the case. Hsieh et al. (2011) identified 43 genes whose imprinting was disrupted in *dme*, *fie*, or *met1* mutants, fitting the known paradigms about the mechanism of gene imprinting. All nine PEGs identified by Hsieh et al. (2011) lost maternal allele silencing when *fie* was inherited maternally, and seven had reduced paternal allele expression when *met1* was inherited paternally. These genes are like *PHE1*. Wolff et al. (2011) independently concluded that repression of maternal alleles of PEGs mostly depends on PcG repression. Paternal alleles might be protected from PcG targeting in the endosperm after fertilization because of the continued presence of DNA methylation at the locus (Weinhofer et al., 2010). Experiments with *met1* and PcG mutants also identify MEGs for which silencing of the paternal allele is regulated by methylation, such as *FWA*, or by the PcG complex, such as *MEA* (Hsieh et al., 2011; Wolff et al., 2011).

Both MEGs and PEGs can be identified that are associated with differentially methylated regions, although this is more common for PEGs (about half of PEGs have DMRs in their flanking sequences) (Gehring et al., 2011). Not surprisingly then, imprinted genes are enriched for TEs in their vicinity, particularly those of the Helitron class (Gehring et al., 2009; Wolf et al., 2011). Interestingly, the MEGs or PEGs associated with differential methylation and TEs are enriched for transcription factors and chromatin-related proteins (Gehring et al., 2011).

Despite the finding that rice endosperm DNA is hypomethylated at transposable element sequences (Zemach et al., 2010b), the 262 imprinted loci identified in rice endosperm do not appear to be enriched in transposons or repeats (Luo et al., 2011). However, TE annotation in rice is less complete than in *A. thaliana*. Therefore, it remains to be determined whether the lessons from the relatively simple *Arabidopsis thaliana* genome, with its low TE density and short distances between genes, will translate to more complex cereal genomes, where most genes are surrounded by TEs. This will be a major challenge for future research.

Outstanding Questions

There has been enormous progress in the study of plant genomic imprinting over the last 40 years. Hundreds of genes are now known to display parentally biased expression during endosperm development. At least some of these are critical for seed viability, although the function of most imprinted genes in the endosperm is unknown. For a significant fraction of genes, imprinted gene expression is established due to genome-wide DNA demethylation that takes place in only one of the four gametes, the central cell, before fertilization. Thus, an epigenetic asymmetry is established that persists after fertilization, leading to imprinted expression either through a direct relationship between DNA methylation and gene expression (e.g., for genes with DNA methylation around the transcriptional start site), or, more frequently, by an indirect mechanism involving PcG targeting and, presumably, as yet other unknown mechanisms.

Critically, DNA demethylation does not appear to be targeted at imprinted genes per se, but at remnants of TEs. For most genes these TE dynamics are neutral, and have little effect on expression (Gehring et al., 2009). However, some of the genes that are imprinted have been swept up in a dynamic process such that their expression state is now tied to the epigenetic status of nearby TEs. This might be adaptive for multiple reasons, not all related to endosperm or seed development. For example, the methylated SINE element 5′ of *FWA* might have been retained because repression of *FWA* expression by DNA methylation during vegetative development—and thus an earlier flowering phenotype—could have been strongly advantageous. Thus, one must be cautious about concluding that a gene has an especially significant function during endosperm development simply because it is imprinted.

The most pressing research question at present is whether or not there is another function to endosperm TE methylation dynamics besides a role in imprinted gene expression. Several groups have suggested that TEs are activated in the central cell or endosperm in order to become small RNA factories for the egg cell or embryo, which carry precious genetic material to the next generation. Transport of small RNAs to the egg cell or embryo could provide extra protection against TE activity in those cells. There is presently extremely little direct experimental evidence to support this hypothesis, but results in these areas are eagerly anticipated.

References

Ahmed, I., Sarazin A., Bowler, C., Colot, V., and Quesneville, H. (2011) Genome-wide evidence for local DNA methylation spreading from small RNA-targeted sequences in *Arabidopsis*. *Nucleic Acids Res. 39*: 6919–6931.

Alleman, M. and Doctor, J. (2000) Genomic imprinting in plants: observations and evolutionary implications. *Plant Mol. Biol. 43*: 147–161.

Barlow, D. (1993) Methylation and imprinting: from host defense to gene regulation? *Science 260*: 309–310.

Baroux, C., Fransz, P., and Grossniklaus, U. (2004) Nuclear fusions contribute to polyploidization of the gigantic nuclei in the chalazal endosperm of *Arabidopsis*. *Planta 220*: 38–46.

Baroux, C., Gagliardini, V., Page, D.R., and Grossniklaus, U. (2006) Dynamic regulatory interactions of Polycomb group genes: MEDEA autoregulation is required for imprinted gene expression in *Arabidopsis*. *Genes Dev. 20*: 1081–1086.

Baroux, C., Pecinka, A., Fuchs, J., Schubert, I., and Grossniklaus, U. (2007) The triploid endosperm genome of *Arabidopsis* adopts a peculiar, parental-, dosage-dependent chromatin organization. *Plant Cell 19*: 1782–1794.

Bartolomei, M.S. and Ferguson-Smith, A.C. (2011) Mammalian genomic imprinting. *Cold Spring Harbor Perspect. Biol. 3*: a002592.

Berger, F. (2003) Endosperm: the crossroad of seed development. *Curr. Opin. Plant Biol. 6*: 42–50.

Boisnard-Lorig, C., Colon-Carmona, A., Bauch, M., Hodge, S., Doerner, P., Bancharel, E., Dumas, C., Haseloff, J., and Berger, F. (2001) Dynamic analyses of the expression of the HISTONE: YFP fusion protein in *Arabidopsis* show that syncytial endosperm is divided in mitotic domains. *Plant Cell 13*: 495–509.

Brandvain, Y., Van Cleve, J., Ubeda, F., and Wilkins, J.F. (2011) Demography, kinship, and the evolving theory of genomic imprinting. *Trends Genet. 27*: 251–257.

Brink, R.A. and Cooper, D.C. (1947) The endosperm in seed development. *Bot. Rev. 13*:423–541.

Brink, R.A., Kermicle, J.L., and Ziebur, N.K. (1970) Derepression in the female gametophyte in relation to paramutant *R* expression in maize endosperms, embryos, and seedlings. *Genetics 66*: 87–96.

Brown, R.C., Lemmon, B.E., Nguyen, H., and Olsen, O.-A. (1999) Development of endosperm in *Arabidopsis thaliana*. *Sex. Plant Reprod. 12*: 32–42.

Cao, X. and Jacobsen, S.E. (2002) Locus-specific control of asymmetric CpNpG methylation by the *DRM* and *CMT3* methyltransferase genes. *Proc. Natl. Acad. Sci. USA 99*: 16491–16498.

Chan, S.W., Zhang, X., Bernatavichute, Y.V., and Jacobsen, S.E. (2006) Two-step recruitment of RNA-directed DNA methylation to tandem repeats. *PLoS Biol. 4*: e363.

Chandler, V.L. (2010) Paramutation's properties and puzzles. *Science 330*: 628–629.

Chaudhury, A.M., Ming, L., Miller, C., Craig, S., Dennis, E.S., and Peacock, W.J. (1997) Fertilization-independent seed development in *Arabidopsis thaliana*. *Proc. Natl. Acad. Sci. USA 94*: 4223–4228.

Chodavarapu, R.K., Feng, S., Bernatavichute, Y.V., Chen, P.Y., Stroud, H., Yu, Y., Hetzel, J.A., Kuo, F., Kim, J., Cokus, S.J., et al. (2010) Relationship between nucleosome positioning and DNA methylation. *Nature 466*: 388–392.

Choi, Y., Gehring, M., Johnson, L., Hannon, M., Harada, J.J., Goldberg, R.B., Jacobsen, S.E., and Fischer, R.L. (2002) DEMETER, a DNA glycosylase domain protein, is required for endosperm gene imprinting and seed viability in *Arabidopsis*. *Cell 110*: 33–42.

Cokus, S.J., Feng, S., Zhang, X., Chen, Z., Merriman, B., Haudenschild, C.D., Pradhan, S., Nelson, S.F., Pellegrini, M., and Jacobsen, S.E. (2008) Shotgun bisulphite sequencing of the *Arabidopsis* genome reveals DNA methylation patterning. *Nature 452*: 215–219.

Costa, L.M., Yuan, J., Rouster, J., Paul, W., Dickinson, H., and Gutierrez-Marcos, J.F. (2012) Maternal control of nutrient allocation in plant seeds by genomic imprinting. *Curr. Biol. 22*: 160–165.

Danilevskaya, O.N., Hermon, P., Hantke, S., Muszynski, M.G., Kollipara, K., and Ananiev, E.V. (2003) Duplicated *fie* genes in maize: expression pattern and imprinting suggest distinct functions. *Plant Cell 15*: 425–438.

Day, R.C., Herridge, R.P., Ambrose, B.A., and Macknight, R.C. (2008) Transcriptome analysis of proliferating *Arabidopsis* endosperm reveals biological implications for the control of syncytial division, cytokinin signaling, and gene expression regulation. *Plant Physiol. 148*: 1964–1984.

Dilkes, B. and Comai, L. (2004) A differential dosage hypothesis for parental effects in seed development. *Plant Cell 16*: 3174–3180.

Feil, R. and Berger, F. (2007) Convergent evolution of genomic imprinting in plants and mammals. *Trends Genet. 23*: 192–199.

Feng. S., Cokus, S.J., Zhang, X., Chen, P.Y., Bostick, M., Goll, M.G., Hetzel, J., Jain, J., Strauss, S.H., Halpern, M.E., et al. (2010a) Conservation and divergence of methylation patterning in plants and animals. *Proc. Natl. Acad. Sci. USA 107*: 8689–8694.

Feng, S., Jacobsen, S.E., and Reik, W. (2010b) Epigenetic reprogramming in plant and animal development. *Science 330*: 622–627.

Feschotte, C. (2008) Transposable elements and the evolution of regulatory networks. *Nature Rev. Genet. 9*: 397–405.

Fransz, P., DeJong, J.H., Lysak, M., Castiglione, M.R., and Schubert, I. (2002) Interphase chromosomes in *Arabidopsis* are organized as well defined chromocenters from which euchromatin loops emanate. *Proc. Natl. Acad. Sci. USA 99*: 14584–14589.

Friedman, W.E. (2006) Embryological evidence for developmental lability during early angiosperm evolution. *Nature 441*: 337–340.

Fujimoto, R., Kinoshita, Y., Kawabe, A., Kinoshita, T., Takashima, K., Nordborg, M., Nasrallah, M.E., Shimizu, K.K., Kudoh, H., and Kakutani, T. (2008) Evolution and control of imprinted *FWA* genes in the genus *Arabidopsis*. *PLoS Genet. 4*: e1000048.

Gehring, M.A. (2005) Mechanisms of gene imprinting. In *Arabidopsis*. (Berkeley: PhD Dissertation. University of California).

Gehring, M., Bubb, K.L., and Henikoff, S. (2009) Extensive demethylation of repetitive elements during seed development underlies gene imprinting. *Science 324*: 1447–1451.

Gehring, M. and Henikoff, S. (2008) DNA methylation and demethylation in *Arabidopsis*. *The Arabidopsis Book 6*: e0102. doi:10.1199/tab.0102.

Gehring, M., Huh, J.H., Hsieh, T.F., Penterman, J., Choi, Y., Harada, J.J., Goldberg, R.B., and Fischer, R.L. (2006) DEMETER DNA glycosylase establishes *MEDEA* polycomb gene self-imprinting by allele-specific demethylation. *Cell 124*: 495–506.

Gehring, M., Missirian, V., and Henikoff, S. (2011) Genomic analysis of parent-of-origin allelic expression in *Arabidopsis thaliana* seeds. *PLoS One 6*: e23687.

Gong, Z., Morales-Ruiz, T., Ariza, R.R., Roldán-Arjona, T., David, L., and Zhu, J.-K. (2002) *ROS1*, a repressor of transcriptional gene silencing in *Arabidopsis*, encodes a DNA glycosylase/lyase. *Cell 111*: 803–814.

Grimanelli, D., Perotti, E., Ramirez, J., and Leblanc, O. (2005) Timing of the maternal-to-zygotic transition during early seed development in maize. *Plant Cell 17*: 1061–1072.

Grossniklaus, U., Vielle Calzada, J.-P., Hoeppner, M.A., and Gagliano, W.B. (1998) Maternal control of embryogenesis by *MEDEA*, a *Polycomb* group gene in *Arabidopsis*. *Science 280*: 446–450.

Gutierrez-Marcos, J.F., Costa, L.M., Dal Pra, M., Scholten, S., Kranz, E., Perez, P., and Dickinson, H.G. (2006) Epigenetic asymmetry of imprinted genes in plant gametes. *Nature Genet. 38*: 876–878.

Haig, D. (2000) The kinship theory of genomic imprinting. *Annu. Rev. Ecol. Syst. 31*: 9–32.

Haig, D. (2004) Genomic imprinting and kinship: How good is the evidence? *Annu. Rev. Genet. 38*: 553–585.

Haig, D. and Westoby, M. (1989) Parent-specific gene expression and the triploid endosperm. *Am. Nat. 134*: 147–155.

Haig, D. and Westoby, M. (1991) Genomic imprinting in endosperm: Its effect on seed development in crosses between species, and between different ploidies of the same species, and its implications for the evolution of apomixis. *Phil. Trans. R. Soc. Lond. B Biol. Sci. 333*: 1–13.

Haun, W.J., Laoueille-Duprat, S., O'Connell M.J., Spillane, C., Grossniklaus, U., Phillips, A.R., Kaeppler, S.M., and Springer, N.M. (2007) Genomic imprinting, methylation and molecular evolution of maize *Enhancer of zeste (Mez) homologs. Plant J. 49*: 325–337.

Hermon, P., Srilunchang, K.O., Zou, J., Dresselhaus, T., and Danilevskaya, O.N. (2007) Activation of the imprinting Polycomb Group *Fie1* gene in maize endosperm requires demethylation of the maternal allele. *Plant Mol. Biol. 64*: 387–395.

Hollister, J.D. and Gaut, B.S. (2007) Population and evolutionary dynamics of *Helitron* transposable elements in *Arabidopsis thaliana. Mol. Biol. Evol. 24*: 2515–2524.

Hollister, J.D. and Gaut, B.S. (2009) Epigenetic silencing of transposable elements: A trade-off between reduced transposition and deleterious effects on neighboring gene expression. *Genome Res. 19*: 1419–1428.

Hollister, J.D., Smith, L.M., Guo, Y.L., Ott, F., Weigel, D., and Gaut, B.S. (2011) Transposable elements and small RNAs contribute to gene expression divergence between *Arabidopsis thaliana* and *Arabidopsis lyrata. Proc. Natl. Acad. Sci. USA 108*: 2322–2327.

Hsieh, T.F., Ibarra, C.A., Silva, P., Zemach, A., Eshed-Williams, L., Fischer, R.L., and Zilberman, D. (2009) Genome-wide demethylation of *Arabidopsis* endosperm. *Science 324*: 1451–1454.

Hsieh, T.F., Shin, J., Uzawa, R., Silva, P., Cohen, S., Bauer, M.J., Hashimoto, M., Kirkbride, R.C., Harada, J.J., Zilberman, D., et al. (2011) Regulation of imprinted gene expression in *Arabidopsis* endosperm. *Proc. Natl. Acad. Sci. USA 108*: 1755–1762.

Ingouff, M., Hamamura, Y., Gourgues, M., Higashiyama, T., and Berger, F. (2007) Distinct dynamics of HISTONE3 variants between the two fertilization products in plants. *Curr. Biol. 19*: 1032–1037.

Jahnke, S. and Scholten, S. (2009) Epigenetic resetting of a gene imprinted in plant embryos. *Curr. Biol. 19*: 1677–1681.

Johnson, M.A. and Bender, J. (2009) Reprogramming the epigenome during germline and seed development. *Genome Biol. 10*: 232.

Johnston, S.A., den Nijs, T.P.M, Peloquin, S.J., and Hanneman, R.E., Jr. (1980) The significance of genic balance to endosperm development in interspecific crosses. *Theor. Appl. Genet. 57*: 5–9.

Jullien, P.E., Katz, A., Oliva, M., Ohad, N., and Berger, F. (2006a) Polycomb group complexes self-regulate imprinting of the Polycomb group gene *MEDEA* in *Arabidopsis. Curr. Biol. 16*: 486–492.

Jullien, P.E., Kinoshita, T., Ohad, N., and Berger, F. (2006b) Maintenance of DNA methylation during the *Arabidopsis* life cycle is essential for parental imprinting. *Plant Cell 18*: 1360–1372.

Jullien, P.E., Mosquna, A., Ingouff, M., Sakata, T., Ohad, N., and Berger, F. (2008) Retinoblastoma and its binding partner MSI1 control imprinting in *Arabidopsis. PLoS Biol. 6*: e194.

Kankel, M.W., Ramsey, D.E., Stokes, T.L., Flowers, S.K., Haag, J.R., Jeddeloh, J.A., Riddle, N.C., Verbsky, M.L., and Richards, E.J. (2003) *Arabidopsis MET1* cytosine methyltransferase mutants. *Genetics 163*: 1109–1122.

Kato, M., Miura, A., Bender, J., Jacobsen, S.E., and Kakutani, T. (2003) Role of CG and non-CG methylation in immobilization of transposons in *Arabidopsis. Curr. Biol. 13*: 421–426.

Kempton, J.H. (1919) Inheritance of spotted aleurone color in hybrids of Chinese maize. *Genetics 4*: 261–274.

Kermicle, J.L. (1970) Dependence of the *R*-mottled aleurone phenotype in maize on mode of sexual transmission. *Genetics 66*: 69–85.

Kermicle, J.L. (1978) Imprinting of gene action in maize endosperm. In *Maize breeding and Genetics*, D.B. Walden, ed. (New York: Wiley), pp. 357–371.

Kinoshita, T., Miura, A., Choi, Y., Kinoshita, Y., Cao, X., Jacobsen, S.E., Fischer, R.L., and Kakutani, T. (2004) One-way control of *FWA* imprinting in *Arabidopsis* endosperm by DNA methylation. *Science 303*: 521–523.

Kinoshita, T., Yadegari, R., Harada, J.J., Goldberg, R.B., and Fischer, R.L. (1999) Imprinting of the *MEDEA* Polycomb gene in the *Arabidopsis* endosperm. *Plant Cell 11*: 1945–1952.

Kiyosue, T., Ohad, N., Yadegari, R., Hannon, M., Dinneny, J., Wells, D., Katz, A., Margossian, L., Harada, J.J., Goldberg, R.B., et al. (1999) Control of fertilization-independent endosperm development by the *MEDEA* polycomb gene in *Arabidopsis. Proc. Natl. Acad. Sci. USA 96*: 4186–4191.

Köhler, C., Hennig, L., Spillane, C., Pien, S., Gruissem, W., and Grossniklaus, U. (2003) The *Polycomb*-group protein MEDEA regulates seed development by controlling expression of the MADS-box gene *PHERES1. Genes Dev. 17*: 1540–1553.

Köhler, C., Page, D.R., Gagliardini, V., and Grossniklaus, U. (2005) The *Arabidopsis* thaliana MEDEA Polycomb group protein controls expression of *PHERES1* by parental imprinting. *Nature Genet. 37*: 28–30.

Kowles, R.V. and Phillips, R.L. (1985) DNA amplification patterns in maize endosperm nuclei during kernel development. *Proc. Natl. Acad. Sci. USA 82*: 7010–7014.

Lauria, M., Rupe, M., Guo, M., Kranz, E., Pirona, R., Viotti, A., and Lund, G. (2004) Extensive maternal DNA hypomethylation in the endosperm of *Zea mays*. *Plant Cell 16*: 510–522.

Law, J.A. and Jacobsen, S.E. (2010) Establishing, maintaining, and modifying DNA methylation patterns in plants and animals. *Nature Rev. Genet. 11*: 204–220.

Le, B.H., Cheng, C., Bui, A.Q., Wagmaister, J.A., Henry, K.F., Pelletier, J., Kwong, L., Belmonte, M., Kirkbride, R.C., Horvath, S., et al. (2010) Global analysis of gene activity during *Arabidopsis* seed development and identification of seed specific transcription factors. *Proc. Natl. Acad. Sci. USA 107*: 8063–8070.

Lin, B.-Y. (1982) Association of endosperm reduction with parental imprinting in maize. *Genetics 100*: 475–486.

Lin, B.-Y. (1984) Ploidy barrier to endosperm development in maize. *Genetics 107*:103–115.

Lippman, Z., Gendrel, A.V., Black. M., Vaughn, M.W., Dedhia, N., McCombie, W.R., Lavine, K., Mittal, V., May, B., Kasschau, K.D., et al. (2004) Role of transposable elements in heterochromatin and epigenetic control. *Nature 430*: 471–476.

Lisch, D. (2009) Epigenetic regulation of transposable elements in plants. *Annu. Rev. Plant Biol. 60*: 43–66.

Lister, R. O'Malley, R.C., Tonti-Filippini, J., Gregory, B.D., Berry, C.C., Millar, A.H., and Ecker, J.R. (2008) Highly integrated single-base resolution maps of the epigenome of *Arabidopsis*. *Cell 133*: 523–536.

Lopes, M.A. and Larkins, B.A. (1993) Endosperm origin, development, and function. *Plant Cell 5*:1383–1399.

Luo, M., Taylor, J.M., Spriggs, A., Zhang H., Wu, X., Russell, S., Singh, M., and Koltunow, A. (2011) A genome-wide survey of imprinted genes in rice seeds reveals imprinting primarily occurs in the endosperm. *PLoS Genet. 7*: e1002125.

Maheshwari, P. (1950) *An introduction to the embryology of angiosperms*. (New York: McGraw-Hill Book Company).

Makarevich, G., Villar, C.B., Erilova, A., and Kohler, C. (2008) Mechanisms of *PHERES1* imprinting in *Arabidopsis*. *J. Cell Sci. 121*: 906–912.

Martienssen, R., Lippman, Z., May, B., Ronemus, M., and Vaughn, M. (2004) Transposons, tandem repeats, and the silencing of imprinted genes. *Cold Spring Harbor Symp. Quant. Biol. 69*: 371–379.

Matzke, M., Kanno, T., Daxinger, L., Huettel, B., and Matzke, A.J. (2009) RNA-mediated chromatin-based silencing in plants. *Curr. Opin. Cell Biol. 21*: 367–376.

McDonald, J.F., Matzke, M.A., and Matzke, A.J. (2005) Host defenses to transposable elements and the evolution of genomic imprinting. *Cytogenet. Genome Res. 110*: 242–249.

Miura, A., Yonebayashi, S., Watanabe, K., Toyama, T., Shimada, H., and Kakutani, T. (2001) Mobilization of transposons by a mutation abolishing full DNA methylation in *Arabidopsis*. *Nature 411*: 212–214.

Mosher, R.A., Melnyk, C.W., Kelly, K.A., Dunn, R.M., Studholme, D.J., and Baulcombe, D.C. (2009) Uniparental expression of PolIV-dependent siRNAs in developing endosperm of *Arabidopsis*. *Nature 460*: 283–286.

Nguyen, H., Brown, R.C., and Lemmon, B.E. (2000) The specialized chalazal endosperm in *Arabidopsis thaliana* and *Lepidium virginicum* (*Brassicaceae*). *Protoplasma 212*: 99–110.

Paldi, A. (2003) Genomic imprinting: Could the chromatin structure be the driving force? *Curr. Topics Dev. Biol. 53*: 115–138.

Pardo-Manuel de Villena, F.D., Casa-Esperon, E.D.L., and Sapienza, C. (2000) Natural selection and the function of genome imprinting: beyond the silenced minority. *Trends Genet. 16*: 573–579.

Pask, A.J., Papenfuss, A.T., Ager, E.I., McColl, K.A., Speed, T.P., and Renfree, M.B. (2009) Analysis of the platypus genome suggests a transposon origin for mammalian imprinting. *Genome Biol. 10*: R1.

Penterman, J., Uzawa, R., and Fischer, R.L. (2007b) Genetic interactions between DNA demethylation and methylation in *Arabidopsis*. *Plant Physiol. 145*: 1549–1557.

Penterman, J., Zilberman, D., Huh, J.H., Ballinger, T., Henikoff, S., and Fischer, R.L. (2007a) DNA demethylation in the *Arabidopsis* genome. *Proc. Natl. Acad. Sci. USA 104*: 6752–6757.

Pillot, M., Baroux, C., Arteaga Vazquez, M., Autran, D., Leblanc, O., Vielle-Calzada, J.P., Grossniklaus, U., and Grimanelli, D. (2010) Embryo and endosperm inherit distinct chromatin and transcriptional states from the female gametes in *Arabidopsis*. *Plant Cell 22*: 307–320.

Renfree, M.B., Hore, T.A., Shaw, G., Marshall Graves, J.A., and Pask, A.J. (2009) Evolution of genomic imprinting: insights from marsupials and monotremes. *Annu. Rev. Genomics Hum. Genet. 10*: 241–262.

Roudier, F., Ahmed, I., Bérard, C., Sarazin, A., Mary-Huard, T., Cortijo, S., Bouyer, D., Caillieux, E., Duvernois-Berthet, E., Al-Shikhley, L., et al. (2011) Integrative epigenomic mapping defines four main chromatin states in *Arabidopsis*. *EMBO J. 30*: 1928–1938.

Sabelli, P.A. and Larkins, B.A. (2009) The development of endosperm in grasses. *Plant Physiol. 149*: 14–26.

Scott, R.J., Spielman, M., Bailey, J., and Dickinson, H.G. (1998) Parent-of-origin effects on seed development in *Arabidopsis thaliana*. *Development 125*: 3329–3341.

Slotkin, R.K. and Martienssen, R. (2007) Transposable elements and the epigenetic regulation of the genome. *Nature Rev. Genet. 8*: 272–285.

Soppe, W.J., Jacobsen, S.E., Alonso-Blanco, C., Jackson, J.P., Kakutani, T., Koornneef, M., and Peeters, A.J. (2000) The late flowering phenotype of *fwa* mutants is caused by gain-of-function epigenetic alleles of a homeodomain gene. *Mol. Cell 6*: 791–802.

Suzuki, S., Ono, R., Narita, T., Pask, A.J., Shaw, G., Wang, C., Kohda, T., Alsop, A.E., Marshall Graves, J.A., Kohara, Y., et al. (2007) Retrotransposon silencing by DNA methylation can drive mammalian genomic imprinting. *PLoS Genet. 3*: e55.

Teixeira, F.X. and Colot, V. (2010) Repeat elements and the *Arabidopsis* DNA methylation landscape. *Heredity 105*: 14–23.

Tycko, B. and Morison, I.M. (2002) Physiological functions of imprinted genes. *J. Cell Phys. 192*: 245–258.

Vielle-Calzada, J.-P., Thomas, J., Spillane, C., Coluccio, A., Hoeppner, M.A., and Grossniklaus, U. (1999) Maintenance of genomic imprinting at the *Arabidopsis medea* locus requires zygotic *DDM1* activity. *Genes Dev. 13*: 2971–2982.

Villar, C.B., Erilova, A., Makarevich, G., Trösch, R., and Köhler, C. (2009) Control of *PHERES1* imprinting in *Arabidopsis* by direct tandem repeats. *Mol. Plant 2*: 654–660.

Walker, E.L. (1998) Paramutation of the *r1* locus of maize is associated with increased cytosine methylation. *Genetics 148*: 1973–1981.

Weinhofer, I., Hehenberger, E., Roszak, P., Hennig, L., and Köhler, C. (2010) H3K27me3 profiling of the endosperm implies exclusion of polycomb group protein targeting by DNA methylation. *PLoS Genet. 6*: e1001152.

Williams, J.H. and Friedman, W.E. (2002) Identification of diploid endosperm in an early angiosperm lineage. *Nature 415*: 522–526.

Wolf, J.B and Hager, R. (2006) A maternal-offspring coadaptation theory for the evolution of genomic imprinting. *PLoS Biol. 4*: e380.

Wolff, P., Weinhofer, I., Seguin, J., Roszak, P., Beisel, C., Donoghue, M.T.A., Spillane, C., Nordborg, M., Rehmsmeier, M., and Köhler, C. (2011) High-resolution analysis of parent-of-origin allelic expression in the *Arabidopsis* endosperm. *PLoS Genet. 7*: e1002126.

Woodhouse, M.R., Freeling, M., and Lisch, D. (2006) The mop1 (mediator of paramutation1) mutant progressively reactivates one of the two genes encoded by the *MuDR* transposon in maize. *Genetics 172*: 579–592.

Xiao, W., Gehring, M., Choi, Y., Margossian, L., Pu, H., Harada, J.J., Goldberg, R.B., Pennell, R.I., and Fischer, R.L. (2003) Imprinting of the *MEA* Polycomb gene is controlled by antagonism between MET1 methyltransferase and DME glycosylase. *Dev. Cell 5*: 891–901.

Zemach, A., Kim, M.Y., Silva, P., Rodrigues, J.A., Dotson, B., Brooks, M.D., and Zilberman, D. (2010b) Local DNA hypomethylation activates genes in rice endosperm. *Proc. Natl. Acad. Sci. USA 107*: 18729–18734.

Zemach, A., McDaniel, I.E., Silva, P., and Zilberman, D. (2010a) Genome-wide evolutionary analysis of eukaryotic DNA methylation. *Science 328*: 916–919.

Zhang, X., Yazaki, A., Sundaresan, S., Cokus, S., Chan, S.W., Chen, H., Henderson, I.R., Shinn, P., Pellegrini, M., Jacobsen, S.E., et al. (2006) Genome-wide high resolution mapping and functional analysis of DNA methylation in *Arabidopsis. Cell 126*: 1189–1201.

Zilberman, D., Gehring, M., Tran, R.K., Ballinger, T., and Henikoff, S. (2007) Genome-wide analysis of *Arabidopsis* DNA methylation uncovers an interdependence between methylation and transcription. *Nature Genet. 39*: 61–69.

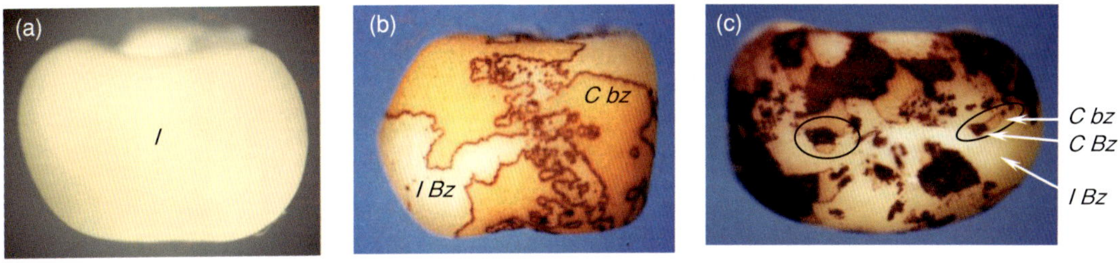

Figure 1.2 Pigment phenotypes of kernels that led to McClintock's discovery of *Ds* transposition. *I*: dominant inhibitory allele of the *C* locus; *C*: full anthocyanin pigmentation when together with the wildtype *Bz* allele of the *Bronze* locus; *bz*: recessive allele of *Bronze* locus. The *Shrunken* and *Waxy* loci are not included in these illustrations for simplicity. The circles highlight twin sectors resulting from formation of dicentric chromosomes and the subsequent unequal breakage and segregation of the dominant alleles.

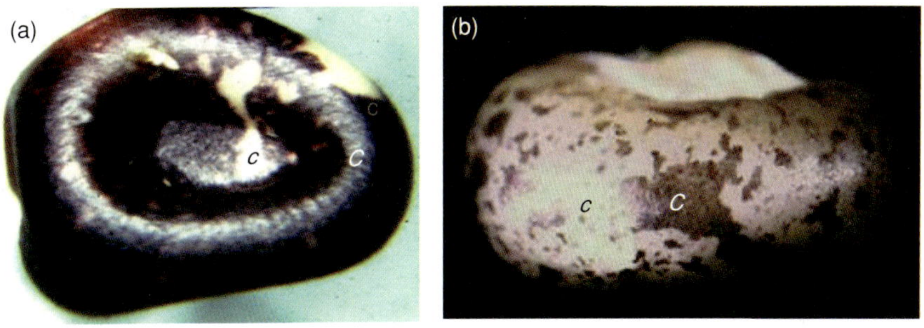

Figure 1.4 A. The phenotype of kernels losing the dominant pigmenting *C* allele to uncover the recessive null *c* due to chromosome breakage at *Ds*. B. The phenotype of a kernel in which *Ds* has inserted at the *C* locus causing loss of pigmentation except in sectors developing from cells in which the *Ds* has transposed out of the *C* locus, restoring a functional *C* allele.

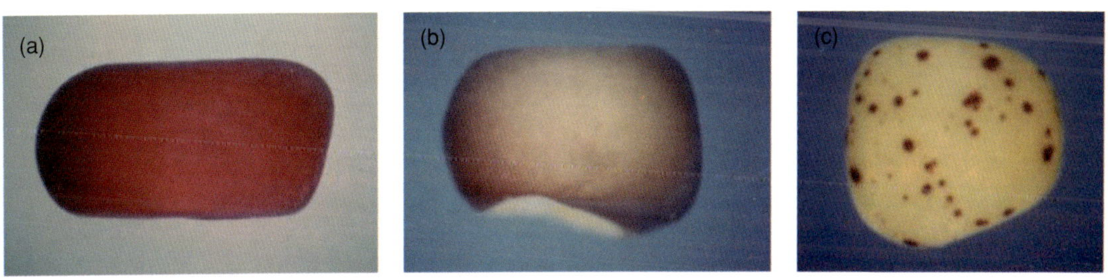

Figure 4.1 Phenotypes of kernels having the genetic constitution: (a) *a1/a1*; (b) *a1-m1/a1*, no *Spm*; (c) *a1-m1/*a*1*, with *Spm*.

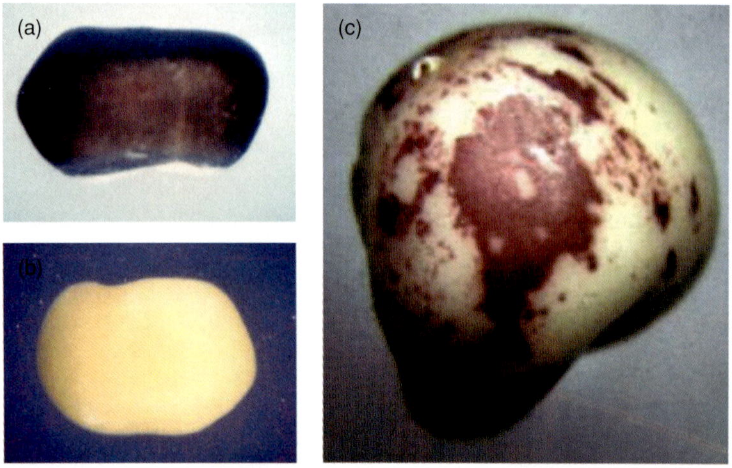

Figure 4.2 Phenotypes of kernels having the genetic constitution *a2-m1* [class II)/a2: (a) without *Spm*; (b) with a fully active *Spm*; (c) with an *Spm-c*].

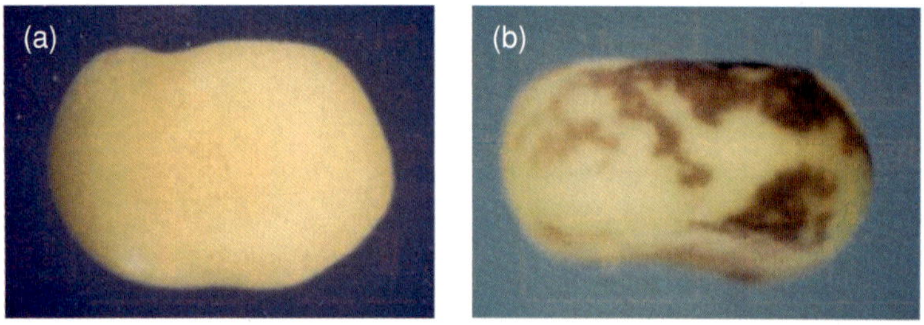

Figure 4.5 (a) The phenotype of the *Spm-dependent a1-m2 7995* allele in the long-term absence of an active *Spm*; (b) the "preset" pattern observed just after meiotic segregation of the *Spm* element.

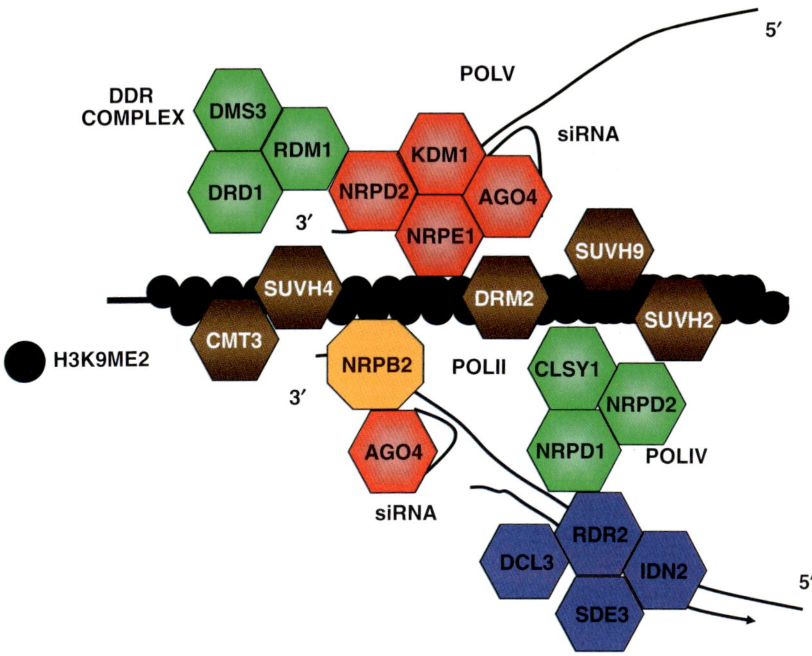

Figure 5.1 A model for RNA-dependent DNA methylation (RdDM) in *Arabidopsis*. Transposable elements, centromeric repeats, and intergenic noncoding repeats are transcribed by PolV, which is recruited to chromatin by the DDR chromatin remodeling complex. AGO4 interacts with GW/WG repeats in the C-terminal domains of the largest subunit of PolV (NRPE1) and the associated Spt5-like protein KDM1. AGO4binds 24nt siRNA and cleaves PolV transcripts. PolII shares many subunits with and can substitute for PolV, and also recruits AGO4 via RDM1. POLIV shares its second subunit (NRPD2) with PolV, but unlike PolII and PolV, PolIV is essential for siRNA biogenesis. The template for RNA synthesis is not known, but PolIV is recruited to chromatin by CLSY1, a homolog of DRD1. PolIV transcripts are converted to double-stranded RNA by RDR2. IDN2, DCL3, and SDE3 are shown associated with Rdr2 by analogy with homologs in *Arabidopsis* (SGS3) and fission yeast (DCR1and HRR1), respectively, and are similarly required for biogenesis of 24nt siRNA. RdDM requires the histone methyltransferases SUVH2 and SUVH9, as well as the DNA methyltransferases DRM2 and CMT3. SUVH2, SUVH9, and SUVH4 have SRA domains that bind methylated cytosines in CHH, CG, and CHG contexts, respectively, while CMT3 has a chromodomain that binds H3K9me2. However, the mechanism by which these methyltransferases are recruited by RNAi is still unclear. By analogy with fission yeast, this mechanism may involve the DNA replication machinery, as mutations in DNA polymerase reduce silencing in *Arabidopsis* and fission yeast.

Figure 8.3 Coordinate suppression of mutant phenotypes during development. Mutations caused by the insertion of nonautonomous *Mu* elements frequently depend on the presence of an autonomous *MuDR* element for their phenotypic effects. A *Mu* element inserted in the 5' end of the maize *high chlorophyll fluorescence106 (hcf106)* gene causes a defect in chlorophyll biosynthesis and sectors of pale green plant color when *MuDR* is present. The dominant *Mutator*-induced *Lesion mimic28* (*Les28)* also requires *MuDR* activity for its phenotypic effects. These phenotypes are gradually and coordinately suppressed in younger (later) leaves as the *Mutator* element is epigenetically silenced during development. The figure shows, from left to right, progressively older (earlier) leaves of an *hcf106/hcf106; Les28/+* double mutant. Sectors that have lost the *hcf* pale-green mutant phenotype are dark green and have also lost the lesion-mimic phenotype (reprinted with permission, from Martienssen and Baron, 1994).

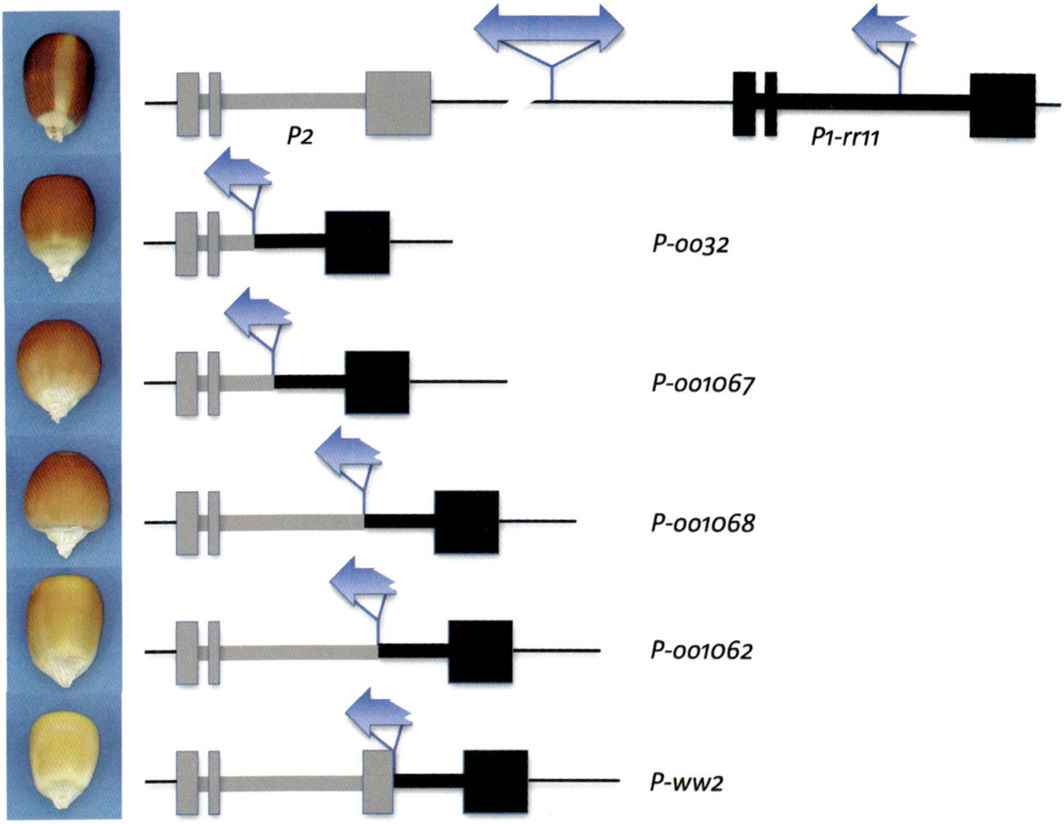

Figure 8.4 Phenotypes and gene structures of four new *P-oo* alleles produced by alternative transposition reactions. The kernel pericarp pigmentation phenotypes specified by the indicated alleles are shown on the left; the corresponding structure of the allele is diagrammed to the right. The structure of the progenitor *P1-rr11* allele is shown at the top. Black and gray boxes are exons 1, 2, and 3 (left to right) of *P1* and *P2*, respectively. Blue arrows indicate an intact *Ac* (left) and a fractured *Ac* (*fAc*) (right) insertion. For the other alleles, the approximate location of the new junction created by an alternative transposition reaction involving the right end of *Ac* and the single end of *fAc* in each *P-oo* allele is shown. *P2* is not expressed in wild-type pericarp tissue; however, in the *P oo* alleles, the *P1* pericarp enhancer (indicated as an asterisk) is located at new positions ranging from 6.2 to 14 kb from the *P2* transcription start site. At these new sites, the *P1* enhancer appears to increase expression of the fusion genes in the pericarp.

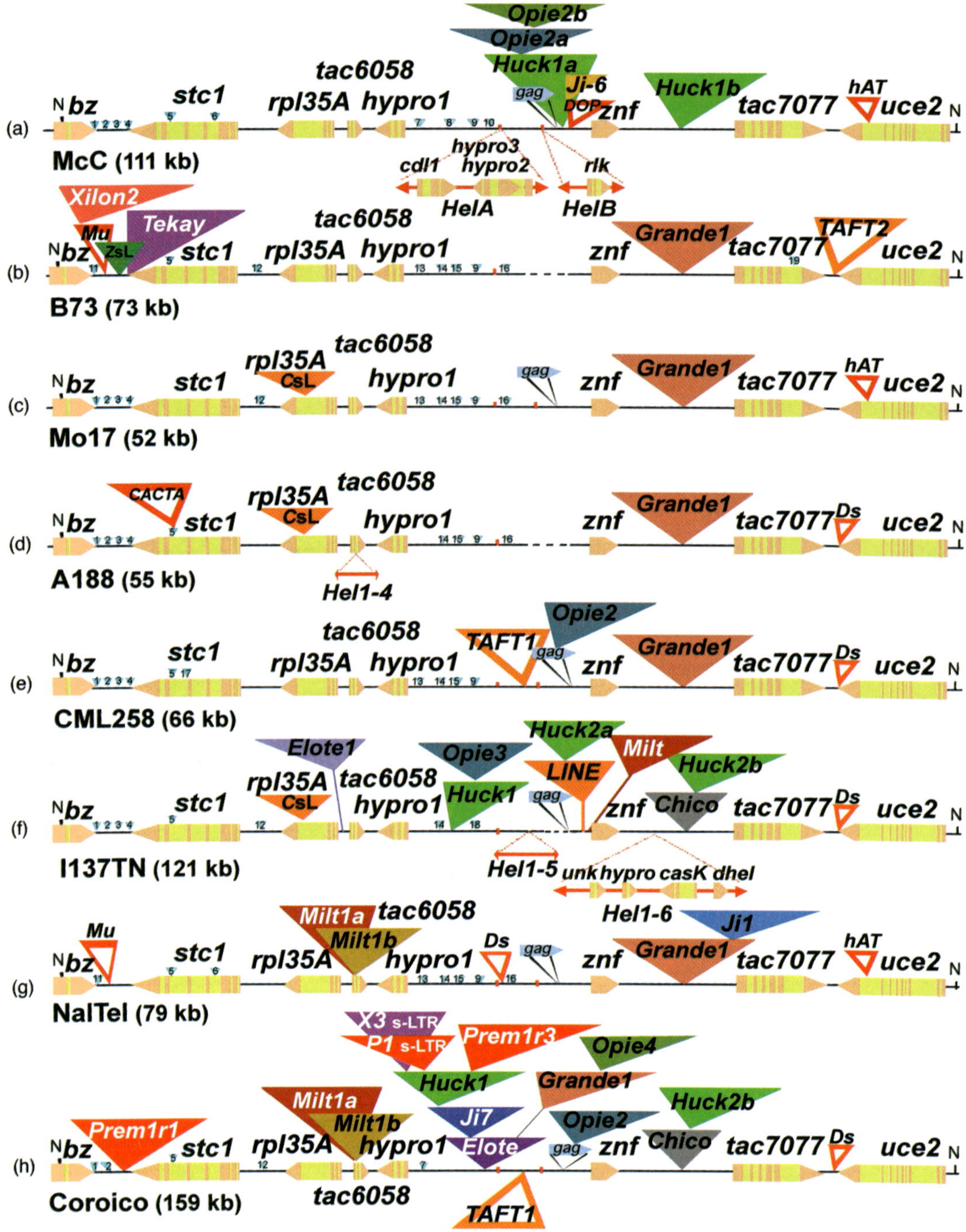

Figure 10.2 Organization of eight *bz* haplotypes. Each haplotype is identified by the name of the genetic line, followed by the size of the cloned NotI fragment in parentheses (NotI sites are indicated by Ns). Genes are shown as pentagons pointing in the direction of transcription; exons are in bronze and introns in yellow (darker and lighter grey). There are eight genes in the region: *bz*, *stc1*, *rpl35A*, *tac6058*, *hypro1*, *znf*, *tac7077*, and *uce2*. The same gene symbols are used for gene fragments carried by helitrons (*Hels*), which are represented as bidirectional arrows below the line for each haplotype. The vacant sites for *HelA* and *HelB* in each haplotype are provided as reference points and marked with short vertical strokes. Dashed lines represent deletions. Retrotransposons are indicated by solid triangles and DNA transposons and TAFTs (probably also DNA transposons), are indicated by open triangles. Small insertions are represented by small triangles and numbered. Only the genes have been drawn to scale (Wang and Dooner, *PNAS 103*: 17644 [2006]. Copyright 2006 National Academy of Sciences, U.S.A, used with permission.).

8 Transposons and Gene Creation

Hugo K. Dooner and Clifford F. Weil

Introduction

It is increasingly evident that the relationship between transposable elements (TE) and their host plant genomes is symbiotic, rather than parasitic. The ability of transposons to restructure genomes has been recognized since their initial discovery as chromosome-breaking elements that can rearrange large chromosomal segments, causing cytologically visible duplications, deletions, and inversions (McClintock, 1952). In fact, in her often-cited Nobel lecture, McClintock (1984) proposed that this restructuring occurred as an imprecisely programmed response of a genome to an unanticipated challenge, such as an accidental interspecific hybridization. However, sequencing of whole or partial plant genomes in the last decade has uncovered unexpected types of interactions between transposons and their host genomes, and rekindled interest in the possibility that TEs may have been involved not just in the evolution of genome structure, but also in the creation of new gene forms. It is this role of transposons that we will address in this review.

How have TEs contributed to gene creation? Taking the word "creation" *sensu lato* to include the generation of new alleles of existing genes, we can group the multiple ways in which transposon and host genes interact to generate genetic material with new properties into the following categories: 1) capture of gene fragments by TEs and formation of chimeric transcripts; 2) co-option of a TE transposase gene by the host; 3) fusion of TE and host genes; 4) alteration of host gene sequences by TE excisions; 5) alterations of host coding sequences by TE insertions; 6) acquisition by a host gene of new regulatory sequences from a TE; 7) alteration of mRNA splicing transcripts by TE insertion; and 8) reshuffling of host sequences by alternative transpositions.

Although both DNA and RNA transposons can participate in the creation of new genes, the former have played a more important role. The reasons for this lie in their respective distributions within the genome and in their different mechanisms of transposition. Retrotransposons constitute the majority of TEs in plant genomes, but are mostly found in pericentromeric and large intergenic regions. In contrast, the less abundant DNA transposons associate preferentially with the euchromatic or genic component of the genome, so they have had a greater opportunity to interact with the genes of the host (Dooner and Weil, 2007). In addition, whereas both DNA and RNA transposons insert by cutting host DNA, only DNA transposons excise, giving them a much greater potential to introduce double-strand breaks (DSBs) into chromosomes. The subsequent repair of these DNA DSBs by the host's enzymatic machinery rarely leaves the DNA in its original condition, but instead leads to a

Plant Transposons and Genome Dynamics in Evolution, First Edition. Edited by Nina V. Fedoroff.

variety of changes, from the addition or deletion of a few base pairs to large-scale chromosomal rearrangements, all of which have the potential to generate new gene forms.

Capture of Gene Fragments by TEs and Formation of Chimeric Genes

The capture of host genomic sequences by a plant transposon was first reported more than 20 years ago (Talbert and Chandler, 1988), but the generality of this property among TEs from monocots and dicots has been only recently appreciated. Both DNA and RNA elements possess the ability to acquire host genomic sequences and move them to new locations, although they do it by altogether different mechanisms. The DNA elements include *MULE*s (*Mutator*-like elements) from maize (Talbert and Chandler, 1988), rice (Jiang et al., 2004), *Lotus* (Jiang et al., 2004), and melon (van Leeuwen et al., 2007), *CACTA* elements from *Sorghum* (Paterson et al., 2009; Wicker et al., 2011), Japanese morning glory (Kawasaki and Nitasaka, 2004), *Antirrhinum* (Roccaro et al., 2007), and soybean (Zabala and Vodkin, 2007) and *Helitron* elements from maize (Lal et al., 2003; Lai et al., 2005; Morgante et al., 2005) and *Sorghum* (Paterson et al., 2009). Peculiarly, and for an as yet unknown reason, some elements in some species, such as the rice *MULEs*—referred to as *Pack-MULE*s (Jiang et al., 2004) or *Anacondas* (Ohtsu et al., 2005)—the maize *Helitrons* (Du et al., 2009; Yang and Bennetzen, 2009), and the *Sorghum CACTA* elements (Paterson et al., 2009) have been highly successful at capturing sequences from multiple genes. Fewer examples of gene capture have been reported for the RNA elements, including retrotransposons of both the LTR (long terminal repeat) and non-LTR class; both of these classes also show species-to-species variation in their ability to capture and/or mobilize host sequences.

Capture of Gene Fragments and Formation of Chimeric Genes by DNA Elements

DNA transposons can capture both exons and introns, making the involvement of an RNA intermediate improbable, although the possibility of capturing unprocessed RNAs has not yet been ruled out. Very likely, host sequences are captured by excisive DNA transposons as a result of DNA replication mistakes that occur during the error-prone repair of the double strand breaks (DSBs) generated by excision. Elements of both the *hAT* (*hobo-Activator-Tam3*) (Rubin and Levy, 1997; Yan et al., 1999) and *Mutator* superfamilies (Raizada et al., 2001) can copy extra sequences from near the excision site (filler DNA) into the transposon during the DNA repair synthesis that occasionally accompanies transposon excision. These filler DNAs are often flanked by short repeats of 2–10 bp; host sequences flanked by short repeats have also been reported within *CACTA* elements (Kawasaki and Nitasaka, 2004). This observation suggests that the transposon and a nearby sequence share sites of microhomology that enable the DNA polymerase to switch templates between the transposon and nontransposon sequence, and vice-versa, during the repair process. Similar template switching errors could also occur as part of *Helitron* excision (Li and Dooner, 2009) or during the rolling-circle replication mechanism by which *Helitrons* presumably transpose (Kapitonov and Jurka, 2001).

Chimeric transcripts containing exons from different genes (exon shuffling) have been detected for the *Pack-MULEs* of rice (Jiang et al., 2004; Hanada et al., 2009), the *Helitrons* of maize (Brunner et al., 2005; Lai et al., 2005; Morgante et al., 2005), and the *Tgm-Express CACTA* transposon of soybean (Zabala and Vodkin, 2007). Many of these transcripts contain premature stop codons in all reading frames and are unlikely to encode functional proteins immediately. It has been argued that the gene fragments carried by these transposons are more likely to play a role in regulating expression of existing genes than in the formation of new genes, possibly through synthesis of antisense RNA fragments (Juretic et al., 2005). However, a reanalysis of ORF length in 2809 rice *Pack-MULEs* found at least 15 cases in which the entire acquired coding region was $\geq$100 bp of uninterrupted ORF (Hanada et al., 2009). Interestingly, this later study also found that chimeric *Pack-MULEs* containing fragments from multiple genes were much more frequently expressed than those derived from a single gene. Furthermore, many *Pack-MULEs* displayed nucleotide substitution rates (Ka/Ks) indicative of purifying selection, suggesting that a subset of them is likely functional.

The gene fragments that transposons can contribute to potential new genes may also help define the character of particular taxa. Comparison among rice genomes has revealed a group of 861 conserved, decidedly more GC-rich genes that appear to be restricted to the Poaceae (Campbell et al., 2007). Strikingly, fragments from this group of genes are nearly three times more likely to be represented in *Pack-MULEs* than gene fragments from the rest of the genome. This bias may also contribute to the increase in number of rice (and maize) genes that show a decreasing GC gradient along their length (discussed further below).

There is suggestive evidence that DNA elements in several species carry fully functional host genes. The *Kaonashi (KI) MULEs* in *Arabidopsis* (Hoen et al., 2006) and similar elements in melon and rice (van Leeuwen et al., 2007) include sequences encoding the protease domain of ubiquitin-like protein-specific proteases (ULPs). These ULP sequences are similar enough to one another to suggest they have evolved under purifying selection, and they show pollen-specific expression. Interestingly, these transposons and their ULP-like genes are silenced by siRNAs and are localized primarily in heterochromatin, suggesting that whatever the functions of the protease fragments might be, they are likely to be in the service of the transposon rather than the plant. In maize, there is a *Helitron* containing a nearly intact cytidine deaminase gene (Xu and Messing, 2006). While there is no evidence to date that this gene is transcribed, the cytidine deaminase ORF lacks only the N-terminal six amino acids and starts with an AUG codon. Similarly, in the wheat A genome, a *CACTA* element containing an acireductone-dioxygenase-like protein (ALP) gene has contributed to a 20-fold acceleration in the duplication of the *ALP* family (Akhunov et al., 2007). While nearly all of these *ALP* genes have no promoter, one of the duplicates appears to have become active using a TE promoter, and is expressed in hexaploid wheat.

The capacity to capture multiple gene fragments suggests that transposons could serve as a caldron in which to mix exons and functional domains to create novel genes. Many evolutionarily young duplications of genes have unique exons in one or both duplicates compared to putative progenitor copies (Katju and Lynch, 2006). Thus, combining fragments of multiple genes may happen frequently, raising the interesting question of how different gene fragments come to be placed right next to one another. Transposon capture could provide just such a mechanism.

Capture of Gene Fragments and Formation of Chimeric Genes by RNA Elements

LTR retrotransposons possess terminal repeats that are longer than those of most DNA elements (usually 0.3-3 kb), and in direct, rather than reverse, orientation. In autonomous elements, the LTRs flank the coding sequences for all of the proteins required for the complex retrotransposition process: gag, protease, reverse transcriptase, and integrase, the latter three being generated by cleavage of a large polyprotein (pol) precursor. In brief, the process of transposition begins with the production of a genomic or full-length transcript of the element. The genomic RNA is transported to the cytoplasm, translated, and assembled with its gene products into a structure called a virus-like particle (VLP), where reverse transcription takes place and where multiple RNAs, including occasional mRNAs from the host, are packaged. The double-stranded DNA product is then imported back to the nucleus and inserted by action of the integrase at a new location in the host genome.

LTR retrotransposons are the most abundant TEs in most plant genomes, accounting for >80% of the nuclear DNA in many angiosperms (Bennetzen, 2005). In maize, the plant species where they have been most carefully annotated, they comprise more than 400 families and account for up to 85% of the genome (Baucom et al., 2009). LTR retrotransposons from several plant species have been reported to acquire host gene fragments and, occasionally, to fuse the coding sequences of two or more genes, creating chimeric genes with the potential to evolve new gene functions. The presence of multiple RNAs within the VLP provides an opportunity for the reverse transcriptase to switch templates and the presence of host mRNA allows the retrotransposon to capture spliced gene fragments (San Miguel and Vitte, 2009).

Evidence that, like retroviruses, an LTR retrotransposon can mobilize a host gene was first reported for the maize *Bs1* retrotransposon, originally isolated in an *adh1* mutation (Johns et al., 1985). *Bs1* is a defective element that has recruited intronless portions of three host genes encoding a proton ATPase (Bureau et al., 1994; Jin and Bennetzen, 1994; Palmgren, 1994), a glucanase, and a xylan endohydrolase (Elrouby and Bureau, 2001). These fragments have been combined to generate a chimeric, new gene that is both transcribed and translated into a BS1 chimeric protein. This gene is expressed during a narrow window in the formation of the ear, suggesting a possible role in reproductive development (Elrouby and Bureau, 2010). Intriguingly, although *Bs1* is present in one to five copies in all maize and teosinte lines tested, a full, uninterrupted ORF encoding the BS1 protein has been found, so far, only in maize.

Bioinformatic analyses of whole genome sequences from rice and maize revealed many cases of gene fragments within LTR retrotransposons, albeit in very different numbers in the two species: 27 in rice (Wang and Brendel, 2006) versus 425 in maize (Baucom et al., 2009). The latter could be traced to a minimum of 189 independent gene fragment captures. Curiously, in light of the *Bs1* example above, no case was identified in maize where a single LTR retrotransposon contained fragments from more than one nuclear gene. Although less extensively documented, cases of host sequences "hitchhiking" in LTR retrotransposons have also been reported in dicots (Wawrzynski et al., 2008).

Non-LTR retrotransposons, sometimes called retroposons, can mobilize gene fragments by a process known as retroposition, which appears to have played a much more important role in animals than in plants. For example, retroposition of transcribed sequences by *L1* (*LINE1*) non-LTR retrotransposons, the most abundant TEs in the human genome, occurs with high frequency (Moran and Gilbert, 2002). Transposition-competent non-LTR retrotransposons contain an internal promoter that initiates RNA synthesis upstream of itself, encode a reverse transcriptase and an endonuclease, end in a poly(A) 3′ tail, and are usually flanked by variable-length target site duplications (TSDs) that range from 3 to 20 bp. They can capture host sequences by two different mechanisms, referred to as cis and trans (Figure 8.1). In the cis (or "read-through") mechanism (Figure 8.1B), transcription proceeds beyond the transposon's poly(A) tract into adjacent sequences, and results in the capture and retrotransposition of sequences derived from the transposon's 3′ flank. The captured sequences could be promoters, enhancers, or exons, and their movement could either lead to the creation of new genes or alter the expression of existing genes (Moran et al., 1999). In the trans mechanism (Figure 8.1A), transcribed DNA that is not adjacent to the retrotransposon is mobilized by a process involving the diversion of the retrotransposon's enzymatic machinery to a host mRNA. The spliced mRNAs are reverse transcribed and inserted into new genomic locations as retrogenes, which differ from genes in that they lack introns and promoters, contain a poly(A) tract, and are flanked by TSDs of variable size (Esnault et al., 2000). Most retrogenes become nonexpressed pseudogenes

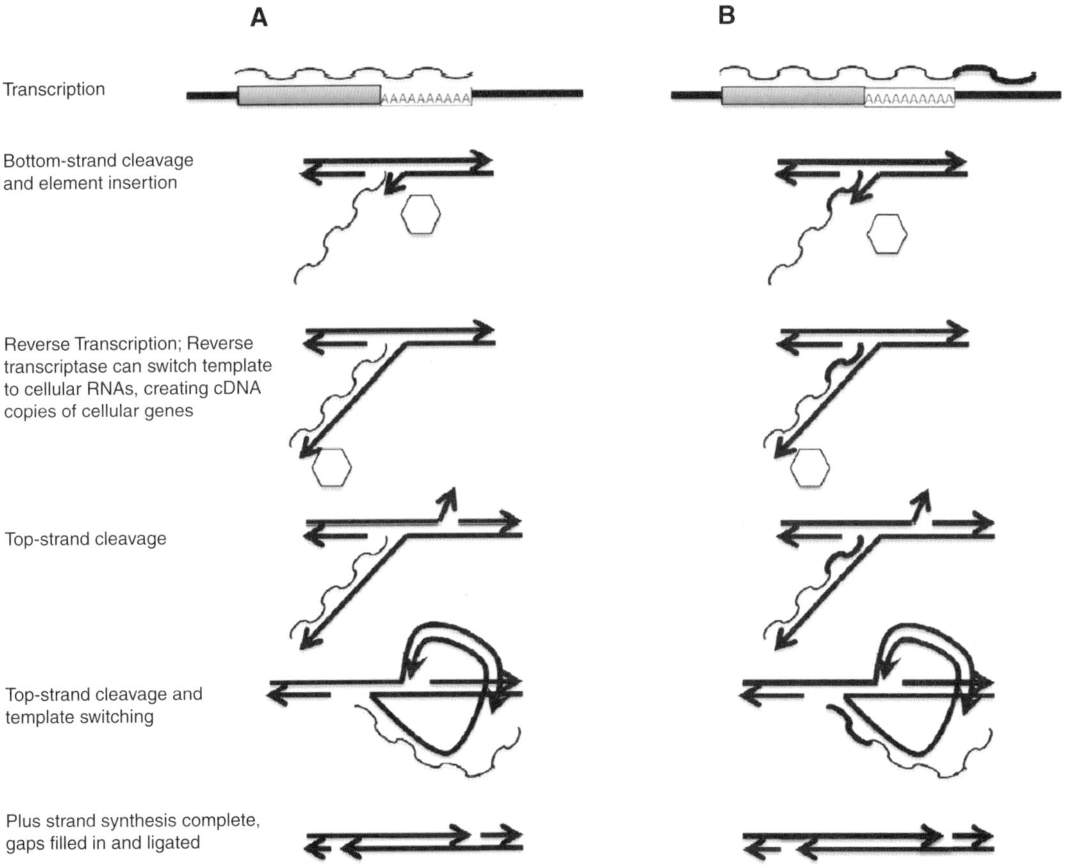

Figure 8.1 Model for nonLTR retrotransposon capture and movement of gene sequence. A) Movement of nonLTR retroelements begins with transcription of the element (shown in blue), export to the cytoplasm and translation of, among other proteins, reverse transcriptase (white hexagon). Following bottom strand cleavage of the target site (shown in gray), element transcripts attach and then serve as templates for minus strand synthesis by reverse transcription. During minus strand synthesis, reverse transcriptase can come off the template and begin reverse transcribing cellular mRNAs, creating pseudogenes that can reinsert into the host genome (the trans model referred to in the text). B) The cis model for nonLTR retrotransposon host sequence capture begins with transcription beyond the 3′ end of the element into host DNA (shown in black). The remainder of the process is essentially the same as in panel A, with the net result that host sequence is moved from one genome position to another, potentially in a position to be joined to other host sequences to create novel genes.

unless, by chance, they are inserted next to a host regulatory sequence, an event that could lead to their expression and potential acquisition of a new function. Retrogenes can also have a chimeric structure in which exons of different genes are combined. In the absence of mutational data, three criteria are conventionally used to assess the likelihood that retrogenes are functional: they must be transcribed (usually meaning found in EST databases), encode ORFs longer than 100 codons, and show patterns of nucleotide substitution indicative of purifying selection (Ka/Ks < 0.5).

Computational searches of whole-genome sequences have shown that the number of retroposed sequences in animals and plants differ dramatically, and that plant retroposition may be lineage specific. Only 69 retrosequences were identified in *Arabidopsis*, of which most were pseudogenes or had unknown functions, and only one was chimeric (Zhang et al., 2005). In contrast, the rice genome has undergone extensive, recent retroposition, with more than 1,200 retrogenes identified (Wang et al., 2006). The vast majority appeared to be functional by the criteria of expression, minimal ORF length, and negative selection (listed previously). Interestingly, 42% of the rice retrogenes had chimeric protein-coding sequences. Most were young, suggesting that potential gene creation by retroposition is a continuing phenomenon in rice. Furthermore, only a quarter of the chimeric retrogenes could also be found in sorghum, indicating that the majority had appeared after the split between the rice and sorghum grass lineages 50 million years ago. Chimeric retrogenes showing typical functional signatures have also been reported in poplar, though in much smaller numbers than in rice (Zhu et al., 2009).

Unexpectedly, intact genes, most probably translocated by a cut-and-paste mechanism, have also been found within LTR-retrotransposons. The maize *fie2* (*fertilization-independent endosperm-like*) gene, composed of 13 exons and 12 introns, is embedded within a *Ji3* LTR retrotransposon in the *orp2* region of chromosome 10 (Du et al., 2006). However, the *fie2* orthologs in neither rice nor sorghum reside in a retrotransposon. Interestingly, the maize *fie2* gene appears to have moved into the retroelement over a short distance on chromosome 10 because it is located very close to its orthologous position in rice and sorghum, presumably its original location. This movement is reminiscent of the short-range translocation of genes by *Ac-Ds* macrotransposons, which has been postulated as a possible mechanism to explain breakdowns of genetic colinearity among close relatives (Huang and Dooner, 2008). In a second, apparent case of non–RNA-mediated capture of a closely linked sequence by an LTR retrotransposon, a *gypsy*-like TE in the soybean wild relative *Glycine tomentella* contains a full-length NBS-LRR disease-resistance-like gene flanked by promoter and terminator sequences similar to those of nearby NBS-LRR sequences (Wawrzynski et al., 2008). These particular LTR retrotransposons are present in single copy in maize and *G. tomentella*, respectively, indicating that they may no longer be capable of transposition.

Co-Option of a TE Gene by the Host

Genome sequencing projects both in plants and in animals have revealed a large number of putative transposase-like genes. These genes are almost all identified on the basis of predicted protein sequence similarity to known transposases (TPases). While a number of these genes have been shown to encode DNA- or RNA-binding proteins, many have yet to be validated as producing transcripts or proteins with annotated functions, and it is likely that the list of TE-derived genes will grow. In many cases, the cis-acting sequences that are important for transposon mobility no longer flank these putative genes, suggesting that, if they are functional, it is not as a part of a TE.

Thus, transposons not only contribute to genome evolution through the restructuring and diversity they create, they literally contribute all or part of their own genes as raw material for genome evolution. TPase proteins include several very useful component domains including efficient DNA binding regions that often recognize specific DNA target sequences, endonuclease activity, and the capacity to form multimers, in some cases while attached to DNA (Sinzelle et al., 2009). Considering the persistence of mobile elements in genomes over evolutionary timescales and the duplication of transposase gene copies that accompanies transposon copy-number increases, it is reasonable to expect that cells would evolve to make valuable use of at least some of these duplicate copies.

Perhaps the best-known example of this comes from vertebrates. The immune system recombinase enzyme, Rag1, which mediates V(D)J rearrangement to create antibody genes, appears likely to be a captured TPase enzyme (Schatz, 1999; Kapitonov and Jurka, 2005). In addition to sequence similarity between Rag1 and TPases, such as the DDE motif (Kim et al., 1999; Landree et al.,

1999; Fugmann et al., 2000), there are mechanistic similarities in the DNA cleavage and rejoining processes between V(D)J rearrangement and *hAT* element transposition (Yu et al., 2004; Zhou et al., 2004). Furthermore, the signal sequences that Rag1 binds to are very likely vestiges of the terminal inverted repeats from a transposon of the *Transib* superfamily (Kapitonov and Jurka, 2005). An even more intriguing twist to this story has emerged in recent years, as the N-terminal part of the Rag1 protein, which had been difficult to align with a transposase, was found to align with a second type of transposon altogether, this time from molluscs (Panchin and Moroz, 2008), suggesting that more than one transposase has gone into making Rag1. Whether some sort of horizontal transfer was involved or whether the origins of Rag1 predate the divergence of vertebrates and molluscs is not yet clear. However, similar use of a modified TPase to carry out such programmed rearrangements has now been extended to ciliates, where a "domesticated" version of the *PiggyBac* TPase called *TPB2* has been shown to play an important part in the programmed deletion of genomic sequences that accompanies macronuclear development (Baudry et al., 2009; Cheng et al., 2010).

To date, these sorts of programmed DNA rearrangements during development have not been described in plants, but there are other examples in which a TPase gene has been "adopted" and modified by the host genome, and now carries out a necessary host function. Bundock and Hooykaas (2005) described an *Arabidopsis* protein, DAYSLEEPER, with strong similarity to a *hAT*-like TPase, which binds upstream of the gene encoding Ku70, a major factor in the repair of DNA DSBs that is required for normal plant development. In addition, a *Mutator-like element (MULE)* TPase appears to have been modified to produce the *FAR1* and *FHY1* proteins that control light signaling (Hudson et al., 2003). Similarly, the *Mustang* family of genes, now found in a wide range of plant species, is derived from the transposase gene of the *Mutator* element (Cowan et al., 2005) and the *gary* genes found in cereals—but not apparently in dicots—are derived from the transposase gene of a *hAT* transposon (Muehlbauer et al., 2006).

An example of a host cell adopting a retrotransposon protein, often cited in the animal literature, is the kinetochore protein CENP-B (Barbosa-Cisneros et al., 1997; Kipling and Warburton, 1997; Casola et al., 2008). CENP-B proteins are also found in plants (Barbosa-Cisneros and Herrera-Esparza, 2002), suggesting either that the capture of this retroelement gene predates the plant-animal divergence or, possibly, that capture of this or a similar retroelement gene has occurred multiple times. The CRR and CRM retroelement families of rice and maize, respectively, may represent a more indirect case of transposon adoption (Aragon-Alcaide et al., 1996; Jiang et al., 1996; Ananiev et al., 1998; Miller et al., 1998; Presting et al., 1998). These elements are specifically targeted to centromeres, and show significant sequence conservation among their LTRs, even between species, which may signal that the sequences are under selective pressure and that they predate the divergence of rice and maize (Zhong et al., 2002; Sharma and Presting, 2008). It has been suggested that their transcription facilitates the replacement of H3 with CEN-H3, a key centromeric histone variant (Zhong et al., 2002). If this proves to be correct, while not representing the creation of a new protein coding sequence per se, it raises interesting questions about how a new gene function is defined. The transposition and targeting functions of these elements are essential to their role at the centromere, suggesting that random transcription units would not be able to serve the same purpose. However, a more heterogeneous mix of centromere-targeted retroelements may suffice, as in barley (Hudakova et al., 2001), and there could be back-up systems that provide centromere function even when the elements are absent entirely (Nasuda et al., 2005). Roles for retroelements in centromere function may extend beyond the grasses, as elements such as *Tat* and *Athila* in *Arabidopsis* and *Beetle1* and *Beetle2* in beet are also targeted specifically to centromeres (Peterson-Burch et al., 2004; Weber and Schmidt, 2009). It will be interesting to see whether, just as transposons have been adopted as telomeres in *Drosophila* (reviewed in Casacuberta and Pardue, 2005), the targeting preferences of these retroelements have allowed them to stand in for the centromeres of some plant species or even, as has been suggested recently, to actively be replacing themselves for centromeres (Wolfgruber et al., 2009).

In a somewhat more indirect way, TE sequences appear to provide the first DNA sequences that initiate a process creating siRNAs that can then suppress nonelement genes (Piriyapongsa and Jordan, 2008; Li et al., 2011). A dramatic example of siRNAs created from TEs is provided by the resurrection plant *Craterostigma plantagineum.* Its desiccation tolerance appears to have arisen from the amplification during its evolution of the dehydration-inducible, non-LTR retrotransposon *CDT1*, which is present today in multiple, surprisingly invariant, copies. This 912-bp element directs the synthesis of a 21-bp siRNA postulated to control the expression of genes involved in the desiccation response (Hilbricht et al., 2008). In rice, many miRNAs are identical or very nearly identical to transposons (TE-MIR), and several TE-MIRs meet the proposed criteria for miRNAs (Meyers et al., 2008). Targets of these TE-MIRs often tend to be non–TE-related genes that contain homologous sequences in their CDS as a result of cognate TE insertions. Interestingly, most of these TE insertions occur at the very edges of the CDS sequences, at the boundaries between protein coding and noncoding sequence, suggesting they have participated in later alterations to splicing or translation start/stop signals. These miRNAs might evolve from being TE related initially to having non-TE, cellular functions.

Fusion of TE and Host Genes

In addition to genes encoding TPases, transposons can contribute other genes to host plant function. These can be fusions of the TPase with adjacent host DNA, either through deletion of one end of the TE or through creation of a new intron that results in splicing together the TPase and the transcript of the adjacent genic DNA, such as the *SETMAR* gene in primates (Cordaux et al., 2006), or the CSB-*PiggyBAC* gene found in Cockayne Syndrome patients (Newman et al., 2008). However, the better-documented route in plants appears to be the capturing of a host gene followed by internal rearrangement of the transposon to create fusions of the captured gene and the TPase gene. These internal rearrangements are common for most transposons as they move from being autonomous elements to being nonautonomous elements. In *Antirrhinum majus* (snapdragon), the product of the *ROSINA* (*RSI*) gene binds upstream of MADS box genes essential for proper development of flower petals and stamens, as well as proper seed morphology and dispersal (Roccaro et al., 2005). The *RSI* gene is multicopy, and sequence analysis revealed it to be part of a *CACTA* family transposon, dubbed *TamRSI*, that also encodes a TPase (Figure 8.2A) (Roccaro et al., 2007). RSI is the product of a second gene that is transcribed inward from the transposon end, in opposite direction to the TPase gene. Many *CACTA* elements have two genes, but this is the first instance in which the two genes are transcribed toward one another. It seems likely that *RSI* was picked up by this transposon, although the mechanism for that remains speculative. Of particular interest is that the spontaneous internal deletions that this element goes through have a special effect on its convergently transcribed genes. For *TamRSI*, such deletions can create novel genes that result in part of the TPase gene, now fused to the 3′ end of the *RSI* gene, being transcribed as antisense RNA; these RNAs are involved in epigenetically silencing *TamRSI* activity (Roccaro et al., 2007). Likewise, antisense transcripts of *RSI* that come from the TPase promoter may play a role in the down-regulation of *RSI* expression in carpels.

An example that appears more like the capture of a transposon by a gene rather than the reverse is seen in maize as part of the *Zmhox1a* gene (Figure 8.2B) (Comelli et al., 1999). In this case,

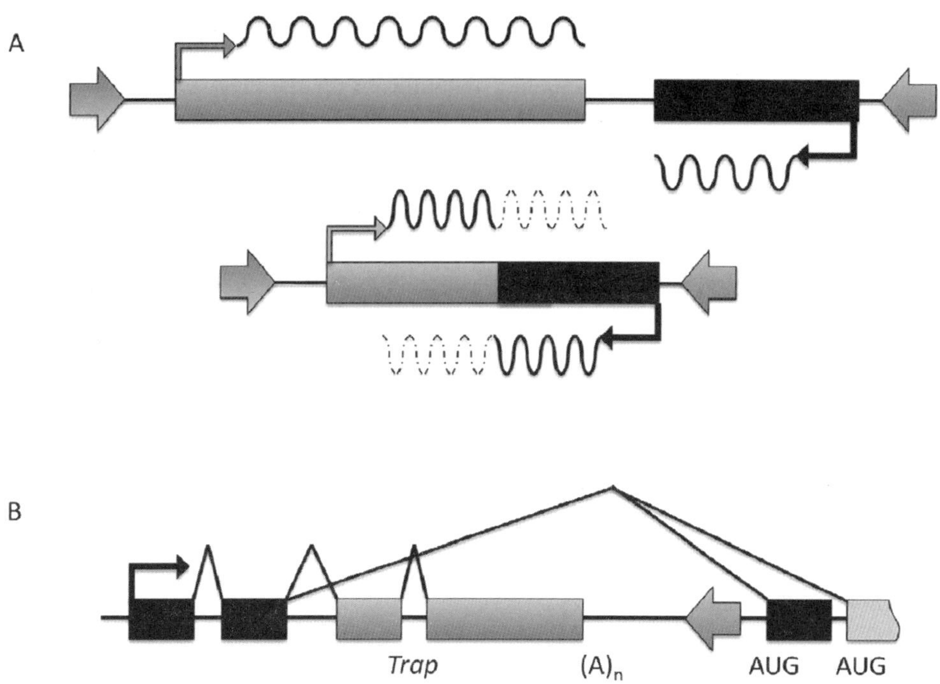

Figure 8.2 A) Top: Diagram of the CACTA *TamRSI* transposon. The transposase gene (grey) and the *RSI* gene (black) are transcribed in opposite orientations. Bottom: A spontaneous internal deletion can create novel genes in which the TPase gene is fused to the 3′ end of *RSI*, creating transcribed antisense transcripts that epigenetically silence *TamRSI* activity and *RSI* expression. B) Organization of part of the *Zmhox1a* gene. *Zmhox1a* exons are represented in black and gray, the *Trap* MULE in grey. An alternative splicing event can join the first two noncoding exons of *Zmhox1a* to an adjacent intron encoding the TRAP protein (see text). The putative TRAP coding region is followed by a *Mutator*-like terminal inverted repeat (grey arrow). The two *Zmhox1a* exons downstream of the TRAP-containing intron are themselves alternatively spliced to the *Zmhox1a* exons upstream of TRAP, resulting in two, alternative start AUG codons. Each of these two proteins appears to have a different function.

an alternative splicing event can join the first two, noncoding exons of *Zmhox1a* to an adjacent intron sequence that is predicted to encode a protein very similar to the C-terminus of the *Mutator* transposase, referred to as Transposase Associated Protein or TRAP. The putative TRAP coding region is followed by a *Mutator*-like terminal inverted repeat. Remarkably, the two *Zmhox1a* exons downstream of the TRAP-containing intron are themselves alternatively spliced to the *Zmhox1a* exons upstream of TRAP, resulting in two alternative start codons. Each of these two proteins appears to have a different function. Whether the transposon-like sequences of the second intron cause this alternative splicing or were inserted into a gene that was already alternatively spliced is not clear.

Alterations of Host Gene Sequences by TE Excisions

The ability of transposon excisions to generate new gene forms and their potential role in the origins of present-day genetic diversity has been recognized for a long time. Insertions of DNA transposons into host DNA generate direct repeats of the host target sequence on either side of the transposon. The size of the flanking TSD is a property of the transposon and generally ranges between 2 and

10 bp (Feschotte and Pritham, 2007). When a DNA transposon excises, it often leaves behind a "footprint" (i.e., a change in the host sequence relative to that found prior to insertion of the transposon). Transposon footprints may include deletions of the flanking host DNA or retention of all or part of the TSD, often with modifications of the central bases or even with filler DNA (e.g., Sommer et al., 1985; Wessler, 1988; Gierl et al., 1989; Scott et al., 1996). These modifications can result in new alleles of a gene, and were detected first as quantitative changes in the amount of pigmentation produced by a series of revertants from mutable alleles that affected anthocyanin pigmentation in the maize seed (McClintock, 1956; Fowler and Peterson, 1974) and *Antirrhinum* flower (Fincham and Harrison, 1967), and later, as qualitative changes in the properties of an enzyme involved in anthocyanin biosynthesis in the maize seed (Dooner and Nelson, 1979).

Clearly, excisions of transposons inserted in coding regions will produce new functional alleles only if they restore the correct reading frame. The new alleles encode proteins with additions or deletions of a few amino acids, which may have different properties compared with their progenitors (Bai et al., 2007). More dramatic changes can occur when transposons excise from promoters of host genes. Novel alleles with altered spatial patterns of gene expression have been reported for excisions of *Tam3* in *Antirrhinum* and *Mu3* in maize. The new *Antirrhinum* alleles resulted from simple deletions within the promoter region that probably altered the spacing between transcription signals (Coen et al., 1986). The new maize allele had a more complex, multibreak DNA rearrangement affecting >400 bp of the promoter region (Kloeckener-Gruissem and Freeling, 1995). The property of transposons to massively rearrange promoter sequences upon excision, creating alleles with novel expression patterns, has been dubbed "promoter scrambling." Regulatory effects of transposons on genes are covered extensively in Chapter 6 in this volume.

Alterations of Host Coding Sequences by TE Insertions

Although TE insertions into coding sequences usually lead to inactivation of the host gene, occasionally they can produce novel functional alleles. Perhaps not surprisingly, one finds examples of this among insertions of retrotransposons, which do not excise, into genes conferring resistance to plant pathogens, a class of genes that evolve under positive selection (Richter and Ronald, 2000).

The *Xa21D* allele of rice, which carries an LTR retrotransposon insertion in the first exon, confers partial resistance to the same six races of *Xanthomonas oryzae* in transgenic plants as *Xa21* alleles lacking the insertion (Wang et al., 1998). The *Xa21D* transcript terminates shortly after the stop codon introduced by the retrotransposon, and encodes a truncated version of the receptor kinase-like protein lacking the transmembrane and kinase domains characteristic of *XA21*. Yet, a comparison of nucleotide substitutions in the LRR coding regions of *Xa21* and *Xa21D* showed that adaptive selection events have played a role in the divergence of these two genes. The evidence suggests that the retrotransposon insertion has created a novel protein, representing a new class of disease-resistance genes.

The *Nicotiana tabaccum N* gene, which confers resistance to tobacco mosaic virus (TMV), contains a *TRIM* (Terminal-repeat Retrotransposon In Miniature) element inserted in the third intron. *N* is transcribed into two mRNA isoforms (NS and NL) and expression of both is required for full resistance to TMV in transgenic plants (Dinesh-Kumar and Baker, 2000). The NL transcript contains a 70-bp alternative exon provided by the *TRIM* element that would encode an amino acid sequence lacking significant similarity to any known protein (Kuang et al., 2009). The predicted product of the NL mRNA is a truncated protein because of a frameshift introduced by the *TRIM*-contributed, alternative exon. Although the mechanism by which the NL transcript contributes to full resistance is unclear, the N alternative exon provides an example of a TE insertion that contributes a coding sequence to a functional alternative-splicing product.

The most numerous DNA-mediated transposons in plants are the Miniature Inverted-repeat Transposable Elements (MITEs), generally nonautonomous elements that are mobilized by the TPase proteins of other elements (reviewed in Feschotte and Pritham, 2007). Given their tremendous copy numbers, it is perhaps not surprising that they have altered the function of a wide range of genes in a wide range of species. Known cases where MITE insertions have provided regulatory signals or altered transcription start sites, polyadenylation signals, splicing sites, and protein-coding sequences, all with the potential to alter gene function, number in the hundreds, if not the thousands (for example, see Xu et al., 2007; Oki et al., 2008).

Acquisition by Host Genes of New Regulatory Sequences from TEs

A more likely outcome of a TE interaction with a gene is the acquisition by the host gene of new regulatory sequences. While this aspect has tremendous importance in the formation of new patterns of gene expression and potential evolutionary consequences for neofunctionalization, we devote only a short section to it here because it is covered in much more detail in Chapter 6 in this volume. TEs carry genes for their own transposition, and many of them retain promoters, even after losing the ability to encode their own transposase (i.e., after becoming nonautonomous). Insertions in host gene regulatory regions or simply upstream of the start codon can create alleles with new tissue specificities (Chen et al., 1987) or place the gene under the same epigenetic regulation as the transposon (Barkan and Martienssen, 1991; Greene et al., 1994; Chatterjee and Martin, 1997). For example, a *Mutator* element inserted in the 5′ end of the maize *high chlorophyll fluorescence106 (hcf106)* gene causes a defect in chlorophyll biosynthesis and sectors of pale, green plant color. This phenotype is gradually suppressed in upper (later) leaves as the *Mutator* element is epigenetically silenced during development (Figure 8.3) (Martienssen and Baron, 1994). Interestingly, in the presence of a second *Mutator*-induced allele at the *lesion mimic28 (les28)* gene, both mutations are coordinately suppressed during development, suggesting that the changes in phenotype are a direct result of epigenetic control of the transposon. The transposon becomes a cis-acting control element for multiple genes that may have shared no common function before, but that now respond as a network to a common signal. In another example, the maize *a1-m2* series of alleles includes one that is *En/Spm* dependent, such that an active *En/Spm* TnpA (the major component of the transposase) must be present to stimulate expression of the *a1* gene (Masson et al., 1987). A nonautonomous member of this transposon family has inserted itself into the promoter region of the *a1* gene. TnpA bound to this element in an *En/Spm*-dependent allele serves as a transcriptional stimulus. Here, the transposon provides both cis- and trans-acting factors in the altered control. In addition, the co-opted gene can be further regulated by any cellular mechanisms that control the transposon.

The effect of transposon insertion on promoter activity is also dependent on the local chromatin structure, some of which may be influenced by sequences within the transposon itself. In *Antirrhinum*, the *hAT* element *Tam3* is inserted into the promoters of several genes, creating what has been termed *Tam3-permissible* gene expression (Uchiyama et al., 2009). In these alleles, sequences in the 3′ end of the *Tam3* element, rich in AA, TA, and TT dinucleotides, may be involved in positioning nucleosomes on the promoter of the target gene.

A TE from the *CACTA* superfamily appears to have been responsible for the acquisition of seed pigmenting ability by the *R-r* allele in maize, one of the most comprehensively studied higher plants genes (Stadler and Nuffer, 1953; Dooner and Kermicle, 1971; Robbins et al., 1991). *R-r* controls

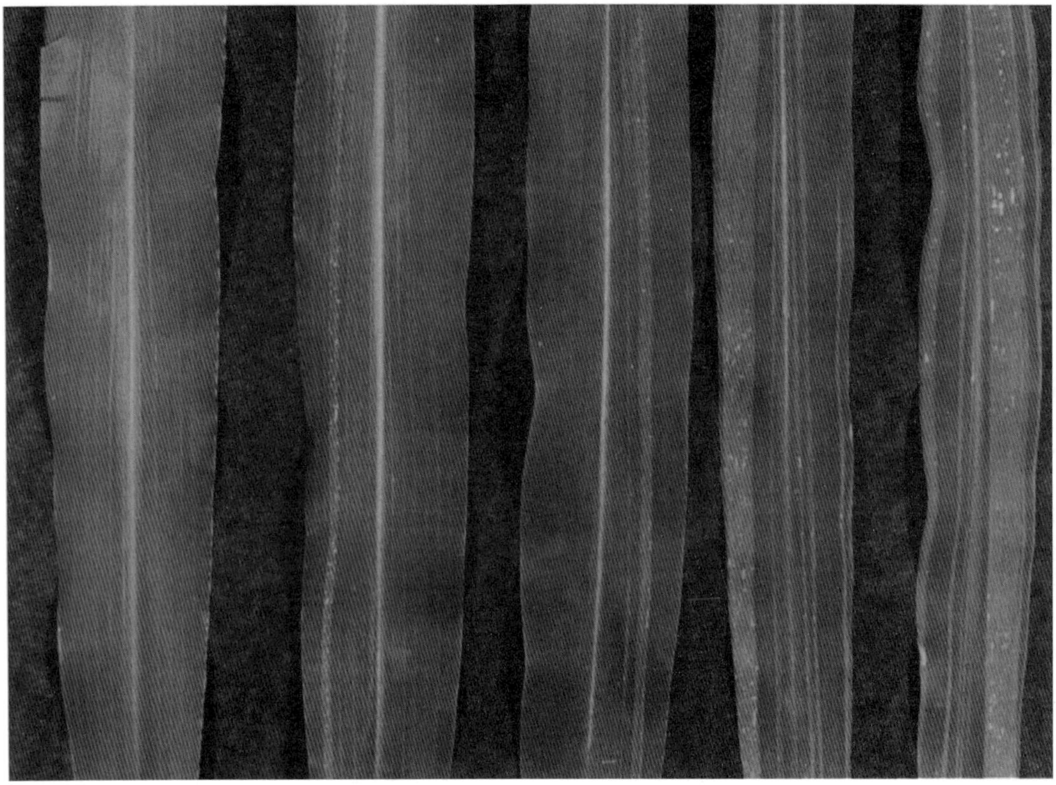

Figure 8.3 Coordinate suppression of mutant phenotypes during development. Mutations caused by the insertion of nonautonomous *Mu* elements frequently depend on the presence of an autonomous *MuDR* element for their phenotypic effects. A *Mu* element inserted in the 5′ end of the maize *high chlorophyll fluorescence106 (hcf106)* gene causes a defect in chlorophyll biosynthesis and sectors of pale green plant color when *MuDR* is present. The dominant *Mutator*-induced *Lesion mimic28* (*Les28*) also requires *MuDR* activity for its phenotypic effects. These phenotypes are gradually and coordinately suppressed in younger (later) leaves as the *Mutator* element is epigenetically silenced during development. The figure shows, from left to right, progressively older (earlier) leaves of an *hcf106/hcf106; Les28/+* double mutant. Sectors that have lost the *hcf* pale-green mutant phenotype are dark green and have also lost the lesion-mimic phenotype (reprinted with permission, from Martienssen and Baron, 1994). For color detail, see color plate section.

anthocyanin pigmentation in vegetative plant tissues and the aleurone layer of the seed through the tissue-specific production of a bHLH transcriptional activator of anthocyanin biosynthetic genes. The aleurone pigmenting capability of *R-r* results from the expression of two divergently expressed genes, *S1* and *S2*, separated by only 707 bp. The promoters for their divergent transcription are included within a 387-bp sequence flanked by inverted repeats that is derived from the ancestral insertion of a *CACTA* element called *Doppia* (Walker et al., 1995). Mutants of *R-r* in which this sequence has been lost also lose their seed pigmenting ability.

In some cases, epigenetic silencing of the transposon subsequently reverses the transcriptional effects initially contributed by a transposon insertion into a promoter region. For example, in rice, a pair of nested MITEs, *Kiddo* and *MDM1*, have inserted into the *ubiquitin2* (*rubq2*) gene promoter and become methylated (Yang et al., 2005). Preventing this methylation increases *rubq2* expression three-fold. Interestingly, promoter analysis of the *rubq2* gene indicates that the *Kiddo* element contributes ~20% of the gene expression, yet a line in which *Kiddo* is absent from this promoter shows the same level of transcription as a line where *Kiddo* is present. These data suggest that the epigenetic control associated with these transposons to control their positive effects on *rubq2* transcription is now part of the regulatory mechanism for this gene.

TE insertions in introns can also lead to changes in gene regulation with important evolutionary consequences. Some *Arabidopsis* accessions are rapid-cycling summer annuals, whereas others are late flowering and vernalization responsive, and thus, behave as winter annuals. One of the mechanisms leading to the evolution of summer-annual flowering is the origination of weak alleles of *FLOWERING LOCUS C* (*FLC*). *FLC* acts as a repressor of flowering, so lines carrying a weak *FLC* allele, such as Landsberg *erecta*, flower early. The *FLC* allele in that accession carries a 1,224-bp *MULE* inserted in the first intron, which causes *FLC-Ler* to be expressed at low levels (Michaels et al., 2003). The intronic *MULE* insertion affects the mRNA steady state level of the *FLC-Ler* host gene by rendering it susceptible to transcriptional gene silencing mediated by siRNAs generated from homologous *MULE*s in the *Arabidopsis* genome (Liu et al., 2004).

Pack-MULE elements have been proposed to influence an interesting trend in the overall structure of genes into which they insert (Jiang et al., 2011), also with important regulatory implications. In monocots, many gene sequences tend to have a higher GC content near their 5′ ends that decreases along the length of the transcript, a gradient seen much less frequently in dicot genes. One idea is that Pack-*MULE*s have accelerated the rate at which genes show this GC gradient. Rice, which has ~3,000 Pack-*MULE*s, and maize, with ~250, show this "negative GC gradient," while *Arabidopsis*, which has only 46 Pack-*MULE*s, does not. Furthermore, the Pack-*MULE*s of maize and rice are significantly more GC rich than the genome average, while *Arabidopsis* Pack-*MULE*s are not. Because Pack-*MULE*s insert preferentially near the 5′ ends of target genes, their GC richness and the GC-rich sequences they can acquire and transduplicate from the 5′ ends of previous targets contribute to the increasing number of genes showing the negative GC gradient. Increasing the GC content near the 5′ end of a gene increases the potential for DNA methylation and epigenetic control, as well as the recombination rate for that gene (Gerton et al., 2000; Drouaud et al., 2006), with significant, potential impact on genome function. It will be interesting in this regard to follow up on the preliminary observations that *Lotus japonicus,* also a dicot, is projected to have ~2,300 Pack-*MULE*s in its genome (Holligan et al., 2006). The completed and annotated genome will need to be in place to compare maize, rice, and *Arabidopsis* Pack-*MULE*s to *Lotus*, but this should occur soon. Pack-*MULE*s have also been observed in the sequenced genome of grapevine (*Vitis vinferas*) (Benjak et al., 2008) and melon (*Cucumis melo*) (van Leeuwen et al., 2007; Gonzalez et al., 2010).

Interaction of TEs with Target Gene mRNA Splicing and Structure

In addition to disrupting gene transcription, insertion of TE sequences into genes can often alter the posttranscriptional processing of mRNAs. Typically these changes affect mRNA splicing, changing the pattern of exons that remain in the finished mRNA, as well as adding some TE sequence to finished, genic mRNA by supplying alternative splice junctions to those regularly used. Unlike the situation in animals, plant introns are generally short (Brown et al., 2002). Insertion of transposon sequences into introns often results in failure to remove those introns, though the mechanistic reasons for this failure remain unclear. Exon skipping can be an additional effect; the exons that are lost from the transcripts can be those flanking the intron, as well as those elsewhere in the gene in cases where the overall pattern of intron splicing for the target gene is disrupted (Varagona et al., 1992). For example, insertion of a retroelement into intron 8 of the maize *waxy* gene (the *wx-G* allele) results in a reduction in the normal splicing pattern and an increase in alternative splicing events that join exon 6 to exon 9, exon 7 to exon 12, and a variety of other patterns. The first step along one route to creation of novel genes by transposons can thus be creation of novel pseudogenes when these novel transcripts are reverse transcribed and then incorporated into the genome.

Transposons can also alter RNA splicing of target genes by masquerading as introns (Giroux et al., 1994); this mechanism certainly contributes to producing new alleles of existing genes, but it is unlikely to create entirely new genes. Mobile elements have survived, in part, because they have evolved mechanisms to minimize their negative impacts on the genomes in which they are found. An example of this strategy is the presence of splice donor signals near the ends of members of the *hAT*, *CACTA*, and *Mutator* superfamilies and of the *Tnt1B* family of retrotransposons, allowing nearly all the transposon sequence to be spliced from the transcripts of genes into which it inserts (Kim et al., 1987; Wessler et al., 1987; Menssen et al., 1990; Ortiz and Strommer, 1990; Okagaki et al., 1992; Leprince et al., 2001). This mechanism reduces the negative selective pressure on transposon insertion, allowing a higher number of insertion events to retain expression of the target gene, though of a mutant allele. Typically, an ectopic splice site located in the gene and near the transposon insertion site is used together with the splice signals within the transposon end. The result is usually replacement of a short stretch of gene sequence by a short stretch of transposon sequence in the finished transcript. Such alleles can retain all or part of the original gene function, but the overall effect is less deleterious to the host than total loss of target gene function.

Another form of posttranscriptional impact that transposons can have comes through RNA editing. In some systems, transposon insertion sets off an increase in adenine to inosine RNA editing (Neeman et al., 2006; Lev-Maor et al., 2007), potentially altering where splice sites form. Whether this occurs regularly in plants for nuclear genes is unclear at present, but recent findings that alternative splicing events in plants may be much more frequent than previously imagined (Wang and Brendel, 2006; Reddy, 2007; Barbazuk et al., 2008; Simpson et al., 2008; Zhang et al., 2010) and that many alternatively spliced transcripts have noncanonical dinucleotide sequences at their intron ends (Filichkin et al., 2010) would be consistent with a model in which transposon activity is correlated directly with increased alternative splicing.

Reshuffling of Host Sequences by Alternative Transpositions

TEs were first identified by McClintock (1947) because of their ability to break chromosomes, and only later were they shown to be mobile (McClintock, 1949). While chromosome rearrangement through aberrant transposition is detailed in Chapter 3 in this volume, these studies are worth mentioning here again because such rearrangements can lead to the creation of novel genes. Recall that the original chromosome-breaking *Ds* (*Dissociation*) element had a complex structure consisting of two identical 2.2-kb *Ds* elements inserted one inside the other in inverse orientation (Doring et al., 1984). This structure, *DoubleDs*, causes chromosome breaks because dicentric chromosomes are formed as a consequence of a transposition reaction involving TE ends on sister chromatids. Transpositions where compatible 5′ and 3′ ends of different TEs, rather than the same TE, are mobilized together are referred to as alternative events (Gray, 2000). Pairs of closely linked transposons, a common outcome of *Ac* and *Ds* transposition, can also cause chromosome breaks by alternative transposition reactions, the frequency of breaks being negatively correlated with the genetic distance between the TEs (Figure 8.4) (Ralston et al., 1989; Dooner et al., 1991). Chromosome-breaking transposon pairs may be in either inverse (Weil and Wessler, 1993) or direct (Zhang and Peterson, 2004; Huang and Dooner, 2008) orientation, so the intertransposon segment can be flanked by either two identical or two opposite TE ends. The tendency of these elements to insert in or near genes

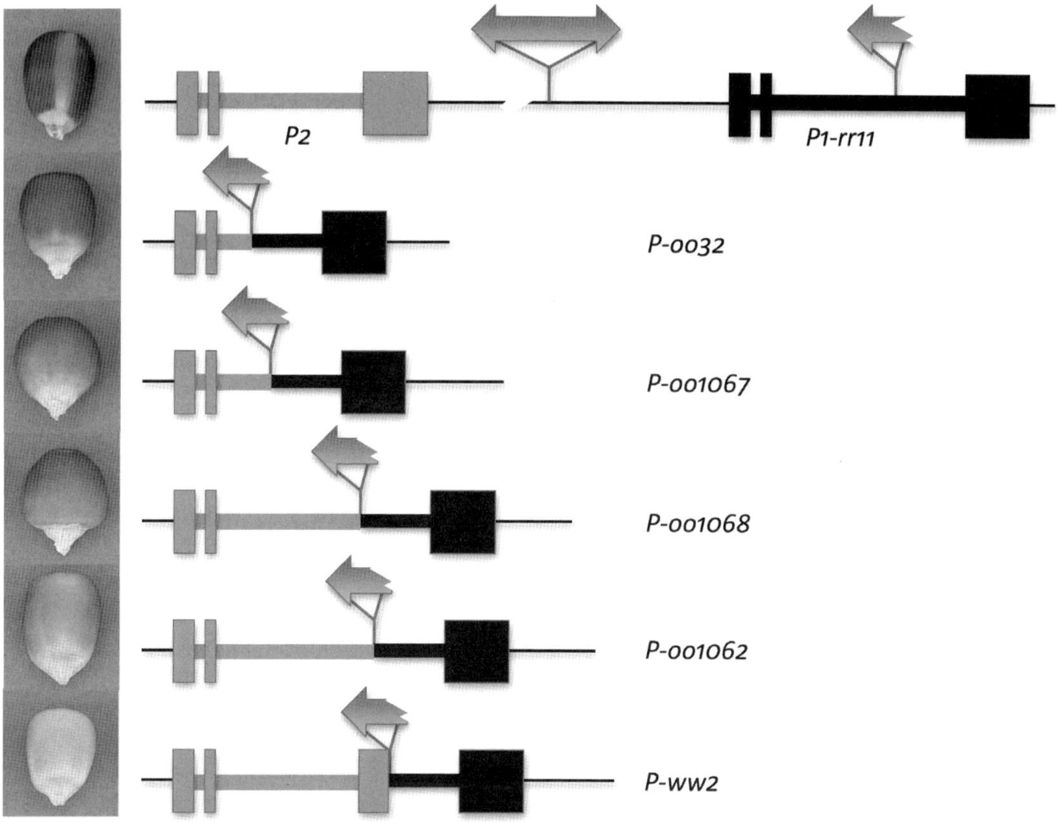

Figure 8.4 Phenotypes and gene structures of four new *P-oo* alleles produced by alternative transposition reactions. The kernel pericarp pigmentation phenotypes specified by the indicated alleles are shown on the left; the corresponding structure of the allele is diagrammed to the right. The structure of the progenitor *P1-rr11* allele is shown at the top. Black and gray boxes are exons 1, 2, and 3 (left to right) of *P1* and *P2*, respectively. Blue arrows indicate an intact *Ac* (left) and a fractured *Ac* (*fAc*) (right) insertion. For the other alleles, the approximate location of the new junction created by an alternative transposition reaction involving the right end of *Ac* and the single end of *fAc* in each *P-oo* allele is shown. *P2* is not expressed in wild-type pericarp tissue; however, in the *P-oo* alleles, the *P1* pericarp enhancer (indicated as an asterisk) is located at new positions ranging from 6.2 to 14 kb from the *P2* transcription start site. At these new sites, the *P1* enhancer appears to increase expression of the fusion genes in the pericarp. For color detail, see color plate section.

makes such rearrangements very likely to put coding sequences near one another that might not otherwise be juxtaposed, leading to potential formation of novel coding regions.

The ability of TE pairs in direct orientation to rearrange chromosomes has also been investigated in detail (Zhang and Peterson, 2004; Holligan et al., 2006; Zhang et al., 2006; Huang and Dooner, 2008; Pulletikurti et al., 2009; Zhang et al., 2009; Yu et al., 2011). In these TE pairs (either an *Ac* and a single-ended *fractured Ac* element or an *Ac* and a *Ds* element), the internal 5′ and 3′ ends from different transposons are separated from each other by host sequence segments of variable size and composition. Transposition reactions involving the two internal ends of the transposon pair can cause deletions, inversions, reshufflings of the intertransposon segment, and chromosome breaks. The rearrangements always include DNA that is adjacent to one or both transposons, and their size can extend up to several hundred kilobases, depending on the location of the reinsertion site (Zhang and Peterson, 2004; Huang and Dooner, 2008).

In a striking report of transposon-induced exon shuffling in maize, four cases were documented of deletions produced by alternative transposition reactions that fused the coding sequences of the two linked, paralogous *P1* and *P2* genes at an intron, producing functional chimeric genes that displayed a new kernel color because of the repositioning of an enhancer sequence (Figure 8.4) (Zhang et al., 2006). This result is a clear demonstration that alternative transposition reactions involving the ends of two separate TEs can rearrange gene segments leading to the creation of new genes expressing novel phenotypes. In another maize study, intertransposon segment reshuffling by alternative transposition reactions involving an *Ac* and a nearby *Ds* element resulted in the formation of ten chimeric genes that combined segments of the adjacent *bz1* and *stc1* genes (Huang and Dooner, 2008). One of the new chimeric genes encodes a continuous ORF that, surprisingly, shares 38% identity over 104 amino acids with a rice ORF of unknown function. Although these genes are most likely nonfunctional, their origin further illustrates the potential of TE pairs to generate gene fusions that can serve as raw material for the evolution of new gene forms.

Conclusion

The extensive surveys of all the transposon and transposon-like sequences within genomes have made the roles of these elements in capturing, moving, reshuffling, and creating new genes and regulatory circuitry clearer. It would seem likely that new roles will continue to emerge the deeper these studies go. The debate over whether or not transposons serve adaptive roles for their hosts would appear to have evolved into a remarkably wide-ranging description of the many ways these elements have created and continue to create genetic diversity. Indeed, it becomes increasingly difficult to imagine the array of modern taxa without them.

References

Akhunov, E.D., Akhunova, A.R., and Dvorak, J. (2007) Mechanisms and rates of birth and death of dispersed duplicated genes during the evolution of a multigene family in diploid and tetraploid wheats. *Mol. Biol. Evol. 24*: 539–550.

Ananiev, E.V., Phillips, R.L., and Rines, H.W. (1998) Chromosome-specific molecular organization of maize (*Zea mays* L.) centromeric regions. *Proc. Natl. Acad. Sci. USA 95*: 13073–13078.

Aragon-Alcaide, L., Miller, T., Schwarzacher, T., Reader, S., and Moore, G. (1996) A cereal centromeric sequence. *Chromosoma 105*: 261–268.

Bai, L., Singh, M., Pitt, L., Sweeney, M., and Brutnell, T.P. (2007) Generating novel allelic variation through *Activator* insertional mutagenesis in maize. *Genetics 175*: 981–992.

Barbazuk, W.B., Fu, Y., and McGinnis, K.M. (2008) Genome-wide analyses of alternative splicing in plants: opportunities and challenges. *Genome Res. 18*: 1381–1392.

Barbosa-Cisneros, O., and Herrera-Esparza, R. (2002) CENP-B is a conserved gene among vegetal species. *Genet. Mol. Res. 1*: 241–245.

Barbosa-Cisneros, O., Fraire-Velazquez, S., Moreno, J., and Herrera-Esparza, R. (1997) CENP-B autoantigen is a conserved protein from humans to higher plants: identification of the aminoterminal domain in *Phaseolus vulgaris*. *Rev. Rhum. Engl. Ed. 64*: 368–374.

Barkan, A., and Martienssen, R.A. (1991) Inactivation of maize transposon-*Mu* suppresses a mutant phenotype by activating an outward-reading promoter near the end of *Mu1*. *Proc. Natl. Acad. Sci. USA. 88*: 3502–3506.

Baucom, R.S., Estill, J.C., Chaparro, C., Upshaw, N., Jogi, A., Deragon, J.M., Westerman, R.P., Sanmiguel, P.J., and Bennetzen, J.L. (2009) Exceptional diversity, non-random distribution, and rapid evolution of retroelements in the B73 maize genome. *PLoS Genet. 5*: e1000732.

Baudry, C., Malinsky, S., Restituito, M., Kapusta, A., Rosa, S., Meyer, E., and Betermier, M. (2009) PiggyMac, a domesticated *PiggyBac* transposase involved in programmed genome rearrangements in the ciliate *Paramecium tetraurelia*. *Genes Dev.* *23*: 2478–2483.

Benjak, A., Forneck, A., and Casacuberta, J.M. (2008) Genome-wide analysis of the "cut-and-paste" transposons of grapevine. *PLoS One 3*: e3107.

Bennetzen, J.L. (2005) Transposable elements, gene creation and genome rearrangement in flowering plants. *Curr. Opin. Genet. Dev. 15*: 621–627.

Brown, J.W., Simpson, C.G., Thow, G., Clark, G.P., Jennings, S.N., Medina-Escobar, N., Haupt, S., Chapman, S.C., and Oparka, K.J. (2002) Splicing signals and factors in plant intron removal. *Biochem. Soc. Trans. 30*: 146–149.

Brunner, S., Pea, G., and Rafalski, A. (2005) Origins, genetic organization and transcription of a family of non-autonomous helitron elements in maize. *Plant J. 43*: 799–810.

Bundock, P., and Hooykaas, P. (2005) An *Arabidopsis hAT*-like transposase is essential for plant development. *Nature 436*: 282–284.

Bureau, T.E., White, S.E., and Wessler, S.R. (1994) Transduction of a cellular gene by a plant retroelement. *Cell 77*: 479–480.

Campbell, M.A., Zhu, W., Jiang, N., Lin, H., Ouyang, S., Childs, K.L., Haas, B.J., Hamilton, J.P., and Buell, C.R. (2007) Identification and characterization of lineage-specific genes within the Poaceae. *Plant Physiol. 145*: 1311–1322.

Casacuberta, E., and Pardue, M.L. (2005) *HeT-A* and *TART*, two *Drosophila* retrotransposons with a bona fide role in chromosome structure for more than 60 million years. *Cytogenet. Genome Res. 110*: 152–159.

Casola, C., Hucks, D., and Feschotte, C. (2008) Convergent domestication of pogo-like transposases into centromere-binding proteins in fission yeast and mammals. *Mol. Biol. Evol. 25*: 29–41.

Chatterjee, M., and Martin, C. (1997) *Tam3* produces a suppressible allele of the *DAG* locus of *Antirrhinum majus* similar to *Mu*-suppressible alleles of maize. *Plant J. 11*: 759–771.

Chen, C.H., Oishi, K.K., Kloeckener-Gruissem, B., and Freeling, M. (1987) Organ-specific expression of maize *Adh1* is altered after a *Mu* transposon insertion. *Genetics 116*: 469–477.

Cheng, C.Y., Vogt, A., Mochizuki, K., and Yao, M.C. (2010) A domesticated *PiggyBac* transposase plays key roles in heterochromatin dynamics and DNA cleavage during programmed DNA deletion in *Tetrahymena thermophila*. *Mol. Biol. Cell 21*: 1753–1762.

Coen, E.S., Carpenter, R., and Martin, C. (1986) Transposable elements generate novel spatial patterns of gene expression in *Antirrhinum majus*. *Cell 47*: 285–296.

Comelli, P., Konig, J., and Werr, W. (1999) Alternative splicing of two leading exons partitions promoter activity between the coding regions of the maize homeobox gene *Zmhox1a* and *Trap* (transposon-associated protein). *Plant Mol. Biol. 41*: 615–625.

Cordaux, R., Udit, S., Batzer, M.A., and Feschotte, C. (2006) Birth of a chimeric primate gene by capture of the transposase gene from a mobile element. *Proc. Natl. Acad. Sci. USA 103*: 8101–8106.

Cowan, R.K., Hoen, D.R., Schoen, D.J., and Bureau, T.E. (2005) *MUSTANG* is a novel family of domesticated transposase genes found in diverse angiosperms. *Mol. Biol. Evol. 22*: 2084–2089.

Dinesh-Kumar, S.P., and Baker, B.J. (2000) Alternatively spliced *N* resistance gene transcripts: their possible role in tobacco mosaic virus resistance. *Proc. Natl. Acad. Sci. USA 97*: 1908–1913.

Dooner, H.K., and Kermicle, J.L. (1971) Structure of the *R-r* tandem duplication in maize. *Genetics 67*: 427–436.

Dooner, H.K., and Nelson, O.E. (1979) Heterogeneous flavonoid glucosyltransferases in purple derivatives from a controlling element-suppressed *bronze* mutant in maize. *Proc. Natl. Acad. Sci. USA 76*: 2369–2371.

Dooner, H.K., and Weil, C.F. (2007) Give-and-take: interactions between DNA transposons and their host plant genomes. *Curr. Opin. Genet. Dev. 17*: 486–492.

Dooner, H.K., Keller, J., Harper, E., and Ralston, E. (1991) Variable patterns of transposition of the maize element *Activator* in tobacco. *Plant Cell 3*: 473–482.

Doring, H.P., Tillman, E., and Starlinger, P. (1984) DNA sequence of the maize transposable element *Dissociation*. *Nature 307*: 127–130.

Drouaud, J., Camilleri, C., Bourguignon, P.Y., Canaguier, A., Berard, A., Vezon, D., Giancola, S., Brunel, D., Colot, V., Prum, B., et al. (2006) Variation in crossing-over rates across chromosome 4 of *Arabidopsis thaliana* reveals the presence of meiotic recombination "hot spots." *Genome Res. 16*: 106–114.

Du, C., Swigonova, Z., and Messing, J. (2006) Retrotranspositions in orthologous regions of closely related grass species. *BMC Evol. Biol. 6*: 62.

Du, C., Fefelova, N., Caronna, J., He, L., and Dooner, H.K. (2009) The polychromatic *Helitron* landscape of the maize genome. *Proc. Natl. Acad. Sci. USA 106*: 19916–19921.

Elrouby, N., and Bureau, T.E. (2001) A novel hybrid open reading frame formed by multiple cellular gene transductions by a plant long terminal repeat retroelement. *J. Biol. Chem. 276*: 41963–41968.

Elrouby, N., and Bureau, T.E. (2010) *Bs1*, a new chimeric gene formed by retrotransposon-mediated exon shuffling in maize. *Plant Physiol. 153*: 1413–1424.

Esnault, C., Maestre, J., and Heidmann, T. (2000) Human LINE retrotransposons generate processed pseudogenes. *Nature Genet. 24*: 363–367.

Feschotte, C., and Pritham, E.J. (2007) DNA transposons and the evolution of eukaryotic genomes. *Annu. Rev. Genet. 41*: 331–368.

Filichkin, S.A., Priest, H.D., Givan, S.A., Shen, R., Bryant, D.W., Fox, S.E., Wong, W.K., and Mockler, T.C. (2010) Genome-wide mapping of alternative splicing in *Arabidopsis thaliana. Genome Res. 20*: 45–58.

Fincham, J.R.S., and Harrison, B. (1967) Instability at the *pal* locus in *Antirrhinum majus*. II. Multiple alleles produced by mutation of one original unstable allele. *Heredity 22*: 211–227.

Fowler, R., and Peterson, P.A. (1974) The *a2m(r-pa-pu)* allele of the *En* controlling element system in maize. *Genetics 76*: 433–446.

Fugmann, S.D., Villey, I.J., Ptaszek, L.M., and Schatz, D.G. (2000) Identification of two catalytic residues in RAG1 that define a single active site within the RAG1/RAG2 protein complex. *Mol. Cell 5*: 97–107.

Gerton, J.L., DeRisi, J., Shroff, R., Lichten, M., Brown, P.O., and Petes, T.D. (2000) Global mapping of meiotic recombination hotspots and coldspots in the yeast Saccharomyces cerevisiae. *Proc. Natl. Acad. Sci. USA 97*: 11383–11390.

Gierl, A., Saedler, H., and Peterson, P.A. (1989) Maize transposable elements. *Annu. Rev. Genet. 23*: 71–85.

Giroux, M.J., Clancy, M., Baier, J., Ingham, L., McCarty, D., and Hannah, L.C. (1994) De novo synthesis of an intron by the maize transposable element *Dissociation. Proc. Natl. Acad. Sci. USA 91*: 12150–12154.

Gonzalez, V.M., Benjak, A., Henaff, E.M., Mir, G., Casacuberta, J.M., Garcia-Mas, J., and Puigdomenech, P. (2010) Sequencing of 6.7 Mb of the melon genome using a BAC pooling strategy. *BMC Plant Biol. 10*: 246.

Gray, Y.H. (2000) It takes two transposons to tango: Transposable-element-mediated chromosomal rearrangements. *Trends Genet. 16*: 461–468.

Greene, B., Walko, R., and Hake, S. (1994) *Mutator* insertions in an intron of the maize *knotted1* gene result in dominant suppressible mutations. *Genetics 138*: 1275–1285.

Hanada, K., Vallejo, V., Nobuta, K., Slotkin, R.K., Lisch, D., Meyers, B.C., Shiu, S.H., and Jiang, N. (2009) The functional role of *Pack-MULEs* in rice inferred from purifying selection and expression profile. *Plant Cell 21*: 25–38.

Hilbricht, T., Varotto, S., Sgaramella, V., Bartels, D., Salamini, F., and Furini, A. (2008) Retrotransposons and siRNA have a role in the evolution of desiccation tolerance leading to resurrection of the plant *Craterostigma plantagineum. New Phytol. 179*: 877–887.

Hoen, D.R., Park, K.C., Elrouby, N., Yu, Z., Mohabir, N., Cowan, R.K., and Bureau, T.E. (2006) Transposon-mediated expansion and diversification of a family of ULP-like genes. *Mol. Biol. Evol. 23*: 1254–1268.

Holligan, D., Zhang, X., Jiang, N., Pritham, E.J., and Wessler, S.R. (2006) The transposable element landscape of the model legume *Lotus japonicus. Genetics 174*: 2215–2228.

Huang, J.T., and Dooner, H.K. (2008) Macrotransposition and other complex chromosomal restructuring in maize by closely linked transposons in direct orientation. *Plant Cell 20*: 2019–2032.

Hudakova, S., Michalek, W., Presting, G.G., ten Hoopen, R., dos Santos, K., Jasencakova, Z., and Schubert, I. (2001) Sequence organization of barley centromeres. *Nucleic Acids Res. 29*: 5029–5035.

Hudson, M.E., Lisch, D.R., and Quail, P.H. (2003) The *FHY3* and *FAR1* genes encode transposase-related proteins involved in regulation of gene expression by the phytochrome A-signaling pathway. *Plant J. 34*: 453–471.

Jiang, J., Nasuda, S., Dong, F., Scherrer, C.W., Woo, S.S., Wing, R.A., Gill, B.S., and Ward, D.C. (1996) A conserved repetitive DNA element located in the centromeres of cereal chromosomes. *Proc. Natl. Acad. Sci. USA 93*: 14210–14213.

Jiang, N., Ferguson, A.A., Slotkin, R.K., and Lisch, D. (2011) *Pack-Mutator-like* transposable elements (*Pack-MULEs*) induce directional modification of genes through biased insertion and DNA acquisition. *Proc. Natl. Acad. Sci. USA 108*: 1537–1542.

Jiang, N., Bao, Z., Zhang, X., Eddy, S.R., and Wessler, S.R. (2004) *Pack-MULE* transposable elements mediate gene evolution in plants. *Nature 431*: 569–573.

Jin, Y.K., and Bennetzen, J.L. (1994) Integration and nonrandom mutation of a plasma membrane proton ATPase gene fragment within the *Bs1* retroelement of maize. *Plant Cell 6*: 1177–1186.

Johns, M.A., Mottinger, J., and Freeling, M. (1985) A low copy number, *copia*-like transposon in maize. *EMBO J. 4*: 1093–1102.

Juretic, N., Hoen, D.R., Huynh, M.L., Harrison, P.M., and Bureau, T.E. (2005) The evolutionary fate of *MULE*-mediated duplications of host gene fragments in rice. *Genome Res. 15*: 1292–1297.

Kapitonov, V.V., and Jurka, J. (2001) Rolling-circle transposons in eukaryotes. *Proc. Natl. Acad. Sci. USA 98*: 8714–8719.

Kapitonov, V.V., and Jurka, J. (2005) RAG1 core and V(D)J recombination signal sequences were derived from *Transib* transposons. *PLoS Biol. 3*: e181.

Katju, V., and Lynch, M. (2006) On the formation of novel genes by duplication in the *Caenorhabditis elegans* genome. *Mol. Biol. Evol. 23*: 1056–1067.

Kawasaki, S., and Nitasaka, E. (2004) Characterization of *Tpn1* family in the Japanese morning glory: *En/Spm*-related transposable elements capturing host genes. *Plant Cell Physiol. 45*: 933–944.

Kim, D.R., Dai, Y., Mundy, C.L., Yang, W., and Oettinger, M.A. (1999) Mutations of acidic residues in RAG1 define the active site of the V(D)J recombinase. *Genes Dev. 13*: 3070–3080.

Kim, H.Y., Schiefelbein, J.W., Raboy, V., Furtek, D.B., and Nelson, O.E., Jr. (1987) RNA splicing permits expression of a maize gene with a defective *Suppressor-mutator* transposable element insertion in an exon. *Proc. Natl. Acad. Sci. USA 84*: 5863–5867.

Kipling, D., and Warburton, P.E. (1997) Centromeres, CENP-B and *Tigger* too. *Trends Genet. 13*: 141–145.

Kloeckener-Gruissem, B., and Freeling, M. (1995) Transposon-induced promoter scrambling: a mechanism for the evolution of new alleles. *Proc. Natl. Acad. Sci. USA 92*: 1836–1840.

Kuang, H., Padmanabhan, C., Li, F., Kamei, A., Bhaskar, P.B., Ouyang, S., Jiang, J., Buell, C.R., and Baker, B. (2009) Identification of miniature inverted-repeat transposable elements (MITEs) and biogenesis of their siRNAs in the Solanaceae: New functional implications for MITEs. *Genome Res. 19*: 42–56.

Lai, J., Li, Y., Messing, J., and Dooner, H.K. (2005) Gene movement by *Helitron* transposons contributes to the haplotype variability of maize. *Proc. Natl. Acad. Sci. USA 102*: 9068–9073.

Lal, S.K., Giroux, M.J., Brendel, V., Vallejos, C.E., and Hannah, L.C. (2003) The maize genome contains a *Helitron* insertion. *Plant Cell 15*: 381–391.

Landree, M.A., Wibbenmeyer, J.A., and Roth, D.B. (1999) Mutational analysis of RAG1 and RAG2 identifies three catalytic amino acids in RAG1 critical for both cleavage steps of V(D)J recombination. *Genes Dev. 13*: 3059–3069.

Leprince, A.S., Grandbastien, M.A., and Christian, M. (2001) Retrotransposons of the *Tnt1B* family are mobile in *Nicotiana plumbaginifolia* and can induce alternative splicing of the host gene upon insertion. *Plant Mol. Biol. 47*: 533–541.

Lev-Maor, G., Sorek, R., Levanon, E.Y., Paz, N., Eisenberg, E., and Ast, G. (2007) RNA-editing-mediated exon evolution. *Genome Biol. 8*: R29.

Li, Y., and Dooner, H.K. (2009) Excision of *Helitron* transposons in maize. *Genetics 182*: 399–402.

Li, Y., Li, C., Xia, J., and Jin, Y. (2011) Domestication of transposable elements into microRNA genes in plants. *PLoS One 6*: e19212.

Liu, J., He, Y., Amasino, R., and Chen, X. (2004) siRNAs targeting an intronic transposon in the regulation of natural flowering behavior in *Arabidopsis*. *Genes Dev. 18*: 2873–2878.

Martienssen, R., and Baron, A. (1994) Coordinate suppression of mutations caused by *Robertson's Mutator* transposons in maize. *Genetics 136*: 1157–1170.

Masson, P., Surosky, R., Kingsbury, J.A., and Fedoroff, N.V. (1987) Genetic and molecular analysis of the *Spm*-dependent *a-m2* alleles of the maize *a* locus. *Genetics 117*: 117–137.

McClintock, B. (1947) Cytogenetic studies of maize and Neurospora. *Carnegie Inst. Wash. Yr. Bk. 46*: 146–152.

McClintock, B. (1949) Mutable loci in maize. *Carnegie Inst. Wash. Yr. Bk. 48*: 142–154.

McClintock, B. (1952) Mutable loci in maize. *Carnegie Inst. Wash. Yr. Bk. 51*: 212–219.

McClintock, B. (1956) Controlling elements and the gene. *Cold Spr. Harbor Symp. Quant. Biol. 21*: 197–216.

McClintock, B. (1984) The significance of responses of the genome to challenge. *Science 226*: 792–802.

Menssen, A., Hohmann, S., Martin, W., Schnable, P.S., Peterson, P.A., Saedler, H., and Gierl, A. (1990) The *En/Spm* transposable element of *Zea mays* contains splice sites at the termini generating a novel intron from a *dSpm* element in the *A2* gene. *EMBO J. 9*: 3051–3057.

Meyers, B.C., Axtell, M.J., Bartel, B., Bartel, D.P., Baulcombe, D., Bowman, J.L., Cao, X., Carrington, J.C., Chen, X., Green, P.J., et al. (2008) Criteria for annotation of plant microRNAs. *Plant Cell 20*: 3186–3190.

Michaels, S.D., He, Y., Scortecci, K.C., and Amasino, R.M. (2003) Attenuation of *FLOWERING LOCUS C* activity as a mechanism for the evolution of summer-annual flowering behavior in *Arabidopsis*. *Proc. Natl. Acad. Sci. USA 100*: 10102–10107.

Miller, J.T., Dong, F., Jackson, S.A., Song, J., and Jiang, J. (1998) Retrotransposon-related DNA sequences in the centromeres of grass chromosomes. *Genetics 150*: 1615–1623.

Moran, J.V., and Gilbert, N. (2002) Mammalian *L1* retrotransposons and related elements. In *Mobile DNA II*, N.L. Craig, R. Craigie, M. Gellert, and A.M. Lambowitz, eds. (Washington, DC: American Society for Microbiology Press), pp. 836–869.

Moran, J.V., DeBerardinis, R.J., and Kazazian, H.H., Jr. (1999) Exon shuffling by *L1* retrotransposition. *Science 283*: 1530–1534.

Morgante, M., Brunner, S., Pea, G., Fengler, K., Zuccolo, A., and Rafalski, A. (2005) Gene duplication and exon shuffling by *helitron*-like transposons generate intraspecies diversity in maize. *Nature Genet. 37*: 997–1002.

Muehlbauer, G.J., Bhau, B.S., Syed, N.H., Heinen, S., Cho, S., Marshall, D., Pateyron, S., Buisine, N., Chalhoub, B., and Flavell, A.J. (2006) A *hAT* superfamily transposase recruited by the cereal grass genome. *Mol. Genet. Genomics 275*: 553–563.

Nasuda, S., Hudakova, S., Schubert, I., Houben, A., and Endo, T.R. (2005) Stable barley chromosomes without centromeric repeats. *Proc. Natl. Acad. Sci. USA 102*: 9842–9847.

Neeman, Y., Levanon, E.Y., Jantsch, M.F., and Eisenberg, E. (2006) RNA editing level in the mouse is determined by the genomic repeat repertoire. *RNA 12*: 1802–1809.

Newman, J.C., Bailey, A.D., Fan, H.Y., Pavelitz, T., and Weiner, A.M. (2008) An abundant evolutionarily conserved CSB-*PiggyBac* fusion protein expressed in Cockayne syndrome. *PLoS Genet. 4*: e1000031.

Ohtsu, K., Hirano, H.Y., Tsutsumi, N., Hirai, A., and Nakazono, M. (2005) *Anaconda*, a new class of transposon belonging to the *Mu* superfamily, has diversified by acquiring host genes during rice evolution. *Mol. Genet. Genomics 274*: 606–615.

Okagaki, R.J., Sullivan, T.D., Schiefelbein, J.W., and Nelson, O.E., Jr. (1992) Alternative 3′ splice acceptor sites modulate enzymic activity in derivative alleles of the maize *bronze1-mutable 13* allele. *Plant Cell 4*: 1453–1462.

Oki, N., Yano, K., Okumoto, Y., Tsukiyama, T., Teraishi, M., and Tanisaka, T. (2008) A genome-wide view of miniature inverted-repeat transposable elements (MITEs) in rice, *Oryza sativa* ssp. japonica. *Genes Genet. Syst. 83*: 321–329.

Ortiz, D.F., and Strommer, J.N. (1990) The *Mu1* maize transposable element induces tissue-specific aberrant splicing and polyadenylation in two *Adh1* mutants. *Mol. Cell Biol. 10*: 2090–2095.

Palmgren, M.G. (1994) Capturing host DNA by a plant retroelement: *Bs1* encodes plasma membrane H^+-ATPase domains. *Plant Mol. Biol. 25*: 137–140.

Panchin, Y., and Moroz, L.L. (2008) Molluscan mobile elements similar to the vertebrate Recombination-Activating Genes. *Biochem. Biophys. Res. Commun. 369*: 818–823.

Paterson, A.H., Bowers, J.E., Bruggmann, R., Dubchak, I., Grimwood, J., Gundlach, H., Haberer, G., Hellsten, U., Mitros, T., Poliakov, A., et al. (2009) The *Sorghum bicolor* genome and the diversification of grasses. *Nature 457*: 551–556.

Peterson-Burch, B.D., Nettleton, D., and Voytas, D.F. (2004) Genomic neighborhoods for *Arabidopsis* retrotransposons: A role for targeted integration in the distribution of the Metaviridae. *Genome Biol. 5*: R78.

Piriyapongsa, J., and Jordan, I.K. (2008) Dual coding of siRNAs and miRNAs by plant transposable elements. *RNA 14*: 814–821.

Presting, G.G., Malysheva, L., Fuchs, J., and Schubert, I. (1998) A *Ty3/gypsy* retrotransposon-like sequence localizes to the centromeric regions of cereal chromosomes. *Plant J. 16*: 721–728.

Pulletikurti, V., Yu, C., Zhang, J., Peterson, T., and Weber, D.F. (2009) Cytological evidence that alternative transposition by *Ac* elements causes reciprocal translocations and in versions in *Zea mays* L. *Maydica 54*: 457–462.

Raizada, M.N., Nan, G.L., and Walbot, V. (2001) Somatic and germinal mobility of the *RescueMu* transposon in transgenic maize. *Plant Cell 13*: 1587–1608.

Ralston, E., English, J., and Dooner, H.K. (1989) Chromosome-breaking structure in maize involving a fractured *Ac* element. *Proc. Natl. Acad. Sci. USA 86*: 9451–9455.

Reddy, A.S. (2007) Alternative splicing of pre-messenger RNAs in plants in the genomic era. *Annu. Rev. Plant Biol. 58*: 267–294.

Richter, T.E., and Ronald, P.C. (2000) The evolution of disease resistance genes. *Plant Mol. Biol. 42*: 195–204.

Robbins, T.P., Walker, E.L., Kermicle, J.L., Alleman, M., and Dellaporta, S.L. (1991) Meiotic instability of the *R-r* complex arising from displaced intragenic exchange and intrachromosomal rearrangement. *Genetics 129*: 271–283.

Roccaro, M., Li, Y., Sommer, H., and Saedler, H. (2007) *ROSINA (RSI)* is part of a CACTA transposable element, *TamRSI*, and links flower development to transposon activity. *Mol. Genet. Genomics 278*: 243–254.

Roccaro, M., Li, Y., Masiero, S., Saedler, H., and Sommer, H. (2005) *ROSINA (RSI)*, a novel protein with DNA-binding capacity, acts during floral organ development in *Antirrhinum majus*. *Plant J. 43*: 238–250.

Rubin, E., and Levy, A.A. (1997) Abortive gap repair: underlying mechanism for *Ds* element formation. *Mol. Cell Biol. 17*: 6294–6302.

San Miguel, P., and Vitte, C. (2009) The LTR-retrotransposons of maize. In *Handbook of maize: Domestication, genetics, and genome*, J.L. Bennetzen and S.C. Hake, eds. (New York: Springer Science), pp. 307–327.

Schatz, D.G. (1999) Transposition mediated by RAG1 and RAG2 and the evolution of the adaptive immune system. *Immunol. Res. 19*: 169–182.

Scott, L., LaFoe, D., and Weil, C.F. (1996) Adjacent sequences influence DNA repair accompanying transposon excision in maize. *Genetics 142*: 237–246.

Sharma, A., and Presting, G.G. (2008) Centromeric retrotransposon lineages predate the maize/rice divergence and differ in abundance and activity. *Mol. Genet. Genomics 279*: 133–147.

Simpson, C.G., Lewandowska, D., Fuller, J., Maronova, M., Kalyna, M., Davidson, D., McNicol, J., Raczynska, D., Jarmolowski, A., Barta, A., et al. (2008) Alternative splicing in plants. *Biochem. Soc. Trans. 36*: 508–510.

Sinzelle, L., Izsvak, Z., and Ivics, Z. (2009) Molecular domestication of transposable elements: from detrimental parasites to useful host genes. *Cell Mol. Life Sci. 66*: 1073–1093.

Sommer, H., Carpenter, R., Harrison, B.J., and Saedler, H. (1985) The transposable element *Tam3* of *Antirrhinum majus* generates a novel type of sequence alterations upon excision. *Molec. Gen. Genet. 199*: 225–231.

Stadler, L.J., and Nuffer, M.G. (1953) Problems of gene structure. II. Separation of *R-r* elements (S) and (P) by unequal crossing over. *Science 117*: 471–472.

Talbert, L.E., and Chandler, V.L. (1988) Characterization of a highly conserved sequence related to *Mutator* transposable elements in maize. *Mol. Biol. Evol. 5*: 519–529.

Uchiyama, T., Fujino, K., Ogawa, T., Wakatsuki, A., Kishima, Y., Mikami, T., and Sano, Y. (2009) Stable transcription activities dependent on an orientation of *Tam3* transposon insertions into Antirrhinum and yeast promoters occur only within chromatin. *Plant Physiol. 151*: 1557–1569.

van Leeuwen, H., Monfort, A., and Puigdomenech, P. (2007) *Mutator*-like elements identified in melon, *Arabidopsis* and rice contain ULP1 protease domains. *Mol. Genet. Genomics 277*: 357–364.

Varagona, M.J., Purugganan, M., and Wessler, S.R. (1992) Alternative splicing induced by insertion of retrotransposons into the maize *waxy* gene. *Plant Cell 4*: 811–820.

Walker, E.L., Robbins, T.P., Bureau, T.E., Kermicle, J., and Dellaporta, S.L. (1995) Transposon-mediated chromosomal rearrangements and gene duplications in the formation of the maize *R-r* complex. *EMBO J. 14*: 2350–2363.

Wang, B.B., and Brendel, V. (2006) Genomewide comparative analysis of alternative splicing in plants. *Proc. Natl. Acad. Sci. USA 103*: 7175–7180.

Wang, G.L., Ruan, D.L., Song, W.Y., Sideris, S., Chen, L., Pi, L.Y., Zhang, S., Zhang, Z., Fauquet, C., Gaut, B.S., et al. (1998) *Xa21D* encodes a receptor-like molecule with a leucine-rich repeat domain that determines race-specific recognition and is subject to adaptive evolution. *Plant Cell 10*: 765–779.

Wang, W., Zheng, H., Fan, C., Li, J., Shi, J., Cai, Z., Zhang, G., Liu, D., Zhang, J., Vang, S., et al. (2006) High rate of chimeric gene origination by retroposition in plant genomes. *Plant Cell 18*: 1791–1802.

Wawrzynski, A., Ashfield, T., Chen, N.W., Mammadov, J., Nguyen, A., Podicheti, R., Cannon, S.B., Thareau, V., Ameline-Torregrosa, C., Cannon, E., et al. (2008) Replication of nonautonomous retroelements in soybean appears to be both recent and common. *Plant Physiol. 148*: 1760–1771.

Weber, B., and Schmidt, T. (2009) Nested *Ty3-gypsy* retrotransposons of a single *Beta procumbens* centromere contain a putative chromodomain. *Chromosome Res. 17*: 379–396.

Weil, C.F., and Wessler, S.R. (1993) Molecular evidence that chromosome breakage by *Ds* elements is caused by aberrant transposition. *Plant Cell 5*: 515–522.

Wessler, S.R. (1988) Phenotypic diversity mediated by the maize transposable elements *Ac* and *Spm*. *Science 242*: 399–405.

Wessler, S.R., Baran, G., and Varagona, M. (1987) The maize transposable element *Ds* is spliced from RNA. *Science 237*: 916–918.

Wicker, T., Mayer, K.F., Gundlach, H., Martis, M., Steuernagel, B., Scholz, U., Simkova, H., Kubalakova, M., Choulet, F., Taudien, S., et al. (2011) Frequent gene movement and pseudogene evolution is common to the large and complex genomes of wheat, barley, and their relatives. *Plant Cell 23*: 1706–1718.

Wolfgruber, T.K., Sharma, A., Schneider, K.L., Albert, P.S., Koo, D.H., Shi, J., Gao, Z., Han, F., Lee, H., Xu, R., et al. (2009) Maize centromere structure and evolution: sequence analysis of centromeres 2 and 5 reveals dynamic loci shaped primarily by retrotransposons. *PLoS Genet. 5*: e1000743.

Xu, J.H., and Messing, J. (2006) Maize haplotype with a *Helitron*-amplified cytidine deaminase gene copy. *BMC Genet. 7*: 52.

Xu, L., Wang, L., Liu, T., Qian, W., Gao, Y., and An, C. (2007) *Triton*, a novel family of miniature inverted-repeat transposable elements (MITEs) in *Trichosanthes kirilowii* Maximowicz and its effect on gene regulation. *Biochem. Biophys. Res. Commun. 364*: 668–674.

Yan, X., Martinez-Ferez, I.M., Kavchok, S., and Dooner, H.K. (1999) Origination of *Ds* elements from *Ac* elements in maize: evidence for rare repair synthesis at the site of *Ac* excision. *Genetics 152*: 1733–1740.

Yang, G., Lee, Y.H., Jiang, Y., Shi, X., Kertbundit, S., and Hall, T.C. (2005) A two-edged role for the transposable element *Kiddo* in the rice *ubiquitin2* promoter. *Plant Cell 17*: 1559–1568.

Yang, L., and Bennetzen, J.L. (2009) Distribution, diversity, evolution, and survival of Helitrons in the maize genome. *Proc. Natl. Acad. Sci. USA 106*: 19922–19927.

Yu, C., Zhang, J., and Peterson, T. (2011) Genome rearrangements in maize induced by alternative transposition of reversed *Ac/Ds* termini. *Genetics 188*: 59–67.

Yu, J., Marshall, K., Yamaguchi, M., Haber, J.E., and Weil, C.F. (2004) Microhomology-dependent end joining and repair of transposon-induced DNA hairpins by host factors in *Saccharomyces cerevisiae*. *Mol. Cell Biol. 24*: 1351–1364.

Zabala, G., and Vodkin, L. (2007) Novel exon combinations generated by alternative splicing of gene fragments mobilized by a CACTA transposon in *Glycine max*. *BMC Plant Biol. 7*: 38.

Zhang, J., and Peterson, T. (2004) Transposition of reversed *Ac* element ends generates chromosome rearrangements in maize. *Genetics 167*: 1929–1937.

Zhang, J., Zhang, F., and Peterson, T. (2006) Transposition of reversed *Ac* element ends generates novel chimeric genes in maize. *PLoS Genet. 2*: e164.

Zhang, J., Yu, C., Pulletikurti, V., Lamb, J., Danilova, T., Weber, D.F., Birchler, J., and Peterson, T. (2009) Alternative *Ac/Ds* transposition induces major chromosomal rearrangements in maize. *Genes Dev. 23*: 755–765.

Zhang, P.G., Huang, S.Z., Pin, A.L., and Adams, K.L. (2010) Extensive divergence in alternative splicing patterns after gene and genome duplication during the evolutionary history of *Arabidopsis*. *Mol. Biol. Evol. 27*: 1686–1697.

Zhang, Y., Wu, Y., Liu, Y., and Han, B. (2005) Computational identification of 69 retroposons in *Arabidopsis*. *Plant Physiol. 138*: 935–948.

Zhong, C.X., Marshall, J.B., Topp, C., Mroczek, R., Kato, A., Nagaki, K., Birchler, J.A., Jiang, J., and Dawe, R.K. (2002) Centromeric retroelements and satellites interact with maize kinetochore protein CENH3. *Plant Cell 14*: 2825–2836.

Zhou, L., Mitra, R., Atkinson, P.W., Hickman, A.B., Dyda, F., and Craig, N.L. (2004) Transposition of *hAT* elements links transposable elements and V(D)J recombination. *Nature 432*: 995–1001.

Zhu, Z., Zhang, Y., and Long, M. (2009) Extensive structural renovation of retrogenes in the evolution of the *Populus* genome. *Plant Physiol. 151*: 1943–1951.

9 Transposons in Plant Speciation

Avraham A. Levy

Introduction

Barbara McClintock described mobilization of "previously silent genomic elements" in response to genomic stress (McClintock, 1984, p. 797). She first described reactivation of *Ac/Ds* transposable elements (TE) following chromosome breakage and hypothesized that various types of genomic shock could trigger similar, programmed responses. At the time, there was no evidence of TE activation as a result of interspecific crosses. Nevertheless, she anticipated that "species crosses are another potent source of genome modification" that could yield TE activation (McClintock, 1984, p. 799). Furthermore, she suggested a direct role of TEs in speciation: "It is difficult to resist concluding that some specific 'genome shock' was responsible for origins of new species." (McClintock, 1984, p. 800). Finally, she challenged the scientific community with the task of determining how cellular processes trigger genomic changes that in turn may affect species formation: "We know nothing however about how the cell senses danger and instigates response to it that often are truly remarkable" (McClintock, 1984, p. 801). Almost three decades later, it has become increasingly harder to resist concluding that TEs are central players in plant speciation. The case for the role of TEs in speciation processes is summarized in this chapter, in the context of the central questions in speciation biology.

Genetic Models of Speciation

Speciation is the evolutionary process of genetic divergence that leads to reproductive incompatibility between two subpopulations of a progenitor species. Reproductive isolation may be either prezygotic or postzygotic. In plants, two-thirds of isolation types are prezygotic and one-third is postzygotic (Lowry et al., 2008). Prezygotic isolation may result from mutations in flower shape, timing of flowering, pollen germination, pollinator attractants, and more. Postzygotic isolation includes embryo abortion, seedling death, hybrid sterility, and other forms of reduced hybrid fitness. A major challenge of evolutionary biology is to understand the genetic basis of reproductive isolation. So far, a leading model in the field is the Dobzhansky-Muller model of hybrid incompatibility (Dobzhansky, 1936). This model proposes that negative genetic interactions between two or more loci that have accumulated mutations in diverging lineages reduce hybrid fitness when brought together in hybrids, hence creating a genetic barrier between these lineages. A review on hybrid necrosis in plants describes 35 examples of 2-locus incompatibilities that reduce hybrid fitness (Bomblies and

---

Plant Transposons and Genome Dynamics in Evolution, First Edition. Edited by Nina V. Fedoroff.

Weigel, 2007). The molecular basis of incompatibility genes in plants has been figured out only for a very few loci, none of these including TEs-induced mutations (Presgraves, 2010). However, considering the large amount of TEs in plant genomes and the broad spectrum of mutations they promote, as detailed in various chapters of this volume, it is very likely that TEs are involved in some of the mutations responsible for generating Dobzhansky-Muller types of genetically incompatible loci, as well as for mutations leading to prezygotic barrier.

In addition to the genic Dobzhansky-Muller model of speciation, it is well established that karyotypic changes can contribute to speciation (Ayala and Coluzzi, 2005). Mutants with chromosome rearrangements, such as inversions, translocations, or fusions, can become reproductively isolated from the rest of the population. When such mutants hybridize with other individuals, this gives rise to hybrids that contain heterozygous chromosome rearrangements. Such hybrids are sterile because they produce dysfunctional gametes upon meiosis. TEs have played a very important role in shaping chromosome structure in plant genomes (Chapters 2, 3, and 10 in this volume). In this chapter, we discuss how TE-induced chromosomal rearrangements may initiate speciation in plants.

Another type of postzygotic genetic isolation that involves TEs as central players is *hybrid dysgenesis*. This phenomenon, described first in *Drosophila*, refers to the lethal unleashing of TEs in hybrid larvae (Malone and Hannon, 2009). Analogous phenomena of TE activation have been reported in plant interspecific hybrids, as detailed, albeit with a less dramatic effect than in *Drosophila*.

Speciation – a Gradual or a Rapid Process?

Another central question in evolutionary biology is whether speciation is a rapid or gradual process. Charles Darwin, in writing on the origin of species (Darwin, 1859), already proposed that speciation can occur through a gradual process of accumulation of mutations, typically in individuals that are geographically isolated and eventually become so genetically different that they become reproductively isolated (Figure 9.1). The role of transposons in generating the mutations that contribute to genetic divergence between genomes is well established and is thoroughly discussed in this volume. It will be discussed here in the context of speciation. An alternative mode of speciation is rapid or instant speciation. The most common mode of rapid speciation in plants is through hybridization and polyploidization, with 35% of vascular plants being recently formed polyploids (Wood et al., 2009). Studies in synthetic hybrids and polyploids, which are analogous to nascent species, have shown that transposons serve as sensitive detectors of genomic shock caused by divergent genome merging. Unleashing transposons in hybrids or polyploids may be analogous to hybrid dysgenesis, prompting reduced hybrid fitness and preventing gene flow between related but diverging species. In such cases, transposons switch roles from being the promoters of diversification and speciation to becoming the keepers of species integrity.

In this chapter we provide evidence for an important role of transposons in both gradual and instant modes of speciation. Finally, we propose a "cellular dysregulation" model that explains how speciation (or other genomic shocks) can activate normally silent transposons, and we discuss the consequences of this activation for the nascent species.

Speciation Through Accumulation of Mutations

When assessing the potential role of TEs in speciation, it is useful to distinguish between the effect of DNA "cut-and-paste" TEs, which induce DNA double-strand breaks (DSBs), and "copy-paste"

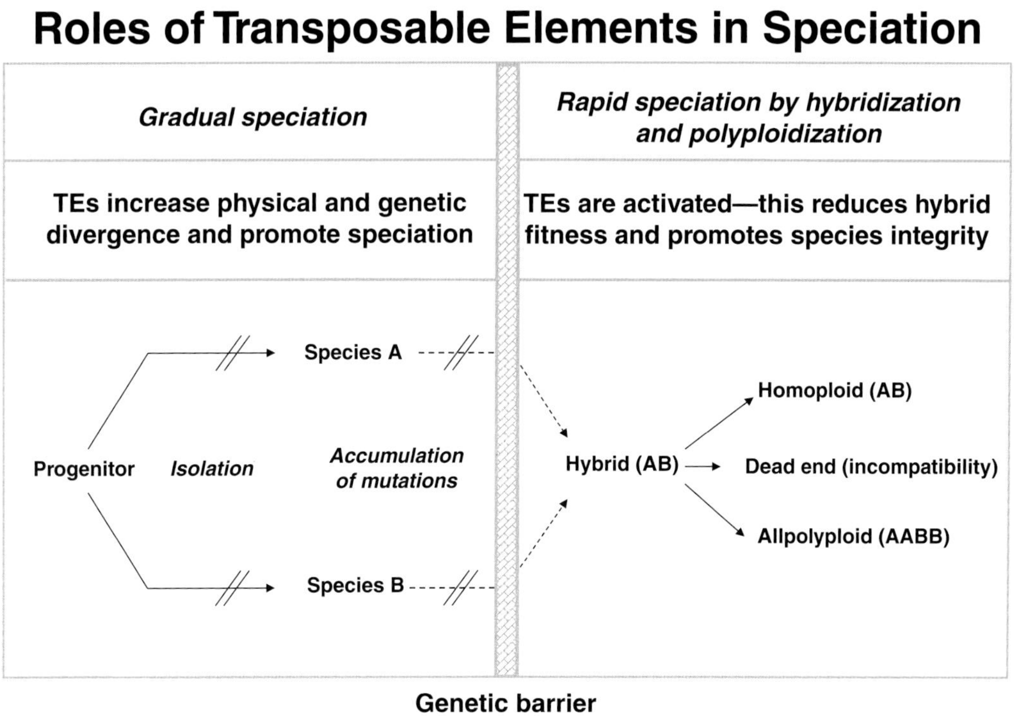

Figure 9.1 The role of transposable elements in speciation. Two modes of speciation are shown: gradual (left part) or rapid, via hybridization and polyploidization (right part). TEs contribute to the divergence between genomes (genetic and physical), generating a reproductive pre- and post-zygotic genetic barrier, and as a result they promote speciation. Occasionally, hybridization between related species can occur. In most cases, this produces a dead-end event due to embryo abortion or reduced fitness of the inter-specific hybrid or hybrid sterility. The role of TEs in promoting chromosomal divergence is well established (Chapters 2, 3 and 10), affecting homeologs pairing and leading to hybrid sterility. The role of TEs in mediating other types of post-zygotic hybrid incompatibility was not yet demonstrated in plants but recent studies suggest that it is a likely possibility ((Kenan-Eichler et al., 2011). Occasionally, interspecific hybridization can give rise to a new hybrid or allopolyploid species which often display heterotic features. Different forms of TEs activation (demethylation, transcription or transposition) are often observed in such nascent species, however, TE silencing can be reestablished after a few generations. In summary, TEs play a dual role in the speciation process: they promote genomic diversification and speciation and they can reduce fitness of inter-specific hybrids, thus maintaining species integrity and hindering speciation by hybridization.

TEs. These elements catalyze different types of mutations, which have different potential impacts on genome structure, gene expression, and speciation.

DNA Cut-and-Paste TEs and Speciation

DNA cut-and-paste elements produce mutations that can have subtle effects, such as point mutations and small indels, or they can have a major effect on chromosome structure, as in the case of inversions and translocations. The early work of McClintock on the role of TEs in inducing a cycle of chromosome breakage and fusions (McClintock, 1946, 1948) is a typical example of how TEs can cause a rapid karyotypic evolution that can lead to genetic isolation and eventually to speciation. McClintock's chromosome-breaking *doubleDs* element turned out to have a peculiar structure, consisting of a *Ds* element nested within another *Ds* element (Ralston et al., 1989; Weil and Wessler, 1993). While this is not a typical element, such *doubleDs* structures are not uncommon, due to the tendency of *Ds* elements to transpose to nearby sites (Greenblatt, 1984; Dooner and Belachew, 1989). TEs usually excise without inducing chromosome rearrangements, creating a DNA DSB that is typically repaired by nonhomologous end joining, leaving behind short footprints (Rinehart et al., 1997). These footprints can contribute to the slow divergence of chromosomes. Such subtle changes accumulating over time can affect the chromosome recombination landscape, which can eventually lead to reproductive isolation. Indeed, DNA mutations even of single nucleotides can significantly reduce the rate of meiotic homologous recombination in a polymorphic region (Emmanuel et al., 2006).

TEs-induced DSBs can also be repaired via homologous recombination (Gorbunova and Levy, 1999), which in turn can mediate a variety of genetic outcomes depending on the nature of the recombination partners (Haber, 2000). For example, intrachromosomal recombination can cause deletions when it occurs between direct repeats and inversions when it occurs between inverted repeats. The work of Peterson on the *P* locus in maize constitutes a thorough documentation of *Ac*-induced intrachromosomal recombination (Chapter 3 in this volume). Recombination between repeats located on different chromosomes can cause translocations, a genetic event that strongly affects gamete viability in heterozygous hybrids. Transgenic plant systems have provided evidence that DNA DSBs induced by the excision of an *Ac* element can induce ectopic recombination (Shalev and Levy, 1997), laying the foundation for translocations. In summary, DNA-breaking TEs promote a broad range of DNA alterations ranging from subtle changes to gross chromosomal rearrangements via both homologous and nonhomologous recombination.

Copy-and-Paste TEs and Speciation

Copy-and-paste TEs that do not induce DSBs upon excision, such as retroelements, can also have far-reaching effects on genome structure, expression, and speciation. Unlike DSB-inducing elements, which can readily catalyze inversions, large deletions, and translocations, retroelements affect genome structure primarily through the accumulation of insertions. Chromosome structure evolution is particularly rapid in centromeres, where retroelements are especially abundant (Arabidopsis-Genome-Initiative, 2000). Recent work in interspecific barley hybrids (Sanei et al., 2011) suggests that coevolution of a centromeric DNA structure together with the centromeric histone variant CEN-H3 can induce the selective loss of chromosomes in embryos of hybrids, leading to the production of haploid plants. It was proposed that centromeric TEs can, in this manner, serve as postzygotic barriers between related species (Chan, 2011).

TE proliferation in both centromeric and noncentromeric regions was shown to lead to genome size doubling during speciation, as seen in *Oryza australiensis* (Piegu et al., 2006). In *Gossypium*, TE proliferation differs among related species, indicating transposition during or after speciation, and contributes to the physical divergence of genomes (Hawkins et al., 2006). At the same time, gene order is maintained, despite extensive physical divergence of chromosomes during speciation (see details in review by Bennetzen, Chapter 10 in this volume). Retroelement proliferation was shown to occur within a species as deduced from intervarietal comparisons of maize (Wang and Dooner, 2006) and of rice (Huang et al., 2008). This proliferation can occur gradually or in bursts, as shown in rice (Naito et al., 2009). Remarkably, intergenic retroelements proliferation in maize generates extensive polymorphic regions where virtually no meiotic recombination occurs, and as a result meiotic recombination is restricted mostly to genic regions (Fu et al., 2002). This suggests that extensive proliferation of TEs within a species and changes in genome size can strongly reduce or even abolish pairing of homologs and their ability to recombine. As a result, a genetically isolated subgroup can be formed, which can become the founder of a new species. In addition, the long terminal repeats (LTRs) of retroelements can drive read-out transcription and affect expression of neighboring genes (Chapters 6 and 8 in this volume). Considering their abundance in plant genomes, this can contribute to gradual genetic divergence and to the Dobzhansky-Muller type of incompatibility.

In summary, there is no direct evidence for TE-induced speciation, but there is plenty of circumstantial evidence that TEs have all the features necessary to promote speciation by chromosomal rearrangements and by alterations in gene expression. Indeed, new patterns of gene expression, mediated by either genetic or epigenetic changes, are likely to contribute to the Dobzhansky-Muller type of incompatibility.

TE-Mediated Speciation – a Likely Scenario?

The experimental evidence for rapid TE-mediated karyotypic restructuring, as well as "archeological" evidence discussed, showing a correlation between TEs proliferation, changes in genome structure and function, and speciation all suggest the likelihood that TEs mediate speciation. However, the evidence available from natural wild populations is quite limited. Recently, the analysis of a marginal population of diploid wheat (*Aegilops speltoides*) has provided some new insight into this question (Belyayev and Raskina, 2010). The marginal population showed extensive variability among individual plants with regard to the expansion or contraction of specific retroelements families. The work of Belyayev et al. (2010) shows that major transposition fluctuations can happen within an isolated population, generating subpopulations of individuals that differ in their genome structure. While this provides in situ support for the role of TEs in speciation, an important question that remains unresolved is whether and how frequently such TEs bursts or TE-mediated chromosome rearrangements can actually lead to speciation. For TE-induced speciation to occur, genetic drift must favor the formation of a subpopulation that differs in its TE profile sufficiently enough that genome restructuring constitutes a reproductive barrier between the subpopulation and its progenitor species. A recent theoretical article that addressed the possibility for such genetic drift lends further plausibility to the hypothesis of TE-driven speciation (Jurka et al., 2011). Future work in the natural habitats will extend our understanding of the TE-mediated speciation processes. Nevertheless, it will remain extremely difficult to catch a TE-mediated speciation event "in the act."

Plant Speciation Through Hybridization and Allopolyploidization

By contrast to gradual speciation through the accumulation of mutations, which can take hundred of thousands or millions of years, speciation can occur "overnight" through hybridization and polyploidization (Leitch and Bennett, 1997; Doyle et al., 2008; Wood et al., 2009). Interspecific hybridization and whole-genome doubling have occurred in the lineage of most, if not all, plants, either long ago (paleopolyploids) or recently (Soltis et al., 2009). Species that are not isolated by prezygotic barriers can readily hybridize, and assuming no lethality of the embryo or of the hybrid plant, interspecific hybrid sterility can be overcome by spontaneous genome doubling through the merging of unreduced gametes. Indeed, plants produce a significant proportion of unreduced gametes (Bretagnolle and Thompson, 1995), which can be as high as 50% in some interspecific wheat hybrids

(Kihara and Lilienfeld, 1949). Despite this high potential for hybrid species formation, only a few hybrid (homoploid) species have been established (Rieseberg and Willis, 2007), and the actual fraction of speciation events directly attributable to polyploidization, remains uncertain (Soltis and Soltis, 2009; Van de Peer et al., 2009). We now discuss the possibility that TEs contribute to reduced hybrid fitness and to postzygotic barriers.

Induction of Transposition upon Hybridization and Polyploidization

Transpositional activation of TEs has been reported following hybridization and polyploidization. A striking example of retrotransposon proliferation is provided by *Helianthus anomalus*, *H. deserticola*, and *H. paradoxus*, which are products of ancient hybridizations between *H. annuus* and *H. petiolaris* sunflower species. The nuclear genomes of these hybrids are at least 50% larger than those of either parental species, despite the fact that both hybrid and parental species are diploids and have the same number of chromosomes. The increased genome size is attributable to the greater abundance of the *Ty3/gypsy*-like LTR retrotransposon in the hybrids than in the parents (Ungerer et al., 2006) and to a lesser extent with the abundance of *Ty1*-like elements (Kawakami et al., 2010). In the foregoing examples, it is difficult to determine whether transposon proliferation occurred gradually following speciation or whether it was triggered by hybridization and polyploidization. Young allopolyploid species, such as wheat, *Spartina*, *Senecio*, and *Tragopogo* (Hegarty and Hiscock, 2009), as well as synthetic polyploids that can be generated in the laboratory, provide conclusive evidence that speciation can occur by hybridization and polyploidization. These nascent or young species exhibit a diversity of transpositional activation responses, including de novo transposition of a synthetic Sunfish element in allotetraploids of *Arabidopsis* (Madlung et al., 2005), the tobacco *Tnt1* retroelement (Petit et al., 2010), and the *Veju* retroelement in synthetic hexaploid wheat (Kraitshtein et al., 2010). However, studies of other species, such as *Spartina* (Parisod et al., 2009), and other TEs, including the wheats *Balduin*, *Apollo*, and *Thalos* (Yaakov and Kashkush, 2011aa) and the wheat *Wis2-1A* retroelement (Kashkush and Yaakov, 2011b), failed to detect transposition bursts following allopolyploidization, despite changes in TE methylation.

TE activation can be lethal, as seen in the case in hybrid dysgenesis in *Drosophila* (Malone and Hannon, 2009). However, massive and lethal transposition as a result of TEs activation in hybrids and polyploids is not well documented in plants. It may be that plants are more robust and resistant to the mutagenic effect of transposition. Alternatively, lethal transposition may occur but escape detection because only seeds or plants where transposition was moderate survive to be studied. By contrast to *Drosophila* larvae, it is difficult to determine the fate of a young embryo embedded within the maternal tissues of a plant.

Epigenetic Alteration of TEs upon Hybridization and Polyploidization

Alterations in TE methylation are a more prevalent response to allopolyploidization than transposition (Parisod et al., 2010; Yaakov and Kashkush, 2011bb). Changes in the methylation status of TEs can be much more extensive than those observed in non-TE genomic sequences. Although some elements are hypermethylated (Yaakov and Kashkush, 2011aa), hypomethylation is more common (Parisod et al., 2010; Yaakov and Kashkush, 2011bb). Hypomethylation is a hallmark of TE activation and seems to be a prerequisite for transposition, even though it is not always sufficient for transposition. Hypomethylation of TEs was first shown in maize to correlate positively with transpositional activity, while inactive elements are hypermethylated (Chandler and Walbot, 1986; Chomet et al., 1987; Banks et al., 1988). Recent work in *Arabidopsis* methylation-deficient mutants *ddm1* (Tsukahara et al., 2009) and *met1* (Mirouze et al., 2009) showed de novo transposition of retroelements.

Hypermethylation of autonomous elements can affect the TE's promoter and its ability to transcribe the transposase gene (Kunze et al., 1988). Hypermethylation can also inhibit binding of transposase to the ends of both autonomous and nonautonomous elements, thereby hindering its ability to catalyze transposition (Kunze and Starlinger, 1989). The methylation status of a TE can, in turn, affect the activity of neighboring genes, as was shown in maize with *Spm* insertions in the *a* locus (Masson et al., 1987) and *Mu* insertion in *hcf106* (Martienssen et al., 1990). In summary, epigenetic changes in TEs triggered by allopolyploidy can affect both genome structure and gene expression. These modifications seem to be a necessary, but not sufficient, to activate transposition.

Transcriptional Activation of TEs upon Hybridization and Polyploidization

Transcriptional activation of the wheat *WIS2-1A* element was reported in the first generation after allopolyploidization (Kashkush et al., 2002). Subsequent work showed that the *WIS2-1A* LTR can drive transcription in a bidirectional manner, affecting the activity of both the element and of nearby genes (Kashkush et al., 2003). Thus, LTR-flanking genes can be either activated or inactivated, depending on their orientation. Interestingly, in the study, *Wis2-1A* transcription was strongly induced and reverse transcription probably occurred, as deduced from the appearance of extrachromosomal cDNA linear putative intermediates of transposition in the synthetic allopolyploid (Kashkush et al., 2003). Nevertheless, transposition of *Wis2-1A* was not detected. Thus, transcriptional activation may be necessary, but not sufficient, for transposition to occur. The necessity of several steps for successful activation of transposition can protect the newly formed species from a lethal transposition burst. Transcriptional activation was also shown for several TEs in newly formed *Arabidopsis* polyploids (Madlung et al., 2005). The *Athila* retroelement was transcriptionally activated in a synthetic *Arabidopsis* interspecific hybrid resulting from crosses between *A. thaliana* and *A. arenosa* (Josefsson et al., 2006). *Athila* activation depended on the dosage and origin of the parental genome. This activation correlated with altered expression of genes affecting seed viability, a trait that may constitute a postzygotic barrier. Considering the large number of retroelements, this mode of activation in hybrids and allopolyploids can cause a genome-wide deregulation of normal gene expression in the newly formed species, the impact of which is likely to be deleterious to the plant (Kashkush and Khasdan, 2007).

Alterations in Small RNAs upon Hybridization and Polyploidization

Small RNAs have been implicated in both gene silencing and RNA-dependent DNA methylation (Matzke and Birchler, 2005). Moreover, small interfering RNAs (siRNAs) play a critical role in heterochromatin maintenance and in transposable element silencing (Zaratiegui et al., 2007). Therefore, small RNAs can function as guardians of the genome, preventing TE demethylation and related changes in transcriptional and transpositional activity (Malone and Hannon, 2009). Recently, it has been shown that small RNAs are involved in TE regulation during pollen development (Slotkin et al., 2009), in female gametophytes (Mosher et al., 2009), and in the endosperm and embryo

(Gehring et al., 2009; Hsieh et al., 2009). Disruption of this regulation in reproductive tissues in hybrids and polyploids can threaten the viability of the newly formed species. Note that embryo abortion, a classical postzygotic reproductive barrier, is very common in interspecific crosses and can be bypassed by embryo rescue, namely by feeding the embryo in culture (Cisneros and Tel-Zur, 2010). This suggests that the embryo is viable, but that the endosperm is not functioning to nourish the embryo. It would be interesting to test whether endosperm dysfunction, possibly as a result of small RNAs alterations, is mediated by TEs in interspecific hybrids. Chapter 7 in this volume suggests that this might be the case.

There are two main classes of small RNAs, microRNAs (miRNA) and small interfering RNAs (siRNAs). While miRNAs are usually targeted to genes, their deregulation might also indirectly affect TEs. In a synthetic *Arabidopsis* allotetraploid, several microRNA species deviated in their expression from the expected midparent values (Ha et al., 2009). In wheat, miR168 was overexpressed in the allohexaploid when compared to the tetraploid and diploid parents (Kenan-Eichler et al., 2011). This may account for the suppression of *ARGONAUTE1* (*AGO1*) (Mallory and Vaucheret, 2009), which could, in turn, reduce the production of siRNAs (Vaucheret, 2008). Such influences on *AGO1*-related regulatory loops may have genome-wide effects on the production of TE-targeting siRNAs. A recent study compared siRNAs matching TEs in *A. thaliana* and in the related *A. lyrata* species, where TE copy numbers are three-fold higher than in *A. thaliana* (Hollister et al., 2011). *A. lyrata* TEs were targeted by a smaller fraction of matching siRNAs, which may explain the difference in TE abundance between the two species. Although siRNAs-matching TEs have not yet been analyzed in the *A. thaliana* x *A. lyrata* hybrid, deregulation of their expression are expected to have a genome-wide impact on TE activity, genome size, and genome stability, as well as gene expression. In wheat, the number of siRNAs corresponding to TEs was significantly lower in the allohexaploid than in its diploid and tetraploid parents (Kenan-Eichler et al., 2011). Moreover, a decrease in the abundance of siRNAs matching the *Veju* retroelement in the nascent allohexaploid was correlated with element hypomethylation (Kenan-Eichler et al., 2011).

A Mechanistic Model for Responses to Genome Shock

Recent comparisons of the transcriptome of plant interspecific hybrids and allopolyploids to that of the parental species shows novel patterns of gene expression, with values of 7%–20% of the genes being expressed beyond the range of their parents (e.g., Wang et al. 2006; Chague et al. (2010); Akhunova et al. 2010). We overview the various molecular mechanisms that can cause rewiring of gene and protein functions in a hybrid. Based on these mechanisms, we propose a general model explaining how TEs controlling genes can be affected by hybridization and polyploidization and how this can activate TEs.

As discussed, speciation via hybridization and polyploidization triggers TE activation leading eventually to transposition. This has been found in many different species and results in the activation of many different TE families. Alterations in TE methylation have been observed in almost all species studied. Therefore, some critical question are: how does the merging of two parental species, each containing silent TEs, almost always results in some form of TE activation? Why is this response to genome merging observed in different genera? We argue that a general mechanism triggering dysregulation of normal cell functions (transcription, protein structure and function, and metabolism) as a result of hybridization and polyploidization, generates different forms of perturbation, all of which result in TE activation (see Figure 9.2).

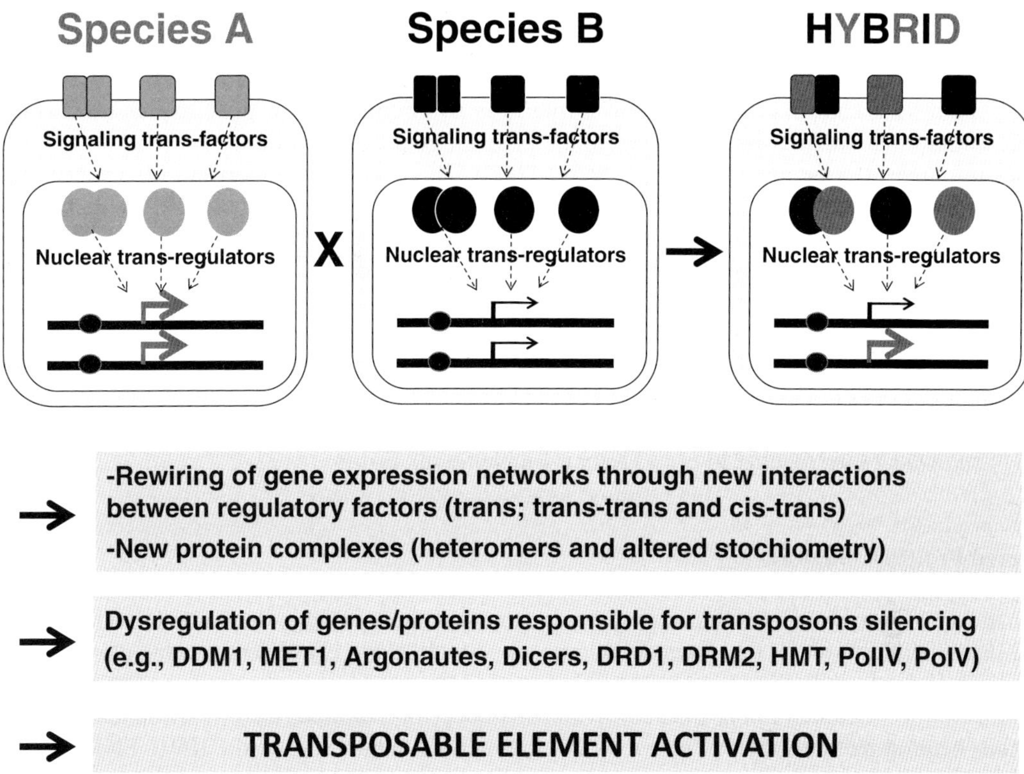

Figure 9.2 Dysregulation of cellular functions in hybrids and polyploids – A model for TE activation Upon the genomic merger of two species, novel interactions occur between cis and trans regulatory factors of the two species. This includes trans factors, such as proteins involved in signal transduction, transcription factors, chromatin remodelers, small RNAs, and cis factors such as promoters (shown here as horizontal arrows on the chromosome) or UTRs. Moreover, protein subunits from the two parental genomes may interact, forming heteromeric complexes. These new protein complexes together with the rewiring of gene expression may alter the genes/proteins involved in TE silencing. This dysregulation may lead TE activation. Considering the large number of genes involved in the various aspects of TE silencing (methylation, chromatin modification and small RNAs-related functions) there is a high probability to dysregulate one or more TE-silencing gene.

Dysregulation of Gene Expression by Novel Interactions Between Regulatory Factors

Upon divergence from a common progenitor, species independently accumulate unique sets of mutations. Some mutations affect coding sequences and therefore protein structure and properties. Other mutations affect regulatory DNA elements that act on genes in cis, such as a promoter or a UTR. Mutations occurring in trans-acting factors, such as small RNAs, transcription factors, chromatin remodelers, or signal transduction genes, can have strong pleiotropic effects (Figure 9.2). Mutations reducing species fitness will be eliminated by natural selection, unless their deleterious effect is abolished by a secondary, compensating mutation (Tirosh et al. 2009; Landry et al. 2007). Therefore, different species contain a different set of mutations, some that reduce fitness and others that increase fitness, some in housekeeping genes and others in regulatory factors. A fascinating question is what happens upon the reunion of two sets of regulatory factors in a hybrid or an allopolyploid? Recent work by Tirosh et al. (2009) addressed this question through global allele-specific expression analysis in a yeast interspecific hybrid. It was shown that only mutations affecting cis regulatory factors had identical influences on both parental and hybrid gene expression. In contrast, mutations in trans-acting factors promote genome-wide rewiring of genetic networks. Merging different trans-acting factors in hybrids provides new cis-trans interactions in both parental genomes (Figure 9.2).

An interesting category of changes in a hybrid is overdominance in gene expression, namely overexpression or suppression compared to the parental genes. It was found to affect ~10% of the genes in a hybrid yeast genome (Tirosh et al., 2009). This can occur as a result of compensatory mutations in the parental species (Tirosh et al., 2009). For example, a strong transcription factor of species A could activate the strong promoter of species B, resulting in overexpression of the gene (Figure 9.2). Thus, new cis-trans interactions in hybrid genomes can lead to large-scale dysregulation of normal gene expression (Tirosh et al., 2009). Such newly formed genetic interactions can lead to a "genomic clash," namely intergenomic interactions that are deleterious to the organism and can lead to hybrid incompatibility, as observed in *Drosophila* (Landry et al., 2007). Similarly, dysregulation of pathogenesis genes is a common feature in plant hybrids and is known as *hybrid necrosis* (Bomblies and Weigel, 2007). Dysregulation of microRNAs was recently detected in *Arabidopsis* (Ng et al., 2011) and wheat interspecific hybrids (Kenan-Eichler et al., 2011).

Altered Protein Complexes

Genetic dysregulation in hybrids and polyploids can also impact proteins (Veitia et al., 2008; Birchler and Veitia, 2010). Proteins that form complexes are more likely to cause deregulation of normal cell function. Hybrids and allopolyploids can form protein heterodimers whose properties may differ from those of the homodimers formed by their parents: for example, there is evidence that some protein heterodimers are not as active as homodimers (Prudhomme et al., 1991; Phillips et al., 1995). In addition, modified protein synthesis or degradation in hybrids or polyploids can induce a stoichiometric imbalance in protein complexes that can lead to the formation of nonproductive subcomplexes (Birchler and Veitia, 2010). Changes in dosage, such as absence or excess of one single protein in a large complex, as could happen through differential gene-copy number variation between the parental species, can also disrupt normal activity of the whole complex and have broad pleiotropic effects on the cell (Veitia et al., 2008; Birchler and Veitia, 2010).

Why TEs Become Activated when Cellular Processes are Dysregulated

The probability that TE activity can be significantly modified in the hybrid or allopolyploid is quite high because tens of genes are involved in regulatory pathways related to silencing TEs (Matzke et al., 2009; MacLean et al., 2010; Haag and Pikaard, 2011). This includes genes involved in cytosine methylation (e.g., *DDM1*, *MET1*) and genes involved in small RNAs synthesis and processing, as well as in RNA-mediated methylation (e.g., DNA-dependent RNA polymerases *PolIV* or *PolV*; *Argonautes* or *Dicer*s; *DRD1*). We reported examples of TE-related siRNAs that were expressed in the parents and completely disappeared in the hybrid (Kenan-Eichler et al., 2011). Moreover, most of the proteins encoded by these genes work in complexes (Haag and Pikaard, 2011). We therefore propose that the dysregulation of cellular regulatory processes explains the genomic shock observed upon hybridization and provides a general framework for understanding the resulting activation of TEs. The repressive machinery that maintains TEs in an inactive, methylated state is complex and can be disrupted at many points. In particular, disruption of the TE-methylation-maintenance mechanisms can occur at various levels, from the formation of multiprotein complexes to the regulation of transcription and abundance of siRNAs. Moreover, the ability of nonautonomous TEs to be mobilized in trans by the proteins encoded by active elements implies that there is considerable potential for amplification of the TE activation response.

There might be some deeper reasons why TEs are so sensitive to dysregulation, probably more so than other genes. We argue that rapidly evolving genes such as TEs are more prone to dysregulation upon hybridization than other genes. Indeed, let's consider the example of a highly conserved gene. By definition, it is quite similar in the diverging species and it has been that way for millions of years, therefore we do not expect a major dysfunction in the interspecific hybrid. By contrast, TEs evolve quickly, and we assume that so does their silencing machinery. Therefore, when two species hybridize, each containing a different set of mutators and silencers, the precarious equilibrium that was reached between mutators and silencers within each parental species becomes disrupted in the hybrid by the mechanisms described above.

While there is no experimental evidence for the hypothesis that rapidly evolving genes are prone to dysregulation in hybrids, there is some support through the analogy with disease resistance genes. These genes are known to evolve quickly as a result of host-pathogen interactions and they are also prone to dysregulation in hybrids—a phenomenon known as *hybrid necrosis* (Bomblies and Weigel, 2007). In addition, a recent study of epigenetic changes in an intraspecific *Arabidopsis* hybrid shows a reduction in the amount of TEs-related small RNAs in the hybrid (Groszmann et al., 2011), similar to what was reported in interspecific wheat hybrids (Kenan-Eichler et al., 2011). This suggests that *Mutators*-silencers coevolution is so rapid that hybrid dysregulation already occurs at the species level.

Conclusions

TEs contribute to speciation in many ways. The mutations they catalyze can stimulate gradual divergence between genomes at the structural level and at the gene expression level, two different effects, both of which promote speciation. On the other hand, they can contribute to rapid genetic isolation through the induction of major chromosome rearrangements, such as translocations and inversions. Rapid speciation can also occur through hybridization and polyploidization. Regardless of TE and species, TE activation is a common response to wide crosses. We have proposed a model that explains TE activation in response to the genomic shock triggered by hybridization and polyploidization. The observed general dysregulation of 10%–20% of genes must often affect genes involved in maintaining TEs in an inactive, highly methylated state. Global dysregulation is clearly deleterious to the nascent hybrid and therefore serves as a genetic barrier between different species, thereby contributing to maintenance of species integrity. The proposal that TEs are activated through dysregulation of TE silencing mechanisms in hybrids and polyploids (Figure 9.2) may explain why TE activation is also induced by various biotic and abiotic stresses that perturb normal gene expression and protein activities. Interestingly, despite the potentially deleterious effect of genome-wide TE activation, there is evidence that the initial shock of speciation can be overcome and that activated TEs return to their silent state within one or two generations (Kenan-Eichler et al., 2011). Moreover, it is remarkable that new hybrids and polyploids often also display heterosis, raising the possibility that TEs are involved in the heterotic response. Moreover, it is possible that the genetic and epigenetic variation induced by TEs as a result of the genomic shock of speciation can open new evolutionary possibilities for a nascent species by producing new mutations and new patterns of gene expression, some of which increase fitness. In this respect, TEs role in speciation may suit Nietzshe's quote: "What does not kill you makes you stronger."

Acknowledgments

I would like to thank Nina Fedoroff for being an intellectually challenging and stimulating editor who contributed much to the evolution of this chapter. This work was supported by a grant from the Israeli Science Foundation. AAL holds the Gilbert de Botton chair of Plant Sciences.

References

Akhunova, A.R., Matniyazov, R.T., Liang, H., and Akhunov, E.D. (2010) Homoeolog-specific transcriptional bias in allopolyploid wheat. *BMC Genomics 11*: 505.

Arabidopsis Genome Initiative. (2000) Analysis of the genome sequence of the flowering plant *Arabidopsis thaliana*. *Nature 408*: 796–815.

Ayala, F.J., and Coluzzi, M. (2005) Chromosome speciation: Humans, *Drosophila*, and mosquitoes. *Proc. Natl. Acad. Sci. USA 102* Suppl 1, 6535–6542.

Banks, J.A., Masson, P., and Fedoroff, N. (1988) Molecular mechanisms in the developmental regulation of the maize *Suppressor-mutator* transposable element. *Genes Dev. 2*: 1364–1380.

Belyayev, A.A., and Raskina, O.M. (2010) Dynamics of highly repetitive DNA fraction as indicator of speciation in species of the family Poaceae. *Russ. J. Genet. 46*: 1122–1124.

Birchler, J.A., and Veitia, R.A. (2010) The gene balance hypothesis: Implications for gene regulation, quantitative traits and evolution. *New Phytol. 186*: 54–62.

Bomblies, K., and Weigel, D. (2007) Hybrid necrosis: Autoimmunity as a potential gene-flow barrier in plant species. *Nature Rev. Genet. 8*: 382–393.

Bretagnolle, F., and Thompson, J.D. (1995) Gametes with the somatic chromosome-number - mechanisms of their formation and role in the evolution of autopolyploid plants. *New Phytol. 129*: 1–22.

Chague, V., Just, J., et al. (2010) Genome-wide gene expression changes in genetically stable synthetic and natural wheat allohexaploids. *New Phytol. 187*: 1181–1194.

Chan, S.W. (2011) In a battle between parental chromosomes, a failure to reload. *Proc. Natl. Acad. Sci. USA 108*: 13361–13362.

Chandler, V.L., and Walbot, V. (1986) DNA modification of a maize transposable element correlates with loss of activity. *Proc. Natl. Acad. Sci. USA 83*: 1767–1771.

Chomet, P.S., Wessler, S., and Dellaporta, S.L. (1987) Inactivation of the maize transposable element *Activator (Ac)* is associated with its DNA modification. *EMBO J. 6*: 295–302.

Cisneros, A., and Tel-Zur, N. (2010) Embryo rescue and plant regeneration following interspecific crosses in the genus *Hylocereus* (Cactaceae). *Euphytica 174*: 73–82.

Darwin, C. (1859) *On the origin of species by means of natural selection* (London: John Murray).

Dobzhansky, T. (1936) Studies on hybrid sterility. II. Localization of sterility factors in *Drosophila* pseudoobscura hybrids. *Genetics 21*: 113–135.

Dooner, H.K., and Belachew, A. (1989) Transposition pattern of the maize element *Ac* from the *bz-m2(Ac)* allele. *Genetics 122*: 447–458.

Doyle, J.J., Flagel, L.E., Paterson, A.H., Rapp, R.A., Soltis, D.E., Soltis, P.S., and Wendel, J.F. (2008) Evolutionary genetics of genome merger and doubling in plants. *Annu. Rev. Genet. 42*: 443–461.

Emmanuel, E., Yehuda, E., Melamed-Bessudo, C., Avivi-Ragolsky, N., and Levy, A.A. (2006) The role of *AtMSH2* in homologous recombination in *Arabidopsis thaliana*. *EMBO Rep. 7*: 100–105.

Fu, H., Zheng, Z., and Dooner, H.K. (2002) Recombination rates between adjacent genic and retrotransposon regions in maize vary by 2 orders of magnitude. *Proc. Natl. Acad. Sci. USA 99*: 1082–1087.

Gehring, M., Bubb, K.L., and Henikoff, S. (2009) Extensive demethylation of repetitive elements during seed development underlies gene imprinting. *Science 324*: 1447–1451.

Gorbunova, V., and Levy, A.A. (1999) How plants make ends meet: DNA double-strand break repair. *Trends Plant Sci. 4*: 263–269.

Greenblatt, I.M. (1984) A chromosome replication pattern deduced from pericarp phenotypes resulting from movements of the transposable element, *Modulator*, in maize. *Genetics 108*: 471–485.

Groszmann, M., Greaves, I.K., Albertyn, Z.I., Scofield, G.N., Peacock, W.J., and Dennis, E.S. (2011) Changes in 24-nt siRNA levels in *Arabidopsis* hybrids suggest an epigenetic contribution to hybrid vigor. *Proc. Natl. Acad. Sci. USA 108*: 2617–2622.

Ha, M., Kim, E.D., and Chen, Z.J. (2009) Duplicate genes increase expression diversity in closely related species and allopolyploids. *Proc. Natl. Acad. Sci. USA 106*: 2295–2300.

Haag, J.R., and Pikaard, C.S. (2011) Multisubunit RNA polymerases IV and V: purveyors of non-coding RNA for plant gene silencing. *Nat. Rev. Mol. Cell Biol. 12*: 483–492.

Haber, J.E. (2000) Partners and pathways repairing a double-strand break. *Trends Genet. 16*: 259–264.

Hawkins, J.S., Kim, H., Nason, J.D., Wing, R.A., and Wendel, J.F. (2006) Differential lineage-specific amplification of transposable elements is responsible for genome size variation in *Gossypium*. *Genome Res. 16*: 1252–1261.

Hegarty, M.J., and Hiscock, S.J. (2009) The complex nature of allopolyploid plant genomes. *Heredity 103*: 100–101.

Hollister, J.D., Smith, L.M., Guo, Y.L., Ott, F., Weigel, D., and Gaut, B.S. (2011) Transposable elements and small RNAs contribute to gene expression divergence between *Arabidopsis thaliana* and *Arabidopsis lyrata*. *Proc. Natl. Acad. Sci. USA 108*: 2322–2327.

Hsieh, T.F., Ibarra, C.A., Silva, P., Zemach, A., Eshed-Williams, L., Fischer, R.L., and Zilberman, D. (2009) Genome-wide demethylation of *Arabidopsis* endosperm. *Science 324*: 1451–1454.

Huang, X., Lu, G., Zhao, Q., Liu, X., and Han, B. (2008) Genome-wide analysis of transposon insertion polymorphisms reveals intraspecific variation in cultivated rice. *Plant Physiol. 148*: 25–40.

Josefsson, C., Dilkes, B., and Comai, L. (2006) Parent-dependent loss of gene silencing during interspecies hybridization. *Curr. Biol. 16*: 1322–1328.

Jurka, J., Bao, W.D., and Kojima, K.K. (2011) Families of transposable elements, population structure and the origin of species. *Biol. Direct. 6*: 44.

Kashkush, K., and Khasdan, V. (2007) Large-scale survey of cytosine methylation of retrotransposons and the impact of readout transcription from long terminal repeats on expression of adjacent rice genes. *Genetics 177*: 1975–1985.

Kashkush, K., Feldman, M., and Levy, A.A. (2002) Gene loss, silencing and activation in a newly synthesized wheat allotetraploid. *Genetics 160*: 1651–1659.

Kashkush, K., Feldman, M., and Levy, A.A. (2003) Transcriptional activation of retrotransposons alters the expression of adjacent genes in wheat. *Nature Genet. 33*: 102–106.

Kashkush, K., and Yaakov, B. (2011) Massive alterations of the methylation patterns around DNA transposons in the first four generations of a newly formed wheat allohexaploid. *Genome 54*: 42–49.

Kawakami, T., Strakosh, S.C., Zhen, Y., and Ungerer, M.C. (2010) Different scales of Ty1/copia-like retrotransposon proliferation in the genomes of three diploid hybrid sunflower species. *Heredity 104*: 341–350.

Kenan-Eichler, M., Leshkowitz, D., Tal, L., Noor, E., Melamed-Bessudo, C., Feldman, M., and Levy, A.A. (2011) Wheat hybridization and polyploidization results in deregulation of small RNAs. *Genetics 188*: 263–272.

Kihara, H., and Lilienfeld, F. (1949) A new synthesized 6x-wheat. In *Proc. Eighth Internat. Congress Genet.*, G.B.a.R. Larsson, ed. (Stockholm, Sweden: Hereditas [Suppl.]), pp. 307–319.

Kraitshtein, Z., Yaakov, B., Khasdan, V., and Kashkush, K. (2010) Genetic and epigenetic dynamics of a retrotransposon after allopolyploidization of wheat. *Genetics 186*: 801–812.

Kunze, R., and Starlinger, P. (1989) The putative transposase of transposable element *Ac* from *Zea mays* L. interacts with subterminal sequences of *Ac*. *EMBO J. 8*: 3177–3185.

Kunze, R., Starlinger, P., and Schwartz, D. (1988) DNA methylation of the maize transposable element *Ac* interferes with its transcription. *Mol. Gen. Genet. 214*: 325–327.

Landry, C.R., Hartl, D.L., and Ranz, J.M. (2007) Genome clashes in hybrids: insights from gene expression. *Heredity 99*: 483–493.

Leitch, I.J., and Bennett, M.D. (1997) Polyploidy in angiosperms. *Trends Plant Sci. 2*: 470–476.

Lowry, D.B., Modliszewski, J.L., Wright, K.M., Wu, C.A., and Willis, J.H. (2008) The strength and genetic basis of reproductive isolating barriers in flowering plants. *Philos. Trans. R. Soc. Lond. B Biol. Sci. 363*: 3009–3021.

MacLean, D., Elina, N., Havecker, E.R., Heimstaedt, S.B., Studholme, D.J., and Baulcombe, D.C. (2010) Evidence for large complex networks of plant short silencing RNAs. *PLoS One 5*: e9901.

Madlung, A., Tyagi, A.P., Watson, B., Jiang, H.M., Kagochi, T., Doerge, R.W., Martienssen, R., and Comai, L. (2005) Genomic changes in synthetic *Arabidopsis* polyploids. *Plant J. 41*: 221–230.

Mallory, A.C., and Vaucheret, H. (2009) ARGONAUTE 1 homeostasis invokes the coordinate action of the microRNA and siRNA pathways. *EMBO Rep. 10*: 521–526.

Malone, C.D., and Hannon, G.J. (2009) Small RNAs as guardians of the genome. *Cell 136*: 656–668.

Martienssen, R., Barkan, A., Taylor, W.C., and Freeling, M. (1990) Somatically heritable switches in the DNA modification of *Mu* transposable elements monitored with a suppressible mutant in maize. *Gene Dev. 4*: 331–343.

Masson, P., Surosky, R., Kingsbury, J.A., and Fedoroff, N.V. (1987) Genetic and molecular analysis of the *Spm*-dependent *a-m2* alleles of the maize *a* locus. *Genetics 117*: 117–137.

Matzke, M.A., and Birchler, J.A. (2005) RNAi-mediated pathways in the nucleus. *Nature Rev. Genet. 6*: 24–35.

Matzke, M., Kanno, T., Daxinger, L., Huettel, B., and Matzke, A.J. (2009) RNA-mediated chromatin-based silencing in plants. *Curr. Opin. Cell Biol. 21*: 367–376.

McClintock, B. (1946) Maize genetics. *Carnegie Inst. Wash. Yr. Bk. 45*: 176–186.

McClintock, B. (1948) Mutable loci in maize. *Carnegie Inst. Wash. Yr. Bk. 47*: 155–169.

McClintock, B. (1984) The significance of responses of the genome to challenge. *Science 226*: 792–801.

Mirouze, M., Reinders, J., Bucher, E., Nishimura, T., Schneeberger, K., Ossowski, S., Cao, J., Weigel, D., Paszkowski, J., and Mathieu, O. (2009) Selective epigenetic control of retrotransposition in *Arabidopsis*. *Nature 461*: 427–430.

Mosher, R.A., Melnyk, C.W., Kelly, K.A., Dunn, R.M., Studholme, D.J., and Baulcombe, D.C. (2009) Uniparental expression of PolIV-dependent siRNAs in developing endosperm of *Arabidopsis*. *Nature 460*: 283–286.

Naito, K., Zhang, F., Tsukiyama, T., Saito, H., Hancock, C.N., Richardson, A.O., Okumoto, Y., Tanisaka, T., and Wessler, S.R. (2009) Unexpected consequences of a sudden and massive transposon amplification on rice gene expression. *Nature 461*: 1130–1134.

Ng, D.W., Zhang, C., Miller, M., Palmer, G., Whiteley, M., Tholl, D., and Chen, Z.J. (2011) *cis*- and *trans*-Regulation of miR163 and target genes confers natural variation of secondary metabolites in two *Arabidopsis* species and their allopolyploids. *Plant Cell 23*: 1729–1740.

Parisod, C., Alix, K., Just, J., Petit, M., Sarilar, V., Mhiri, C., Ainouche, M., Chalhoub, B., and Grandbastien, M.A. (2010) Impact of transposable elements on the organization and function of allopolyploid genomes. *New Phytol. 186*: 37–45.

Parisod, C., Salmon, A., Zerjal, T., Tenaillon, M., Grandbastien, M.A., and Ainouche, M. (2009) Rapid structural and epigenetic reorganization near transposable elements in hybrid and allopolyploid genomes in *Spartina*. *New Phytol. 184*: 1003–1015.

Petit, M., Guidat, C., Daniel, J., Denis, E., Montoriol, E., Bui, Q.T., Lim, K.Y., Kovarik, A., Leitch, A.R., Grandbastien, M.A., and Mhiri, C. (2010) Mobilization of retrotransposons in synthetic allotetraploid tobacco. *New Phytol. 186*: 135–147.

Phillips J.P., Tainer J.A., Getzoff E.D., Boulianne G.L., Kirby K., Hilliker A.J. (1995) Subunit-destabilizing mutations in *Drosophila* copper/zinc superoxide dismutase: Neuropathology and a model of dimer dysequilibrium. *Proc. Natl. Acad. Sci. USA 92*: 8574–8578.

Piegu, B., Guyot, R., Picault, N., Roulin, A., Sanyal, A., Kim, H., Collura, K., Brar, D.S., Jackson, S., Wing, R.A., et al. (2006) Doubling genome size without polyploidization: Dynamics of retrotransposition-driven genomic expansions in *Oryza australiensis*, a wild relative of rice. *Genome Res. 16*: 1262–1269.

Presgraves, D.C. (2010) The molecular evolutionary basis of species formation. *Nature Rev. Genet. 11*: 175–180.

Prudhomme M., Mejean V., Martin B., Claverys J.P. (1991) Mismatch repair genes of *Streptococcus pneumoniae*: *HexA* confers a mutator phenotype in *Escherichia coli* by negative complementation. *J. Bacteriol. 173*: 7196–7203.

Ralston, E., English, J., and Dooner, H.K. (1989) Chromosome-breaking structure in maize involving a fractured *Ac* element. *Proc. Natl. Acad. Sci. USA 86*: 9451–9455.

Rieseberg, L.H., and Willis, J.H. (2007) Plant speciation. *Science 317*: 910–914.

Rinehart, T.A., Dean, C., and Weil, C. (1997) Comparative analysis of non-random DNA repair following *Ac* transposon excision in maize and *Arabidopsis*. *Plant J. 12*: 1419–1427.

Sanei, M., Pickering, R., Kumke, K., Nasuda, S., and Houben, A. (2011) Loss of centromeric histone H3 (CENH3) from centromeres precedes uniparental chromosome elimination in interspecific barley hybrids. *Proc. Natl. Acad. Sci. USA 108*: E498–E505.

Shalev, G., and Levy, A.A. (1997) The maize transposable element *Ac* induces recombination between the donor site and an homologous ectopic sequence. *Genetics 146*: 1143–1151.

Slotkin, R.K., Vaughn, M., Borges, F., Tanurdzic, M., Becker, J.D., Feijo, J.A., and Martienssen, R.A. (2009) Epigenetic reprogramming and small RNA silencing of transposable elements in pollen. *Cell 136*: 461–472.

Soltis, D.E., Buggs, R.J., Barbazuk, W.B., Schnable, P.S., and Soltis, P.S. (2009) On the origins of species: does evolution repeat itself in polyploid populations of independent origin? *Cold Spring Harbor Symp. Quant. Biol. 74*: 215–223.

Soltis, P.S., and Soltis, D.E. (2009) The role of hybridization in plant speciation. *Annu. Rev. Plant Biol. 60*: 561–588.

Tirosh, I., Reikhav, S., Levy, A.A., and Barkai, N. (2009) A yeast hybrid provides insight into the evolution of gene expression regulation. *Science 324*: 659–662.

Tsukahara, S., Kobayashi, A., Kawabe, A., Mathieu, O., Miura, A., and Kakutani, T. (2009) Bursts of retrotransposition reproduced in *Arabidopsis*. *Nature 461*: 423–426.

Ungerer, M.C., Strakosh, S.C., and Zhen, Y. (2006) Genome expansion in three hybrid sunflower species is associated with retrotransposon proliferation. *Curr. Biol. 16*: R872–R873.

Van de Peer, Y., Maere, S., and Meyer, A. (2009) The evolutionary significance of ancient genome duplications. *Nat. Rev. Genet. 10*: 725–732.

Vaucheret, H. (2008) Plant ARGONAUTES. *Trends Plant Sci. 13*: 350–358.

Veitia, R.A., Bottani, S., and Birchler, J.A. (2008) Cellular reactions to gene dosage imbalance: genomic, transcriptomic and proteomic effects. *Trends Genet. 24*: 390–397.

Wang, J., Tian, L., et al. (2006) Genomewide nonadditive gene regulation in *Arabidopsis* allotetraploids. *Genetics 172*: 507–517.

Wang, Q., and Dooner, H.K. (2006) Remarkable variation in maize genome structure inferred from haplotype diversity at the *bz* locus. *Proc. Natl. Acad. Sci. USA 103*: 17644–17649.

Weil, C.F., and Wessler, S.R. (1993) Molecular evidence that chomosome breakage by *Ds* elements is caused by aberrant transposition. *Plant Cell 5*: 515–522.

Wood, T.E., Takebayashi, N., Barker, M.S., Mayrose, I., Greenspoon, P.B., and Rieseberg, L.H. (2009) The frequency of polyploid speciation in vascular plants. *Proc. Natl. Acad. Sci. USA 106*: 13875–13879.

Yaakov, B., and Kashkush, K. (2011a) Massive alterations of the methylation patterns around DNA transposons in the first four generations of a newly formed wheat allohexaploid. *Genome 54*: 42–49.

Yaakov, B., and Kashkush, K. (2011b) Methylation, transcription, and rearrangements of transposable elements in synthetic allopolyploids. *Int. J. Plant Genomics 2011*, 1–7.

Zaratiegui, M., Irvine, D.V., and Martienssen, R.A. (2007) Noncoding RNAs and gene silencing. *Cell 128*: 763–776.

10 Transposons, Genomic Shock, and Genome Evolution

Nina V. Fedoroff and Jeffrey L. Bennetzen

How Transposons Came to be Called "Selfish" DNA

The invention and perfection of DNA sequencing techniques in the late 1970s underlies the contemporary explosion of knowledge on the structure, gene content, and organization of genomes. The 1960s had seen the development of nucleic acid reassociation techniques whose application revealed the presence of much repetitive DNA in eukaryotic genomes (Britten and Davidson, 1976a, 1976b). As DNA sequencing became a reality, there was a good deal of discussion about the value of sequencing entire genomes in view of the growing evidence that only a small fraction of eukaryotic DNA comprised genes in the sense of protein and structural RNA-coding sequences.

A pair of papers published back to back in *Nature* in 1980 solidified Ohno's view that much of eukaryotic DNA, including transposons, was nonfunctional DNA—labeled "junk"—because it was thought to make no contribution to phenotype and existed only by virtue of its ability to increase by duplication (Doolittle and Sapienza, 1980; Ohno, 1972; Orgel and Crick, 1980). The objective of both papers was to get beyond the then–still-prevalent view that every bit of an organism's DNA has a specific function as a result of selection. Thus, both papers pushed Dawkin's concept of "selfish DNA" in order to focus attention on the fact that DNA capable of proliferating within a genome, as transposable elements do, may need no other explanation for its survival. Orgel and Crick asserted: "The spread of selfish DNA sequences within the genome can be compared to the spread of a not-too-harmful parasite within its host" (Orgel and Crick, 1980).

Fortunately, genome sequencing raced forward, and today we have a vastly more complex understanding of genome structure and organization than we might have had if sequencing had been restricted to protein-coding sequences and RNA-coding sequences, but we also find ourselves neck-deep in transposons. The selfish DNA concept offered an explanation for the long-standing C-value paradox that organisms of similar evolutionary complexity differ vastly in their DNA content (Gall, 1981). The C-value, which is the DNA content per haploid genome, varies widely among closely related organisms, and has been known for some time to be attributable to the repetitive portion of the genome (Rosbash et al., 1974). This is especially striking in angiosperms, with C-values ranging from 0.0648 pg for *Genlisea margaretae* to 132.5 pg for a hexaploid hybrid *Trillium* × *hagae*, a 2,000-fold range (Bennett and Leitch, 2011; Bennett and Smith, 1976). The explanation of the C-value paradox, as described in more detail in Chapter 2 in this volume, resides in huge

Plant Transposons and Genome Dynamics in Evolution, First Edition. Edited by Nina V. Fedoroff.

differences among genomes in the abundance of transposons and retrotransposons, even as gene numbers remain relatively constant.

The "Selfish DNA" Label Stuck to Transposons

Ideas about noncoding DNA have evolved substantially over the past decades, and it is no longer dismissed as junk (Zuckerkandl, 1992; Zuckerkandl and Cavalli, 2007). But the transposon monikers from the selfish DNA papers have persisted. Transposons are still referred to as "invaders" (Kidwell and Lisch, 1998) and "parasites" (Martienssen, 1998; Slotkin and Martienssen, 2007; Zeh et al., 2009).

When epigenetic mechanisms began to be taken seriously in the late 1990s, the idea was advanced that they evolved precisely to "control" the destructive potential of "parasitic sequences," mostly transposons (Bestor, 1998; Martienssen, 1998; Slotkin and Martienssen, 2007; Yoder et al., 1997). Today we know that there is a set of interacting epigenetic feedback mechanisms that regulate the chromosomal organization, as well as the recombinational and transcriptional activity of repetitive sequences, in general, and not just transposons. But the conviction that epigenetic gene regulatory mechanisms originated from transposons has persisted (Lisch and Bennetzen, 2011). Here we ask whether the notion that they evolved to control transposons still fits the facts in the light of the many that have accumulated since this hypothesis was first advanced.

Transposons Coevolved with Eukarotic Genomes

The difficulty with the idea that transposons are invaders and parasites begins when one asks where such parasites might have come from. It turns out that genes encoding transposases, which all have certain common structural motifs in their catalytic cores, are present throughout eukaryotes (Yuan and Wessler, 2011), and can be traced back into prokaryotic organisms that do not have the elaborate epigenetic superstructure of eukaryotes (Hickman et al., 2010). That is, transposons were around long before the eukaryotic lifestyle, with its bloated genomes, appeared on the evolutionary scene. So there's no "where" whence these parasites might have invaded eukaryotes: they coevolved.

Moreover, the movement of prokaryotic transposons is regulated by a variety of mechanisms that are not dissimilar to those in eukaryotes, including a requirement for hemi-methylation of specific sites for both transcription and transposition of certain elements (Casadesus and Low, 2006; Kleckner, 1990; Roberts et al., 1985). It is true that the ability of eukaryotic transposons to accumulate in large numbers, together with their highly generic transposition mechanisms, means that the proliferation of a transposon introduced into a genome lacking it, whether by a genetic cross or a virus, makes it resemble an invader (Engels, 1992; Evgen'ev et al., 2000). Indeed, there is growing appreciation that transposons are subject to horizontal transfer in eukaryotes, in some cases through host-parasite interactions (Gilbert et al., 2012; Gilbert et al., 2010; Kidwell, 1992; Silva and Kidwell, 2000; Thomas et al., 2010). However, new transposons can arise within a genome and silent transposons can be mobilized anew by a variety of physiological and genetic stresses, undergoing "bursts" of transposition and expanding genomes before being silenced and decaying (Baucom et al., 2009b).

Sequence Duplication: The Real Innovation

There are central differences between prokaryotic and eukaryotic genome organization: most prokaryotic genomes are small, not extensively endoreduplicated, and do not accumulate the extraordinary numbers of transposons and retrotransposons characteristic of eukaryotic genomes. So there's a deeper issue than the ancient origin of transposons. What distinguishes eukaryotic genome organization from that of prokaryotes is the presence of vast amounts of duplicated sequences. This is something that we take for granted, but should not.

Prokaryotes can readily duplicate genome segments by virtue of small stretches of homology, but tandem duplications are rapidly lost in the absence of selection or survive by dispersing (Achaz et al., 2002; Anderson and Roth, 1977; Andersson and Hughes, 2009). Indeed, absent selection or a reduction in homology, tandem duplications are inevitably eliminated by the mechanics of homologous unequal crossing over between duplicated sequences, generating one-copy organisms (an absorbing state) and organisms with increasing numbers of copies that, in turn, throw off singletons (Achaz et al., 2002; Koch, 1979). Organisms with many copies are eventually eliminated either by virtue of the small energetic drag of the extra DNA or by a population bottleneck. This is borne out by the observation that duplicate genes in prokaryotic organisms are generally acquired by horizontal gene transfer rather than arising by tandem duplication (Isambert and Stein, 2009; Treangen and Rocha, 2011).

The essence of the selfish DNA argument is that there is a category of self-reproducing DNA within the genome that that has little or no phenotypic impact, hence it is not subject to selective pressure. This is not an unreasonable thesis, given the evidence that eukaryotic genomes are full of repetitive DNA, but that it continues to pass unremarked is remarkable and profoundly misses the point: how did eukaryotes tip the balance between duplication and deletion that keeps genome size small in organisms in which homology-dependent recombination mechanisms predominate? How can transposons and retrotransposons, whose duplicative mechanisms create dispersed repetitive sequences, build up in large numbers in spite of the fact that homologous recombination among them has the potential to cause major, even catastrophic, chromosomal rearrangements? What were the critical genetic inventions that enabled the evolution of the contemporary eukaryotic genome organization and, perhaps, multicellularity, as well?

The Facilitator: Epigenetic Control of Homologous Recombination

The answer to these questions lies precisely in the epigenetic mechanisms that eukaryotes have elaborated to a much greater extent than prokaryotes. Repressive protein complexes, histone methylation, RNA interference (RNAi), and RNA-directed DNA methylation, described in detail in Chapter 5 of this volume, as well as recombinational regulatory complexes (Moldovan et al., 2012), are among the epigenetic mechanisms that have so far surfaced. These serve a variety of structural and regulatory functions, but perhaps the critical one for understanding the evolution of eukaryotic genomes is the minimization of ectopic recombination between homologous sequences in the course of the DNA break-repair processes that maintain genome and chromosome stability.

Heterochromatin, the highly compacted chromosome regions rich in repetitive DNA, is recombinationally inert (Peng and Karpen, 2008; Vanderlyn, 1949). Maize genes are hotspots of recombination, with little recombinational activity in the intergenic, retrotransposons-rich regions (Fu et al., 2001; Fu et al., 2002; Nelson, 1959; Okagaki and Weil, 1997). While not all eukaryotes use all of the known epigenetic mechanisms, even lower eukaryotes with relatively small genomes generally use RNAi to stabilize repetitive DNAs, such as ribosomal RNA genes and centromeric repeats (Peng and Karpen, 2007, 2008; Torres-Rosell et al., 2007). In fission yeast, noncoding transcripts of repetitive sequences initiate a process that generates small RNAs that, in turn, target further transcripts for degradation and attract protein complexes that induce heterochromatization through histone modification (Bayne et al., 2010; Verdel et al., 2004). Disruption of the RNAi machinery disturbs the repair of double-strand breaks in centromeric repeats, stimulating repair by homologous recombination (Zaratiegui et al., 2011). Recent studies on DNA damage-repair pathways in *Arabidopsis* and humans identify a crucial role for small RNAs, although precisely how they do so is not yet clear (Francia et al., 2012; Wei et al., 2012).

Plants have a more complex and redundant array of epigenetic mechanisms than animals, exhibiting three different systems of DNA methylation, as well as posttranscriptional mechanisms, all of which have small noncoding RNAs at their heart (Law and Jacobsen, 2010; Simon and Meyers, 2011; Zaratiegui et al., 2007). Plants do not set aside a germline early in development, as higher animals do, which imposes more stringent requirements for maintaining genetic integrity because differentiated genomes must eventually be reprogrammed for reproduction. Indeed, the elaboration of epigenetic mechanisms may have made possible the indeterminate lifestyle of many higher plants and reproduction cued to environmental signals. Epigenetic mechanisms, for example, determine such environmentally responsive developmental transitions as vernalization and flowering time (Groszmann et al., 2011a; Yaish et al., 2011). The other side of the coin is that the very redundancy and efficiency of plant epigenetic regulatory systems in all likelihood facilitates the accumulation of very large numbers of transposons and retrotransposons in higher plant genomes.

Although identified early as epigenetically silenced, transposons are far from the only targets of epigenetic regulation in plants (Banks et al., 1988; Fedoroff et al., 1984). In 1994, it was reported that an artificial gene comprising a viroid cDNA became methylated and transcriptionally inactive in the tobacco genome, but only if viroid RNA replication had occurred, suggesting a feedback mechanism initiated by transcript overabundance (Wassenegger et al., 1994). In the early days of plant molecular modification, it was discovered that an introduced transgene encoding an enzyme in the pigment biosynthetic pathway was subject to silencing (Jorgensen, 1990; Jorgensen, 1994; Que and Jorgensen, 1998). In time, silencing was found to entail both transcriptional and posttranscriptional mechanisms that share characteristics with those plants use to control viral pathogens and are mediated by the production of small interfering RNAs (siRNAs) (Chen, 2010a; Kooter et al., 1999; Ratcliff et al., 1997). Gene duplication underlies paramutation, one of the very first described plant-epigenetic silencing phenomena (Arteaga-Vazquez and Chandler, 2010; Brink, 1958; Panavas et al., 1999). Detailed analysis of how introduced repeats are silenced has made it increasingly evident that the repetitive character of the sequence is a common trigger for siRNA-mediated gene silencing and methylation (Chan et al., 2006). Several recent reports link epigenetic mechanisms with DNA replication and repair in plants (Liu et al., 2010; Wei et al., 2012; Yin et al., 2009). Thus, epigenetic mechanisms are integral components of the systems that maintain genome structure and regulate gene expression in plants.

Ectopic recombination between transposons in different locations has the potential to profoundly disrupt genomes. So the very existence of epigenetic mechanisms that suppress recombination must necessarily favor the accumulation of transposons in view of the amplification mechanisms built into both cut-and-paste transposons and copy-and-paste retrotransposons (Chapter 2 in this volume). Since unequal crossing over between long repeats can eliminate duplicated sequences more readily than illegitimate recombination can generate them, mechanisms that suppress homologous recombination are precisely what is needed to tip the balance between duplication and deletion. As well, the fact that small diffusible RNA molecules are at the heart of the silencing machinery means that duplications, including new copies of transposons, do not evade regulation by moving to new locations. Thus, it seems most likely that it was the elaboration of epigenetic mechanisms in eukaryotes that made it possible for genomes to grow by both endoreduplication and transposon proliferation.

Epigenetic Mechanisms, Duplication and Genome Evolution

The elaboration of the epigenetic mechanisms is arguably also a critical step in the evolution of multicellular organisms, underpinning the ability to keep duplicate genes for subsequent diversification for expression in specific cells and tissues and at different developmental moments (Adams, 2007). Equally key is the ability to program genes for differential expression by a variety of mechanisms, among which are the relatively stable mechanisms involving DNA and histone modification, as well as the more labile small–RNA-mediated mechanisms, such as microRNA, all of which are interconnected (Chapter 5 in this volume). It is also increasingly evident that the information that determines the epigenetic program of a sequence resides in the sequence itself (Lienert et al., 2011), a recent observation that has antecedents in the genetic and molecular analysis of maize transposons. For example, the *Suppressor-mutator* (*Spm*) transposon identified by McClintock contains a highly GC-rich sequence in its first exon whose presence promotes silencing, and the transposon encodes a positive regulatory protein that can overcome the methylation-associated silencing, comprising an element-encoded, epigenetic autoregulatory loop (Chapter 4 in this volume).

Thus, it appears vastly more likely that contemporary eukaryotic genomes evolved in the context of epigenetic mechanisms than that they invented them as an afterthought to combat a plague of parasitic transposons. Eukaryotic DNA methylases evolved from bacterial restriction-modification methylases, while small–RNA-based regulatory systems may have originated in a bacterial surveillance system that uses an RNAi-like mechanism, both of which serve bacteria to distinguish between their own DNA and that of DNA acquired through horizontal gene transfer of plasmids and from bacteriophage infection (Barrangou et al., 2007; Iyer et al., 2011).

Contemporary eukaryotic genome organization makes eminent sense in the light of epigenetic mechanisms that minimize homologous recombination and transposon and retrotransposon activity. However, while such mechanisms slow the scrambling of genomes to an evolutionary time scale, they do not eliminate it. Comparative analysis of related genomes has revealed the fingerprints of homologous and illegitimate recombination and transposition at every level of genome organization. Whether examining the results of transposition events involving a single transposon (Chapter 3 in this volume) or viewing the contribution of transposition to the evolution of chromosomes, described below, the centrality of transposition and both homologous and illegitimate recombination mechanisms to contemporary genome organization is inescapable.

At an even finer level of resolution, transposons contribute to creating genes, modifying them, programming and reprogramming them, as described in detail in Chapters 6–9 in this volume. The traffic in genes and regulatory sequences is bidirectional: transposons pick up bits and pieces of genes that code for proteins other than transposases, and transposase genes are pressed into services other than transposition. Transposons provide the telomeres of some organisms and jump in to replace them in others (Belfort et al., 2011; McEachern and Haber, 2006). Centromeres contain and are often surrounded by transposons (Baucom et al., 2009a; Kanizay and Dawe, 2009; Ma and Bennetzen, 2006; Slotkin, 2010). The human immune system uses recombinase proteins that evolved from transposases to generate sequence diversity through V(D)J recombination (Hiom et al., 1998; Kapitonov and Jurka, 2005), a phenomenon with antecedents in the diversity-generating retrotransposons of bacteria (Medhekar and Miller, 2007). Transposons are central to the epigenetic phenomenon of imprinting that imbues genes with different expression patterns depending on whether they were transmitted through male or female gametes (Chapter 7 in this volume), differences that arise during the major epigenetic reprogramming events of gametogenesis itself (Chapter 5 in this volume).

Plant Genome Organization: Gene Islands in a Sea of Repetitive DNA

The genomes of most flowering plants are large and complex: the median size of *Poaceae* genomes is ~4,700 Mb (http://data.kew.org/cvalues/). Plant genome studies, however, tend to focus on the genomes of species such as *Arabidopsis* (~140 Mb), peach (~270 Mb), *Brachypodium* (~360 Mb), rice (~400 Mb), and *Populus* (~480 Mb) with the unusually small nuclear DNA contents that make them amenable to comprehensive DNA sequence studies. However, sample sequence analysis and comprehensive full-genome studies of soybean (~1,100 Mb), maize (~2,400 Mb), barley (~5,500 Mb), and other more typically sized plant genomes have revealed a common genome organization that is shared by all angiosperms studied to date. Genes are arranged on islands amidst seas of repetitive DNA. In the small genome species, the gene islands are quite large and the seas are few, mostly comprising pericentromeric heterochromatin, but the opposite is true for genomes of angiosperms with more typical genome sizes. In maize and wheat, for instance, the median gene island contains only 1–2 genes covering less than 10 kb, while most interisland seas are 20–100 kb (Bennetzen et al., 1994; Devos et al., 2005; Schnable et al., 2009; Wicker et al., 2001).

It is not known at what genome size the typical angiosperm genome organization becomes standard or whether a precise size threshold exists. For instance, the ~700 Mb genomes of sorghum has most of its genes in gene-rich mega-islands, such as *Arabidopsis* and rice, but with larger regions of pericentromeric repeats than seen in the other two species (Paterson et al., 2009). When more genomes in the C-value range of 1,000–2,000 Mb are investigated, it will be interesting to see whether there is some particular DNA content that is correlated with a tipping point at which lagoons of repetitive DNA appear inside genic regions. Because the island structure of a plant genome can be an outcome of either or both differential rates of TE amplification and DNA removal, it seems likely that cases will be found where repetitive stretches have arisen recently in genic regions even in some small-genome plants, while some large-genome plants will have the great majority of their repetitive DNA only in pericentromeric and satellite regions, with genic areas having been cleansed of most TEs and other repeats by recombinational processes.

Early studies on small regions of the maize genome revealed that the intergenic seas of repetitive DNA are composed of LTR (long terminal repeat) retrotransposons (SanMiguel et al., 1998; SanMiguel et al., 1996). A large percentage of the LTR retrotransposons was found as nested insertions, with older elements often containing multiple insertions of younger, more recently inserted LTR retrotransposons (Figure 10.1). The insertion times of the retrotransposons within retrotransposons in a 240-kb region containing the *adh1* gene were dated by analyzing the divergence between the terminal LTRs of each element inserted within the previous one, leading to the conclusion that the insertions had all occurred within the last 6 million years, most in the preceding 3 million (SanMiguel et al., 1998). Generalization of this observation implies that the maize genome doubled in size by retrotransposition over just a few million years. A similar estimate of rapid

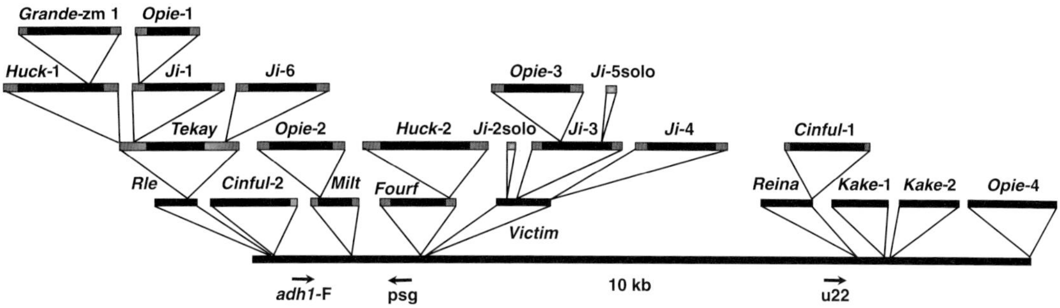

Figure 10.1 The arrangement of retrotransposons in the maize *adh1*-F region. The short lines represent retrotransposons, with the internal domains represented in black and the LTRs in grey; triangles below the retrotransposons converge on the insertion sites of retrotransposons, and the small arrows show the direction of transcription of the genes, shown under the long black line that represents the sequence near the *adh1* gene into which the transposons inserted. Redrawn from SanMiguel et al., *Nature Genetics 20*: 43–45 (1998).

genome expansion by retrotransposon proliferation has been reported for a wild rice genome (Piegu et al., 2006).

Based on the draft whole-genome sequence of the maize B73 inbred line, it is now known that close to 85% of the contemporary 2.3-gigabase genome comprises transposons, more than 75% of which are LTR retrotransposons (Schnable et al., 2009). Its roughly 40,000 genes, averaging about 3.3 kb in length, exist in a sea of more than 1 million transposons and retrotransposons belonging to almost 1,300 different gene families. A detailed analysis of a relatively gene-rich 22-Mb region revealed a gene density of 25 genes/Mb, with the distance between genes varying from less than 1 kilobase to more than 100 kb, with half exceeding 20 kb (Wei et al., 2009). However, by contrast to retrotransposons, which comprise long, relatively gene-free regions, Class II *CACTA*s, *MULE*s, MITEs, and *Helitrons* and *hATs,* as well as the small SINE retroelements, are in or near genes, and many transposons and retroelements contain captured gene fragments (Wei et al., 2009).

Examples of similar genome organization have now been identified by sample sequence and full genome sequence analysis in every angiosperm investigated. Differences in the abundances of LTR retrotransposons account for most of the >1,000-fold variation in genome size that differentiates angiosperm species (Bennetzen et al., 2005). For instance, the tiny *Arabidopsis* genome contains only 20–25 Mb of LTR retrotransposons (Liu and Bennetzen, 2008), while the maize genome contains >1,800 Mb of this TE class (Baucom et al., 2009a; Schnable et al., 2009).

Transposon Neighborhoods and Insertion Site Selection

Beyond their very different abundances in different species, TEs also show differences in their distribution patterns. These differences are most obvious in intraspecific comparisons. In maize, for instance, the *gypsy* superfamily of LTR retrotransposons is mostly pericentromeric, while the *copia* superfamily is underrepresented in the pericentromeric domain. In contrast, maize CACTA transposons are relatively evenly distributed, but other DNA transposons such as the *Mutator* elements are greatly enriched inside or near genes (Cresse et al., 1995; Schnable et al., 2009). Maize *Helitron* transposons accumulate mainly near but not inside each other, with a strong preference for members of the same subfamily as nearest neighbors and in the same orientation as the nearby

Helitron (Yang and Bennetzen, 2009)(Figure 10.3). Such clustering may reflect the propensity of some TEs to move to nearby sites, long documented for the *Ac-Ds* transposon family of maize (Greenblatt and Brink, 1962; Huang et al., 2003; Raina et al., 2002).

Few and perhaps none of the insertion sites for different element types and families exhibit a primary DNA sequence pattern that suggests how insertions are targeted. Moreover, not all of the transposon and retrotransposon distribution biases may be attributable to preferential insertion, because some de novo insertions may be selected against, such as those that disrupt gene function, or be in places in the genome that are more prone to be active in elimination of elements. However, insertion biases are the simplest explanation when differences are observed between elements with the same structure but different accumulations. A good example of this is the LTR retrotransposon superfamilies: *copia* and *gypsy* elements have the same basic structure, and thus are expected to be removed by the same processes, including recombination, natural selection, and illegitimate recombination, yet they have very different distributions in many genomes (Paterson et al., 2009; Schnable et al., 2009). It seems more likely that differences in insertion site selection than differences in DNA removal rates account for the differential distribution of LTR retrotransposon families in complex plant genomes.

In bacteria, diversity-generating retroelements target specific sequences through a combination of sequence and structural features (Guo et al., 2008; Guo et al., 2011). Targeting can also be achieved through the interaction of retrotransposon-specific proteins and chromatin proteins, which are themselves preferentially associated with certain types of sequences. An example is provided by the interactions of *Sir4p*, a structural component of heterochromatin, with a 6-amino acid motif of the yeast *Ty5* integrase protein that targets it to telomeric heterochromatin (Fuerst and Voytas, 2003; Gai and Voytas, 1998). Insertion of an *Arabidopsis lyrata* centromeric retrotransposon into *Arabidopsis thaliana* centromeres was recently reported (Tsukahara et al., 2012). In view of the fact that the centromeric sequences are quite different in the two species, it appears likely that targeting occurs through conserved centromere-specific structural proteins. Maize *Mu* transposons, by contrast, target open chromatin, which is recombinationally active (Liu et al., 2009). Thus, there is reason to believe that both chromatin protein-targeted insertion site selection and DNA-based target site selection influence the genomic distribution of transposons.

Because each TE family has a somewhat different distribution pattern, it is possible that every TE family has a unique insertion bias. Analysis of retrotransposon distribution across the maize chromosomes showed that the chromosomal regions with the most abundant LTR retrotransposons, generally in pericentromeric regions, have the least LTR retrotransposon diversity, perhaps reflecting both a common insertion site selection mechanism and a low elimination rate (Baucom et al., 2009a). It has been known for some time that the most abundant TEs in maize, such as the high–copy-number LTR retrotransposons, are less likely to cause mutations than are lower–copy-number TEs, such as low–copy-number LTR retrotransposons and DNA elements of the *Ac/Ds*, *Spm/En*, or *Mutator* type (Bennetzen, 2000b). With completion of the maize genome sequence, insertion biases associated with LTR retrotransposons family copy number have become very clear (Baucom et al., 2009a). Low–copy-number TEs are mostly found near or inside genes, while those with a high copy number are rarely observed either near or inside genes.

As in other organisms, TEs comprise a larger fraction of the total DNA in gene-poor pericentromeric regions than elsewhere in the maize genome. Insertion dates of intact retrotransposons, as inferred from the divergence of their LTRs, vary with distance from the centromere, with the younger elements more abundant in euchromatin and older elements more abundant near centromeres. Low–copy-number retrotransposons are, on average, older than high–copy-number transposons, which are concentrated in the pericentromeric regions where recombination is suppressed and therefore less effective in eliminating retrotransposons (Baucom et al., 2009a).

No sequence-based hypothesis has yet been advanced to explain why some LTR retrotransposon attain high copy numbers and others do not. It is possible that this property is intrinsic to the retrotransposon itself, perhaps a consequence of the molecular mechanism underlying insertion site selection. In some systems, it has been observed that a low–copy-number LTR retrotransposon can suddenly, at least on an evolutionary timescale, explode into a high–copy-number LTR retrotransposons (Piegu et al., 2006). Many TE families may have the potential to reach high copy numbers, but perhaps require either an internal mutation or a specific host cue for release from epigenetic repression.

Genome Evolution: Colinearity and Its Erosion

Genome divergence through transposon activity and recombination are ongoing processes that occur within species and at surprisingly high rates in some. For example, a comparison among maize inbred lines, illustrated in Figure 10.2, revealed remarkable differences in retrotransposon abundance and composition within a region of just 150 kb surrounding the *bronze* gene (Fu and Dooner, 2002; Wang and Dooner, 2006).

It has long been known that inbreeding is detrimental, a phenomenon termed "inbreeding depression," and that certain intraspecific crosses and even some interspecific and intergeneric crosses produce progeny that are superior to either parent (Chen, 2010b; Hochholdinger and Hoecker, 2007). The latter is known as "hybrid vigor" or "heterosis" and was first described by Darwin, who observed that progeny of crosspollinated maize plants were substantially taller than the progeny of inbred plants (Darwin, 1786). Heterosis was rediscovered by George Harrison Shull and Edward M. East early in the twentieth century, laying the groundwork for today's extraordinarily productive hybrid corn industry (Hochholdinger and Hoecker, 2007). However, despite a long history of active investigation, the molecular basis of heterosis is not yet understood (Smith and Weigel, 2012). As well, it is observed that segregating hybrids can exhibit phenotypes that are either novel or more extreme in both positive and negative directions than those characteristic of the parental lines, a phenomenon called "transgressive segregation" (Rieseberg et al., 2003). The observation that both gene and TE content diverge rapidly suggests that at least one of the factors contributing to heterosis is the ability of diverging genomes to complement each other's lost genes (Fu and Dooner, 2002). Recent efforts to understand heterosis and transgressive segregation have focused on changes in the regulation of genes by small RNAs (Groszmann et al., 2011b; Kenan-Eichler et al., 2011; Shivaprasad et al., 2012).

Intraspecific *Arabidopsis* hybrids that exhibit a strong heterotic effect show marked reduction in the overall abundance of 24-nt small RNAs (sRNA) and specific differences generally at genomic locations that show markedly different small RNA levels in the parental lines (Groszmann et al., 2011b). The greater the difference between the parental lines, the more likely is the sRNA level to be altered in the hybrid. Dissimilar sRNA levels in parental lines correlate with altered gene methylation and gene expression levels (Groszmann et al., 2011b). A recent study on stable transgressive phenotypes in the progeny of crosses between cultivated and wild tomatoes suggests that at least in some cases, the suppressed and methylated genes are the source of secondary 21-nt siRNAs (Shivaprasad et al., 2012).

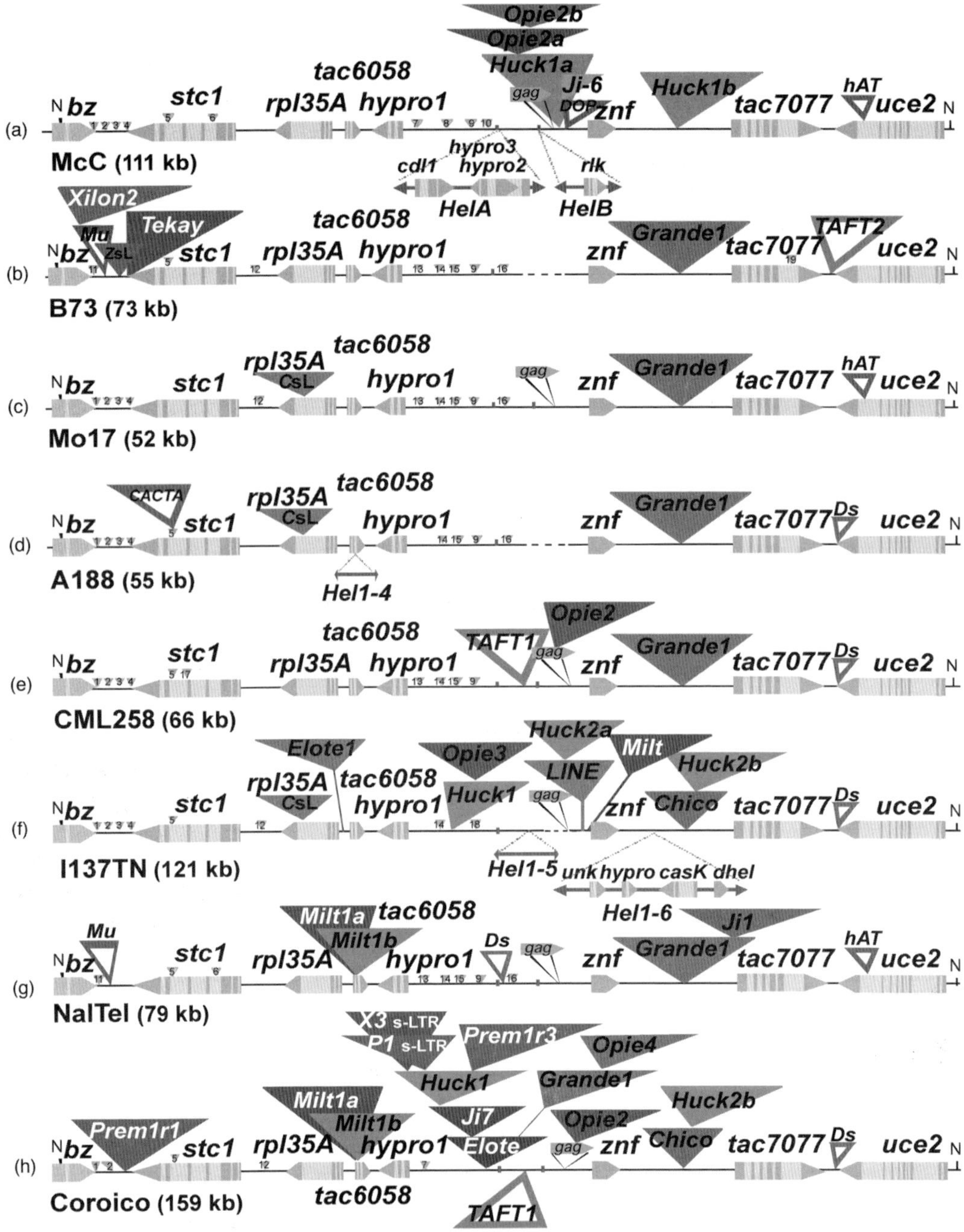

Figure 10.2 Organization of eight *bz* haplotypes. Each haplotype is identified by the name of the genetic line, followed by the size of the cloned *Not*I fragment in parentheses (NotI sites are indicated by Ns). Genes are shown as pentagons pointing in the direction of transcription; exons are in a darker grey than introns. There are eight genes in the region: *bz*, *stc1*, *rpl35A*, *tac6058*, *hypro1*, *znf*, *tac7077*, and *uce2*. The same gene symbols are used for gene fragments carried by helitrons (*Hels*), which are represented as bidirectional arrows below the line for each haplotype. The vacant sites for *HelA* and *HelB* in each haplotype are provided as reference points and marked with short vertical strokes. Dashed lines represent deletions. Retrotransposons are indicated by solid triangles and DNA transposons and TAFTs (probably also DNA transposons), are indicated by open triangles. Small insertions are represented by small triangles and numbered. Only the genes have been drawn to scale (Wang and Dooner, *PNAS 103*: 17644 [2006]. Copyright 2006 National Academy of Sciences, U.S.A, used with permission.). For color detail, see color plate section.

Speciation is the process by which subgroups of a reproductively compatible population become reproductively isolated and give rise to a new species. As explored in detail in Chapter 9 in this volume, this occurs by a variety of mechanisms, some of which involve both transposon mobilization and active genome restructuring. Whether plant speciation occurs by a significant genome restructuring or by a genetic change that simply alters some aspect of reproductive timing, the genomes of newly formed plant species are necessarily similar and the gene order is largely colinear. Analysis of orthologous genomic regions in related species reveals that as the evolutionary distance increases, the colinearity declines, although the number and nature of the genes remains more or less constant (Wicker et al., 2010; Wicker et al., 2011).

It has long been known that genes change their chromosome locations and it has been speculated that transposons mobilize large DNA segments because they are often found at the ends of inverted or transposed sequences. (Bennetzen, 2000a; Bennetzen and Freeling, 1997; Tikhonov et al., 1999; Wicker et al., 2005). McClintock's studies on the *Ac/Ds* transposon family showed that chromosome breaks at the site of insertion of a nonautonomous *Ds* element could be resolved with attendant duplications, deletions, inversions, and translocations of large chromosomal segments, as described in Chapter 1 of this volume (McClintock, 1950a, 1950b). Early molecular analysis of a *Ds* transposition-associated mutation derived by McClintock revealed that its origin was associated with a large duplication of the short arm of maize chromosome 9 (Courage-Tebbe et al., 1983). Genetic analysis of *Ac/Ds*-associated rearrangements at the P locus, described in detail in Chapter 3 in this volume, identified transposition events that initiate at the 5′ end of one transposon and terminate at the 3′ end of a nearby transposon (Chapter 3 in this volume). This has been designated "alternative transposition," and can generate a variety of rearrangements, depending on the relative orientation of the transposon ends, and translocate large segments of intervening DNA.

The results of three-way comparisons among the genomes of *Brachypodium*, rice, and sorghum suggested that the apparent movement of genes is often mediated by a process that duplicates the gene and flanking sequences, leaving a copy at the original insertion site (Wicker et al., 2011). Since translocated genome segments containing such genes are often flanked by transposons, this observation suggests that the movement of a chromosome segment is likely to be initiated by a double-strand break at the new insertion site of a transposon and may be repaired through one of several known processes that repair double-strand breaks in plants, including synthesis-dependent strand annealing, template slippage, or unequal crossing over (Puchta, 2005; Wicker et al., 2010). Such duplications may also arise during mitotic chromosome replication when transposition is known to occur, transferring long genomic segments between two copies of the same transposon by the alternative transposition mechanism, resulting in the translocation of a large chromosome segment to either the sister chromatid or a different chromosome and leaving a copy of the original sequence behind at the donor site on the other sister chromatid. The well-known tendency of transposons of the *Ac/Ds* family to undergo short-range transpositions from only one sister chromatid to an unreplicated site nearby gives rise to genic regions flanked by copies of the same transposon (Greenblatt and Brink, 1962), facilitating subsequent mobilization of the intervening chromosome segment (Chapter 3 in this volume).

Genome Contraction and Divergence of Intergenic Sequences

Genomes expand by sequence duplication, transposition, and retrotransposition and contract by deletion mediated by a variety of homologous and illegitimate recombination events (Buchmann et al., 2012; Devos et al., 2002; Ma et al., 2004; Vitte and Bennetzen, 2006; Xu et al., 2012). Solo LTR generation by unequal homologous recombination between the LTRs at the ends of a single retroelement is frequent in some species, particularly near genes and at the kinetochore (Ma et al., 2004). Such unequal events can also occur between adjacent elements, leading to the deletion of the DNA between two TE copies (Devos et al., 2002; Ma et al., 2004). In most regions where TEs are abundant, homologous recombination is dramatically inhibited, a likely causal factor in TE accumulation, as noted earlier (Fu et al., 2001; Fu et al., 2002). Indeed, retrotransposons elimination by unequal and illegitimate recombination is most frequent in recombinationally active genome regions (Tian et al., 2009).

Autonomous DNA transposons commonly decay by internal deletions that reveal the operation of double-strand break repair mechanism that duplicates genetic information, now called "filler DNA," by the invasion of a single strand into a duplex nearby and copying of a short sequence that is then inserted at the deletion breakpoint (Puchta, 2005). This was originally reported for nonautonomous, internally deleted *Spm* transposons and later for the origin of new nonautonomous *Ds* transposons from the autonomous *Ac* transposon (Masson et al., 1987; Yan et al., 1999). McClintock identified and investigated a series of *Spm* derivatives that originated from an internally deleted, nonautonomous *Spm*, all of which were the result of further deletions within the same element at the same location (Masson et al., 1987; Schwarz-Sommer et al., 1985). These further deletions arose at a fairly high frequency, but only in the presence of an autonomous element, strongly suggesting that they were initiated either directly by the double-strand cleavage activity of the transposase encoded by the autonomous element or indirectly by secondary double-strand breaks incurred during the transposition reaction (McClintock, 1968).

Large-scale comparisons in related species have shown that turnover of intergenic regions is very rapid on an evolutionary time scale. Fine-grained analysis of the sequence dynamics shows that the intergenic volatility is mediated by transposons, which both delete adjacent DNA sequences and insert filler DNA from elsewhere in the process of repairing double-strand breaks in the DNA catalyzed by transposases (Buchmann et al., 2012). Deletions commonly occur adjacent to a transposon end, but can remove entire transposons. Statistically significant clustering of such deletions in the vicinity of transposons suggests that they arise as a result of the double-strand breaks that initiate transposition.

Transposases Sculpt Genomes

Transposases, the enzymes encoded by transposons and required for their mibility, hold a special place in the pantheon of genome sculptors because of how they bind, cut, and relegate DNA sequences. The essential elements of the transposase mechanism are: 1) strict sequence recognition and cutting at the donor site; and 2) either a relaxed sequence specificity or no sequence specificity at the target site. That is, transposases, arguably the most abundant genes on earth (Aziz et al., 2010), cut transposon ends at the donor site and attach them to a new sequence elsewhere in the genome.

Although the transposition mechanism has been investigated in much greater detail for the prokaryotic *Mu* transposase than for any eukaryotic transposon, the *Mu* mechanism appears to be paradigmatic and common to the DDE superfamily (Craig, 1997; Craigie and Mizuuchi, 1985, 1987; Dreyfus, 2006; Kaufman and Rio, 1992; Keith et al., 2008; Mitra et al., 2008; Montano and Rice, 2011; Nesmelova and Hackett, 2010). Multimers of the transposase form a transpososome complex that brings transposon ends together with the target insertion site (Haapa-Paananen et al., 2002; Mizuuchi and Craigie, 1986). The transposon ends are brought into close juxtaposition with each other at the donor site for a coupled cleavage/joining reaction that cleaves the transposon termini, introduces a staggered cleavage at the target site and transfers the 3′ ends of the transposon to the overhanging 5′ ends at the target (Savilahti et al., 1995; Williams et al., 1999). The gapped duplexes at the two element ends are then repaired to generate the target site duplication whose length is a TE family characteristic determined by the transposase. Retroviral and retrotransposons integrases, which belong to the same superfamily of nucleotidyltransferases as transposases, use a similar integration mechanism, and are likely to have a common evolutionary origin (Capy et al., 1997; Cherepanov et al., 2011; Polard and Chandler, 1995; Yuan and Wessler, 2011).

From the perspective of genome evolution, the crucial characteristics of the transposition complex are a strict recognition and precise cleavage of the transposon donor site sequences and relaxed sequence specificity at the target site. In both bacteria and plants, the mechanism that generally ensures genome integrity and confines transposition to just one of the two newly replicated daughter strands or sister chromatids is the recognition and cleavage of the two hemi-methylated terminal inverted repeats of a single transposon (Roberts et al., 1985; Ros and Kunze, 2001). But like other aspects of transposition, recombination, and DNA repair, it is both error prone and can be fooled by such transposition events as the insertion of one *Ds* transposon in inverted orientation into the center of itself, giving the chromosome-breaking *doubleDs* transposon that led McClintock to her momentous discoveries about how transposons move and restructure chromosomes (Doring et al., 1990; English et al., 1995). The bottom line for genomes is that the cleavage and resection of DNA by transposases virtually guarantees genome scrambling and the appearance of transposons at rearrangement breakpoints. Simply put, transposases drive genome evolution.

Small Regulatory RNAs from Transposons

Transposons and retrotransposons can give rise to both autoregulatory siRNAs and trans-acting siRNAs and miRNAs by the transposition and recombination processes described in preceding sections. Isolation and characterization of a maize *Mu* killer locus, known to silence the highly active maize *Mu* transposon family, showed it to be an inverted duplication of a partially deleted autonomous *Mu* element (Slotkin et al., 2003, 2005). The duplicated structure is transcribed and processed into small RNAs, which are in turn amplified to induce the heritable silencing of *Mu* transposons throughout the genome (Slotkin et al., 2005).

A substantial number of rice and *Arabidopsis* miRNAs have been reported to be homologous to transposons and retrotransposons (Li et al., 2011; Piriyapongsa and Jordan, 2008). Of 92 TE-homologous rice miRNAs analyzed, 80% exhibit homology to MITEs, 10% to retrotransposons, and 10% to other families of DNA transposons (Li et al., 2011). Some TE-homologous rice miRNA loci also encode siRNAs (Piriyapongsa and Jordan, 2008). More generally, there is growing evidence that miRNAs arise from inverted duplications, many representing duplicated segments of the target genes that eventually come under their regulatory control, possibly evolving through siRNA-producing intermediates by the accumulation of internal mutations in their precursors that convert them to suitable substrates for DCL1, the RNAseIII family enzyme that processes miRNA precursors (Allen et al., 2004; Fahlgren et al., 2007; Kurihara and Watanabe, 2004). Whether transposon-mediated rearrangements contribute to the origin of regulatory RNA loci more broadly has not yet been determined.

Genome Shocks

McClintock left us a final genome puzzle. She used the term genomic "shock" to refer to unpredictable stresses, such as irradiation, mutagenesis or the introduction of a broken chromosome, that elicit what she viewed as a highly programmed response from the genome, including transposon activation.

She says in her Nobel Prize lecture (McClintock, 1984) that her maize experiments revealed to her:

> . . . how a genome may react to conditions for which it is unprepared, but to which it responds in a totally unexpected manner. Among these is the extraordinary response of the maize genome to entrance of a single ruptured end of a chromosome into a telophase nucleus. It was this event that, basically, was responsible for activations of potentially transposable elements that are carried in a silent state in the maize genome. The mobility of these activated elements allows them to enter different gene loci and to take over control of action of the gene wherever one may enter. Because the broken end of a chromosome entering a telophase nucleus can initiate activations of a number of different potentially transposable elements, the modifications these elements induce in the genome may be explored readily. In addition to modifying gene action, these elements can restructure the genome at various levels, from small changes involving a few nucleotides, to gross modifications involving large segments of chromosomes, such as duplications, deficiencies, inversions, and other more complex reorganizations.
>
> A cell capable of repairing a ruptured end of a chromosome must sense the presence of this end in its nucleus. This sensing activates a mechanism that is required for replacing the ruptured end with a functional telomere. That such a mechanism must exist was revealed by a mutant that arose in my stocks. When homozygous, this mutant would not allow the repair mechanism to operate in the cells of the plant. Entrance of a newly ruptured end of a chromosome into the zygote is followed by the chromatid type of breakage-fusion-bridge cycle throughout mitoses in the developing plant. This suggests that the repair mechanism in the maize strains I had been using is repressed in cells producing the male and female gametophytes and also in the endosperm, but is activated in the embryo.

In the years since her pioneering observations more than half a century ago, it has been amply documented that plant transposons are activated in response to a variety of DNA-damaging agents and both biotic and abiotic stresses, as well as pathogen infection and the passage of plant cells through tissue culture (Chapter 6 in this volume). Other sources of natural chromosomal disturbance are provided by interspecific hybridization and allopolyploidization, both of which trigger the activation of transposons (Chapter 9 in this volume). This appears to be true, as well, in other eukaryotes, from yeast to flies to humans, and the common denominator is, of course, DNA damage. Telomerases are relatives of retrotransposon-encoded reverse transcriptases, and transposons either comprise or can fill in for missing telomeres in flies and yeast, respectively (Belfort et al., 2011; Curcio and Belfort, 2007; Eickbush, 1997; Pardue et al., 1997). Telomere loss activates transposons, precisely where McClintock's extraordinary experiments began.

Chapter 9 in this volume proposes that the repressive machinery maintaining transposons and other repetitive sequences in a genetically silent state is dysregulated by "genomic shocks," all of which are associated with DNA damage. Evidence in support of this hypothesis comes from the analysis of *Drosophila* hybrid dysgenesis (Khurana et al., 2011). In a dysgenic cross between a male containing P transposons and a female lacking them, both the novel P elements and resident elements are massively mobilized, resulting in hybrid sterility. However, dysgenic females recover fertility with age. Several proteins required for the biogenesis and activity of the transposon-silencing piRNAs are mislocalized in young ovaries, suggesting a general disruption of the silencing machinery.

Destruction of Vasa, a DEAD-box protein required for piRNA biogenesis, appears to be triggered by the DNA damage caused by P element mobilization (Khurana et al., 2011). It is proposed that activation of Chk2, a cell-cycle checkpoint kinase, links DNA damage to down-regulation of the silencing machinery and blocks stem cell proliferation in oogenesis to cause early sterility. Fertility restoration requires silencing of both P and resident elements, which in turn is mediated by their transposition into the heteromeric transposon-rich piRNA clusters that are the source of piRNAs (Khurana et al., 2011).

Genome Evolvability

The division of genomes into selfish DNA bins and, by implication, unselfish DNA bins appears less and less useful as we understand more about genome organization and evolution. The same types of epigenetic mechanisms underpin the expansion of eukaryotic genomes, the management of transcription, repair of DNA damage, and the division of genomes into chromosomes, as well as the stability, replication, and segregation of the genetic material in both mitosis and meiosis. The accumulation of TEs, particularly of retrotransposons, is likely to be a consequence of epigenetic mechanisms that inhibit the homology-dependent recombination events that would otherwise eliminate them either directly by intrachromosomal deletions within or between tandem insertions or indirectly by creating nonviable chromosomes through ectopic recombination events. The propensity of TEs to accumulate is perhaps most apparent in higher plants, some of whose genomes comprise small, recombinationally active gene islands in a vast sea of recombinationally inert retrotransposons.

Yet, although epigenetic mechanisms slow the pace of genome restructuring, the evolutionary impact of TEs is inescapable. Indeed, their ability to move and to move sequences has shaped higher eukaryotic genomes from the structuring and restructuring of genes and their regulatory sequences to the shaping and reshaping of the genomic landscape. It is difficult to escape the conclusion that eukaryotic genome evolution is driven from within not just by the gentle breeze of the genetic mechanisms that replicate and repair DNA, but by a strong wind of transposon activity, vastly accelerating the generation of material for natural selection. This is perhaps the essence of eukaryotic evolvability, the ability of larger and larger genomes to maintain evolutionary flexibility.

It appears likely that eukaryotic epigenetic mechanisms have their roots in prokaryotic genetic systems that discriminate self from non-self DNA in an environment dominated by horizontal transfer of genetic material through plasmids and bacteriophages. In prokaryotes, such systems damp, but do not prevent genome evolution through horizontal gene exchanges. But while both DNA methylation and small RNA regulatory mechanisms still function to recognize pathogens in eukaryotes, they have acquired a panoply of new functions. And linking the sequence specificity of the latter to the heritability of the former was a stroke of pure evolutionary genius, one we've barely begun to understand.

Finally, we begin to confront the challenge McClintock left us. She concluded her Nobel lecture with the following:

> In the future, attention undoubtedly will be centered on the genome, with greater appreciation of its significance as a highly sensitive organ of the cell that monitors genomic activities and corrects common errors, senses unusual and unexpected events, and responds to them, often by restructuring the genome. We know about the components of genomes that could be made available for such restructuring. We know nothing, however, about how the cell senses danger and instigates responses to it that often are truly remarkable.

How and why is it that many different types of genetic and environmental stresses activate transposons and retrotransposons? Just as McClintock reported that broken chromosome ends can "heal," so do transposition bursts subside, over both short and long runs, but how and why?

References

Achaz, G., Rocha, E.P., Netter, P., and Coissac, E. (2002) Origin and fate of repeats in bacteria. *Nucleic Acids Res. 30*: 2987–2994.

Adams, K.L. (2007) Evolution of duplicate gene expression in polyploid and hybrid plants. *J. Hered. 98*: 136–141.

Allen, E., Xie, Z., Gustafson, A.M., Sung, G.H., Spatafora, J.W., and Carrington, J.C. (2004) Evolution of microRNA genes by inverted duplication of target gene sequences in *Arabidopsis thaliana*. *Nat. Genet. 36*: 1282–1290.

Anderson, R.P., and Roth, J.R. (1977) Tandem genetic duplications in phage and bacteria. *Annu. Rev. Microbiol. 31*: 473–505.

Andersson, D.I., and Hughes, D. (2009) Gene amplification and adaptive evolution in bacteria. *Annu. Rev. Genet. 43*: 167–195.

Arteaga-Vazquez, M.A., and Chandler, V.L. (2010) Paramutation in maize: RNA mediated trans-generational gene silencing. *Curr. Opin. Genet. Dev. 20*: 156–163.

Aziz, R.K., Breitbart, M., and Edwards, R.A. (2010) Transposases are the most abundant, most ubiquitous genes in nature. *Nucleic Acids Res. 38*: 4207–4217.

Banks, J.A., Masson, P., and Fedoroff, N. (1988) Molecular mechanisms in the developmental regulation of the maize *Suppressor-mutator* transposable element. *Genes Dev. 2*: 1364–1380.

Barrangou, R., Fremaux, C., Deveau, H., Richards, M., Boyaval, P., Moineau, S., Romero, D.A., and Horvath, P. (2007) CRISPR provides acquired resistance against viruses in prokaryotes. *Science 315*: 1709–1712.

Baucom, R.S., Estill, J.C., Chaparro, C., Upshaw, N., Jogi, A., Deragon, J.M., Westerman, R.P., Sanmiguel, P.J., and Bennetzen, J.L. (2009a) Exceptional diversity, non-random distribution, and rapid evolution of retroelements in the B73 maize genome. *PLoS Genet. 5*: e1000732.

Baucom, R.S., Estill, J.C., Leebens-Mack, J., and Bennetzen, J.L. (2009b) Natural selection on gene function drives the evolution of LTR retrotransposon families in the rice genome. *Genome Res. 19*: 243–254.

Bayne, E.H., White, S.A., Kagansky, A., Bijos, D.A., Sanchez-Pulido, L., Hoe, K.L., Kim, D.U., Park, H.O., Ponting, C.P., Rappsilber, J., et al. (2010) Stc1: a critical link between RNAi and chromatin modification required for heterochromatin integrity. *Cell 140*: 666–677.

Belfort, M., Curcio, M.J., and Lue, N.F. (2011) Telomerase and retrotransposons: reverse transcriptases that shaped genomes. *Proc. Natl. Acad. Sci. USA 108*: 20304–20310.

Bennett, M.D., and Leitch, I.J. (2011) Nuclear DNA amounts in angiosperms: Targets, trends and tomorrow. *Ann. Bot. 107*: 467–590.

Bennett, M.D., and Smith, J.B. (1976) Nuclear DNA amounts in angiosperms. *Philos. Trans. R. Soc. Lond. B Biol. Sci. 274*: 227–274.

Bennetzen, J.L. (2000a) Comparative sequence analysis of plant nuclear genomes: Microcolinearity and its many exceptions. *Plant Cell 12*: 1021–1029.

Bennetzen, J.L. (2000b) Transposable element contributions to plant gene and genome evolution. *Plant Mol. Biol. 42*: 251–269.

Bennetzen, J.L., and Freeling, M. (1997) The unified grass genome: synergy in synteny. *Genome Res. 7*: 301–306.

Bennetzen, J.L., Ma, J., and Devos, K.M. (2005) Mechanisms of recent genome size variation in flowering plants. *Ann. Bot. 95*: 127–132.

Bennetzen, J.L., Schrick, K., Springer, P.S., Brown, W.E., and SanMiguel, P. (1994) Active maize genes are unmodified and flanked by diverse classes of modified, highly repetitive DNA. *Genome 37*: 565–576.

Bestor, T.H. (1998) The host defence function of genomic methylation patterns. *Novartis Found. Symp. 214*: 187–195; discussion 195–189, 228–132.

Brink, R.A. (1958) Paramutation at the R locus in maize. *Cold Spring Harbor Symp. Quant. Biol. 23*: 379–391.

Britten, R.J., and Davidson, E.H. (1976a) DNA sequence arrangement and preliminary evidence on its evolution. *Fed. Proc. 35*: 2151–2157.

Britten, R.J., and Davidson, E.H. (1976b) Studies on nucleic acid reassociation kinetics: Empirical equations describing DNA reassociation. *Proc. Natl. Acad. Sci. USA 73*: 415–419.

Buchmann, J.P., Matsumoto, T., Stein, N., Keller, B., and Wicker, T. (2012) Interspecies sequence comparison of *Brachypodium* reveals how transposon activity corrodes genome colinearity. *Plant J. 71*:550–563.

Capy, P., Langin, T., Higuet, D., Maurer, P., and Bazin, C. (1997) Do the integrases of LTR-retrotransposons and class II element transposases have a common ancestor? *Genetica 100*: 63–72.

Casadesus, J., and Low, D. (2006) Epigenetic gene regulation in the bacterial world. *Microbiol. Mol. Biol. Rev. 70*: 830–856.

Chan, S.W., Zhang, X., Bernatavichute, Y.V., and Jacobsen, S.E. (2006) Two-step recruitment of RNA-directed DNA methylation to tandem repeats. *PLoS Biol. 4*: e363.

Chen, X. (2010a) Small RNAs: Secrets and surprises of the genome. *Plant J. 61*: 941–958.

Chen, Z.J. (2010b) Molecular mechanisms of polyploidy and hybrid vigor. *Trends Plant Sci. 15*: 57–71.

Cherepanov, P., Maertens, G.N., and Hare, S. (2011) Structural insights into the retroviral DNA integration apparatus. *Curr. Opin. Struct. Biol. 21*: 249–256.

Courage-Tebbe, U., Doring, H.P., Fedoroff, N., and Starlinger, P. (1983) The controlling element *Ds* at the *Shrunken* locus in *Zea mays*: Structure of the unstable *sh-m5933* allele and several revertants. *Cell 34*: 383–393.

Craig, N.L. (1997) Target site selection in transposition. *Annu. Rev. Biochem. 66*: 437–474.

Craigie, R., and Mizuuchi, K. (1985) Mechanism of transposition of bacteriophage *Mu*: Structure of a transposition intermediate. *Cell 41*: 867–876.

Craigie, R., and Mizuuchi, K. (1987) Transposition of *Mu* DNA: Joining of *Mu* to target DNA can be uncoupled from cleavage at the ends of *Mu*. *Cell 51*: 493–501.

Cresse, A.D., Hulbert, S.H., Brown, W.E., Lucas, J.R., and Bennetzen, J.L. (1995) *Mu1*-related transposable elements of maize preferentially insert into low copy number DNA. *Genetics 140*: 315–324.

Curcio, M.J., and Belfort, M. (2007) The beginning of the end: Links between ancient retroelements and modern telomerases. *Proc. Natl. Acad. Sci. USA 104*: 9107–9108.

Darwin, C.R. (1786) *The effects of cross- and self-fertilization in the vegetable kingdom*, 1st edition (London: John Murray).

Devos, K.M., Brown, J.K., and Bennetzen, J.L. (2002) Genome size reduction through illegitimate recombination counteracts genome expansion in *Arabidopsis*. *Genome Res. 12*: 1075–1079.

Devos, K.M., Ma, J., Pontaroli, A.C., Pratt, L.H., and Bennetzen, J.L. (2005) Analysis and mapping of randomly chosen bacterial artificial chromosome clones from hexaploid bread wheat. *Proc. Natl. Acad. Sci. USA 102*: 19243–19248.

Doolittle, W.F., and Sapienza, C. (1980) Selfish genes, the phenotype paradigm and genome evolution. *Nature 284*: 601–603.

Doring, H.P., Pahl, I., and Durany, M. (1990) Chromosomal rearrangements caused by the aberrant transposition of double *Ds* elements are formed by *Ds* and adjacent non-*Ds* sequences. *Mol. Gen. Genet. 224*: 40–48.

Dreyfus, D.H. (2006) The DDE recombinases: Diverse roles in acquired and innate immunity. *Ann. Allergy Asthma Immunol. 97*: 567–576; quiz 576–568, 602.

Eickbush, T.H. (1997) Telomerase and retrotransposons: Which came first? *Science 277*: 911–912.

Engels, W.R. (1992) The origin of *P* elements in *Drosophila melanogaster*. *BioEssays 14*: 681–686.

English, J.J., Harrison, K., and Jones, J.D.G. (1995) Aberrant transpositions of maize double *Ds*-like elements usually involve *Ds* ends on sister chromatids. *Plant Cell 7*: 1235–1247.

Evgen'ev, M., Zelentsova, H., Mnjoian, L., Poluectova, H., and Kidwell, M.G. (2000) Invasion of *Drosophila virilis* by the *Penelope* transposable element. *Chromosoma 109*: 350–357.

Fahlgren, N., Howell, M.D., Kasschau, K.D., Chapman, E.J., Sullivan, C.M., Cumbie, J.S., Givan, S.A., Law, T.F., Grant, S.R., Dangl, J.L., et al. (2007) High-throughput sequencing of *Arabidopsis* microRNAs: evidence for frequent birth and death of MIRNA genes. *PLoS One 2*: e219.

Fedoroff, N., Furtek, D., and Nelson, O. (1984) Cloning of the *Bronze* locus in maize by a simple and generalizable procedure using the transposable controlling element *Ac*. *Proc. Natl. Acad. Sci. USA 81*: 3825–3829.

Francia, S., Michelini, F., Saxena, A., Tang, D., de Hoon, M., Anelli, V., Mione, M., Carninci, P., and di Fagagna, F.D. (2012) Site-specific DICER and DROSHA RNA products control the DNA-damage response. *Nature 488*: 231–235.

Fu, H., and Dooner, H.K. (2002) Intraspecific violation of genetic colinearity and its implications in maize. *Proc. Natl. Acad. Sci. USA 99*: 9573–9578.

Fu, H., Park, W., Yan, X., Zheng, Z., Shen, B., and Dooner, H.K. (2001) The highly recombinogenic *bz* locus lies in an unusually gene-rich region of the maize genome. *Proc. Natl. Acad. Sci. USA 98*: 8903–8908.

Fu, H., Zheng, Z., and Dooner, H.K. (2002) Recombination rates between adjacent genic and retrotransposon regions in maize vary by 2 orders of magnitude. *Proc. Natl. Acad. Sci. USA 99*: 1082–1087.

Fuerst, P.G., and Voytas, D.F. (2003) CEN plasmid segregation is destabilized by tethered determinants of *Ty5* integration specificity: a role for double-strand breaks in CEN antagonism. *Chromosoma 112*: 58–65.

Gai, X., and Voytas, D.F. (1998) A single amino acid change in the yeast retrotransposon *Ty5* abolishes targeting to silent chromatin. *Mol. Cell 1*: 1051–1055.

Gall, J.G. (1981) Chromosome structure and the C-value paradox. *J. Cell Biol. 91*: 3s–14s.

Gilbert, C., Hernandez, S.S., Flores-Benabib, J., Smith, E.N., and Feschotte, C. (2012) Rampant horizontal transfer of *SPIN* transposons in squamate reptiles. *Mol. Biol. Evol. 29*: 503–515.

Gilbert, C., Schaack, S., Pace, J.K., 2nd, Brindley, P.J., and Feschotte, C. (2010) A role for host-parasite interactions in the horizontal transfer of transposons across phyla. *Nature 464*: 1347–1350.

Greenblatt, I.M., and Brink, R.A. (1962) Twin mutations in medium variegated pericarp maize. *Genetics 47*: 489–501.

Groszmann, M., Greaves, I.K., Albert, N., Fujimoto, R., Helliwell, C.A., Dennis, E.S., and Peacock, W.J. (2011a) Epigenetics in plants-vernalisation and hybrid vigour. *Biochim. Biophys. Acta 1809*: 427–437.

Groszmann, M., Greaves, I.K., Albertyn, Z.I., Scofield, G.N., Peacock, W.J., and Dennis, E.S. (2011b) Changes in 24-nt siRNA levels in *Arabidopsis* hybrids suggest an epigenetic contribution to hybrid vigor. *Proc. Natl. Acad. Sci. USA 108*: 2617–2622.

Guo, H., Tse, L.V., Barbalat, R., Sivaamnuaiphorn, S., Xu, M., Doulatov, S., and Miller, J.F. (2008) Diversity-generating retroelement homing regenerates target sequences for repeated rounds of codon rewriting and protein diversification. *Mol. Cell 31*: 813–823.

Guo, H., Tse, L.V., Nieh, A.W., Czornyj, E., Williams, S., Oukil, S., Liu, V.B., and Miller, J.F. (2011) Target site recognition by a diversity-generating retroelement. *PLoS Genet. 7*: e1002414.

Haapa-Paananen, S., Rita, H., and Savilahti, H. (2002) DNA transposition of bacteriophage *Mu*. A quantitative analysis of target site selection *in vitro*. *J. Biol. Chem. 277*: 2843–2851.

Hickman, A.B., Chandler, M., and Dyda, F. (2010) Integrating prokaryotes and eukaryotes: DNA transposases in light of structure. *Crit. Rev. Biochem. Mol. Biol. 45*: 50–69.

Hiom, K., Melek, M., and Gellert, M. (1998) DNA transposition by the RAG1 and RAG2 proteins: A possible source of oncogenic translocations. *Cell 94*: 463–470.

Hochholdinger, F., and Hoecker, N. (2007) Towards the molecular basis of heterosis. *Trends Plant Sci. 12*: 427–432.

Huang, H., Zhang, S., Raina, S., Li, H., Li, J., Dec, E., Ma, H., and Fedoroff, N.V. (2003) Resources for targeted insertional and deletional mutagenesis in *Arabidopsis*. *Plant Mol. Biol. 53*: 133–150.

Isambert, H., and Stein, R.R. (2009) On the need for widespread horizontal gene transfers under genome size constraint. *Biol. Direct. 4*: 28.

Iyer, L.M., Abhiman, S., and Aravind, L. (2011) Natural history of eukaryotic DNA methylation systems. *Prog. Mol. Biol. Transl. Sci. 101*: 25–104.

Jorgensen, R. (1990) Altered gene expression in plants due to *trans* interactions between homologous genes. *Trends Biotechnol. 8*: 340–344.

Jorgensen, R. (1994) Developmental significance of epigenetic impositions on the plant genome: A paragenetic function for chromosomes. *Dev. Genet. 15*: 53–76.

Kanizay, L., and Dawe, R.K. (2009) Centromeres: Long intergenic spaces with adaptive features. *Funct. Integr. Genomics 9*: 287–292.

Kapitonov, V.V., and Jurka, J. (2005) RAG1 core and V(D)J recombination signal sequences were derived from *Transib* transposons. *PLoS Biol. 3*: e181.

Kaufman, P.D., and Rio, D.C. (1992) *P* element transposition *in vitro* proceeds by a cut-and-paste mechanism and uses GTP as a cofactor. *Cell 69*: 27–39.

Keith, J.H., Schaeper, C.A., Fraser, T.S., and Fraser, M.J., Jr. (2008) Mutational analysis of highly conserved aspartate residues essential to the catalytic core of the *piggyBac* transposase. *BMC Mol. Biol. 9*: 73.

Kenan-Eichler, M., Leshkowitz, D., Tal, L., Noor, E., Melamed-Bessudo, C., Feldman, M., and Levy, A.A. (2011) Wheat hybridization and polyploidization results in deregulation of small RNAs. *Genetics 188*: 263–272.

Khurana, J.S., Wang, J., Xu, J., Koppetsch, B.S., Thomson, T.C., Nowosielska, A., Li, C., Zamore, P.D., Weng, Z., and Theurkauf, W.E. (2011) Adaptation to *P* element transposon invasion in *Drosophila melanogaster*. *Cell 147*: 1551–1563.

Kidwell, M.G. (1992) Horizontal transfer of *P* elements and other short inverted repeat transposons. *Genetica 86*: 275–286.

Kidwell, M.G., and Lisch, D.R. (1998) Hybrid genetics: Transposons unbound. *Nature 393*: 22–23.

Kleckner, N. (1990) Regulation of transposition in bacteria. *Annu. Rev. Cell Biol. 6*: 297–327.

Koch, A.L. (1979) Selection and recombination in populations containing tandem multiple genes. *J. Mol. Evol. 14*: 273–285.

Kooter, J.M., Matzke, M.A., and Meyer, P. (1999) Listening to the silent genes: transgene silencing, gene regulation and pathogen control. *Trends Plant Sci. 4*: 340–347.

Kurihara, Y., and Watanabe, Y. (2004) *Arabidopsis* micro-RNA biogenesis through Dicer-like 1 protein functions. *Proc. Natl. Acad. Sci. USA 101*: 12753–12758.

Law, J.A., and Jacobsen, S.E. (2010) Establishing, maintaining and modifying DNA methylation patterns in plants and animals. *Nat. Rev. Genet. 11*: 204–220.

Li, Y., Li, C., Xia, J., and Jin, Y. (2011) Domestication of transposable elements into microRNA genes in plants. *PLoS One 6*: e19212.

Lienert, F., Wirbelauer, C., Som, I., Dean, A., Mohn, F., and Schubeler, D. (2011) Identification of genetic elements that autonomously determine DNA methylation states. *Nature Genet. 43*: 1091–1097.

Lisch, D., and Bennetzen, J.L. (2011) Transposable element origins of epigenetic gene regulation. *Curr. Opin. Plant Biol. 14*: 156–161.

Liu, Q., Wang, J., Miki, D., Xia, R., Yu, W., He, J., Zheng, Z., Zhu, J.-K., and Gong, Z. (2010) DNA replication factor C1 mediates genomic stability and transcriptional gene silencing in *Arabidopsis*. *Plant Cell 22*: 2336–2352.

Liu, R., and Bennetzen, J.L. (2008) Enchilada redux: How complete is your genome sequence? *New Phytol. 179*: 249–250.

Liu, S., Yeh, C.T., Ji, T., Ying, K., Wu, H., Tang, H.M., Fu, Y., Nettleton, D., and Schnable, P.S. (2009) *Mu* transposon insertion sites and meiotic recombination events co-localize with epigenetic marks for open chromatin across the maize genome. *PLoS Genet. 5*: e1000733.

Ma, J., and Bennetzen, J.L. (2006) Recombination, rearrangement, reshuffling, and divergence in a centromeric region of rice. *Proc. Natl. Acad. Sci. USA 103*: 383–388.

Ma, J., Devos, K.M., and Bennetzen, J.L. (2004) Analyses of LTR-retrotransposon structures reveal recent and rapid genomic DNA loss in rice. *Genome Res. 14*: 860–869.

Martienssen, R. (1998) Transposons, DNA methylation and gene control. *Trends Genet. 14*: 263–264.

Masson, P., Surosky, R., Kingsbury, J.A., and Fedoroff, N.V. (1987) Genetic and molecular analysis of the *Spm*-dependent *a-m2* alleles of the maize *a* locus. *Genetics 117*: 117–137.

McClintock, B. (1950a) Mutable loci in maize. *Carnegie Inst. Wash. Yr. Bk. 49*: 157–167.

McClintock, B. (1950b) The origin and behavior of mutable loci in maize. *Proc. Natl. Acad. Sci. USA 36*: 344–355.

McClintock, B. (1968) The states of a gene locus in maize. *Carnegie Inst. Wash. Yr. Bk. 66*: 20–28.

McClintock, B. (1984) The significance of responses of the genome to challenge. *Science 226*: 792–801.

McEachern, M.J., and Haber, J.E. (2006) Break-induced replication and recombinational telomere elongation in yeast. *Annu. Rev. Biochem. 75*: 111–135.

Medhekar, B., and Miller, J.F. (2007) Diversity-generating retroelements. *Curr. Opin. Microbiol. 10*: 388–395.

Mitra, R., Fain-Thornton, J., and Craig, N.L. (2008) piggyBac can bypass DNA synthesis during cut and paste transposition. *EMBO J. 27*: 1097–1109.

Mizuuchi, K., and Craigie, R. (1986) Mechanism of bacteriophage mu transposition. *Annu. Rev. Genet. 20*: 385–429.

Moldovan, G.L., Dejsuphong, D., Petalcorin, M.I., Hofmann, K., Takeda, S., Boulton, S.J., and D'Andrea, A.D. (2012) Inhibition of homologous recombination by the PCNA-interacting protein PARI. *Mol. Cell 45*: 75–86.

Montano, S.P., and Rice, P.A. (2011) Moving DNA around: DNA transposition and retroviral integration. *Curr. Opin. Struct. Biol. 21*: 370–378.

Nelson, O.E. (1959) Intracistron recombination in the *Wx/wx* region in maize. *Science 130*: 794–795.

Nesmelova, I.V., and Hackett, P.B. (2010) DDE transposases: Structural similarity and diversity. *Adv. Drug Deliv. Rev. 62*: 1187–1195.

Ohno, S. (1972) So much "junk" DNA in our genome. *Brookhaven Symp. Biol. 23*: 366–370.

Okagaki, R.J., and Weil, C.F. (1997) Analysis of recombination sites within the maize *waxy* locus. *Genetics 147*: 815–821.

Orgel, L.E., and Crick, F.H.C. (1980) Selfish DNA: The ultimate parasite. *Nature 284*: 604–607.

Panavas, T., Weir, J., and Walker, E.L. (1999) The structure and paramutagenicity of the *R-marbled* haplotype of *Zea mays*. *Genetics 153*: 979–991.

Pardue, M.L., Danilevskaya, O.N., Traverse, K.L., and Lowenhaupt, K. (1997) Evolutionary links between telomeres and transposable elements. *Genetica 100*: 73–84.

Paterson, A.H., Bowers, J.E., Bruggmann, R., Dubchak, I., Grimwood, J., Gundlach, H., Haberer, G., Hellsten, U., Mitros, T., Poliakov, A., et al. (2009) The *Sorghum bicolor* genome and the diversification of grasses. *Nature 457*: 551–556.

Peng, J.C., and Karpen, G.H. (2007) H3K9 methylation and RNA interference regulate nucleolar organization and repeated DNA stability. *Nature Cell Biol. 9*: 25–35.

Peng, J.C., and Karpen, G.H. (2008) Epigenetic regulation of heterochromatic DNA stability. *Curr. Opin. Genet. Dev. 18*: 204–211.

Piegu, B., Guyot, R., Picault, N., Roulin, A., Sanyal, A., Kim, H., Collura, K., Brar, D.S., Jackson, S., Wing, R.A., et al. (2006) Doubling genome size without polyploidization: Dynamics of retrotransposition-driven genomic expansions in *Oryza australiensis*, a wild relative of rice. *Genome Res. 16*: 1262–1269.

Piriyapongsa, J., and Jordan, I.K. (2008) Dual coding of siRNAs and miRNAs by plant transposable elements. *RNA 14*: 814–821.

Polard, P., and Chandler, M. (1995) Bacterial transposases and retroviral integrases. *Mol. Microbiol. 15*: 13–23.

Puchta, H. (2005) The repair of double-strand breaks in plants: mechanisms and consequences for genome evolution. *J. Exp. Bot. 56*: 1–14.

Que, Q., and Jorgensen, R.A. (1998) Homology-based control of gene expression patterns in transgenic petunia flowers. *Dev. Genet. 22*: 100–109.

Raina, S., Mahalingam, R., Chen, F., and Fedoroff, N. (2002) A collection of sequenced and mapped *Ds* transposon insertion sites in *Arabidopsis thaliana*. *Plant Mol. Biol. 50*: 93–110.

Ratcliff, F., Harrison, B.D., and Baulcombe, D.C. (1997) A similarity between viral defense and gene silencing in plants. *Science 276*: 1558–1560.

Rieseberg, L.H., Widmer, A., Arntz, A.M., and Burke, J.M. (2003) The genetic architecture necessary for transgressive segregation is common in both natural and domesticated populations. *Philos. Trans. R. Soc. Lond. B Biol. Sc. 358*: 1141–1147.

Roberts, D., Hoopes, B.C., McClure, W.R., and Kleckner, N. (1985) *IS10* transposition is regulated by DNA adenine methylation. *Cell 43*: 117–130.

Ros, F., and Kunze, R. (2001) Regulation of *activator/dissociation* transposition by replication and DNA methylation. *Genetics 157*: 1723–1733.

Rosbash, M., Ford, P.J., and Bishop, J.O. (1974) Analysis of the C-value paradox by molecular hybridization. *Proc. Natl. Acad. Sci. USA 71*: 3746–3750.

SanMiguel, P., Gaut, B.S., Tikhonov, A., Nakajima, Y., and Bennetzen, J.L. (1998) The paleontology of intergene retrotransposons of maize. *Nature Genet. 20*: 43–45.

SanMiguel, P., Tikhonov, A., Jin, Y.K., Motchoulskaia, N., Zakharov, D., Melake-Berhan, A., Springer, P.S., Edwards, K.J., Lee, M., Avramova, Z., et al. (1996) Nested retrotransposons in the intergenic regions of the maize genome. *Science 274*: 765–768.

Savilahti, H., Rice, P.A., and Mizuuchi, K. (1995) The phage *Mu* transpososome core: DNA requirements for assembly and function. *EMBO J. 14*: 4893–4903.

Schnable, P.S., Ware, D., Fulton, R.S., Stein, J.C., Wei, F., Pasternak, S., Liang, C., Zhang, J., Fulton, L., Graves, T.A., et al. (2009) The B73 maize genome: Complexity, diversity, and dynamics. *Science 326*: 1112–1115.

Schwarz-Sommer, Z., Gierl, A., Berndtgen, R., and Saedler, H. (1985) Sequence comparison of "states" of *a1-m1* suggest a model of *Spm* (*En*) action. *EMBO J. 4*: 2439–2443.

Shivaprasad, P.V., Dunn, R.M., Santos, B.A., Bassett, A., and Baulcombe, D.C. (2012) Extraordinary transgressive phenotypes of hybrid tomato are influenced by epigenetics and small silencing RNAs. *EMBO J. 31*: 257–266.

Silva, J.C., and Kidwell, M.G. (2000) Horizontal transfer and selection in the evolution of *P* elements. *Mol. Biol. Evol. 17*: 1542–1557.

Simon, S.A., and Meyers, B.C. (2011) Small RNA-mediated epigenetic modifications in plants. *Curr. Opin. Plant Biol. 14*: 148–155.

Slotkin, R.K. (2010) The epigenetic control of the *Athila* family of retrotransposons in *Arabidopsis*. *Epigenetics 5*: 483–490.

Slotkin, R.K., Freeling, M., and Lisch, D. (2003) *Mu* killer causes the heritable inactivation of the *Mutator* family of transposable elements in *Zea mays*. *Genetics 165*: 781–797.

Slotkin, R.K., Freeling, M., and Lisch, D. (2005) Heritable transposon silencing initiated by a naturally occurring transposon inverted duplication. *Nature Genet. 37*: 641–644.

Slotkin, R.K., and Martienssen, R. (2007) Transposable elements and the epigenetic regulation of the genome. *Nature Rev. Genetics 8*: 272–285.

Smith, L.M., and Weigel, D. (2012) On epigenetics and epistasis: Hybrids and their non-additive interactions. *EMBO J. 31*: 249–250.

Thomas, J., Schaack, S., and Pritham, E.J. (2010) Pervasive horizontal transfer of rolling-circle transposons among animals. *Genome Biol. Evol. 2*: 656–664.

Tian, Z., Rizzon, C., Du, J., Zhu, L., Bennetzen, J.L., Jackson, S.A., Gaut, B.S., and Ma, J. (2009) Do genetic recombination and gene density shape the pattern of DNA elimination in rice long terminal repeat retrotransposons? *Genome Res. 19*: 2221–2230.

Tikhonov, A.P., SanMiguel, P.J., Nakajima, Y., Gorenstein, N.M., Bennetzen, J.L., and Avramova, Z. (1999) Colinearity and its exceptions in orthologous *adh* regions of maize and sorghum. *Proc. Natl. Acad. Sci. USA 96*: 7409–7414.

Torres-Rosell, J., Sunjevaric, I., De Piccoli, G., Sacher, M., Eckert-Boulet, N., Reid, R., Jentsch, S., Rothstein, R., Aragon, L., and Lisby, M. (2007) The Smc5-Smc6 complex and SUMO modification of Rad52 regulates recombinational repair at the ribosomal gene locus. *Nature Cell Biol. 9*: 923–931.

Treangen, T.J., and Rocha, E.P. (2011) Horizontal transfer, not duplication, drives the expansion of protein families in prokaryotes. *PLoS Genet. 7*: e1001284.

Tsukahara, S., Kawabe, A., Kobayashi, A., Ito, T., Aizu, T., Shin-i, T., Toyoda, A., Fujiyama, A., Tarutani, Y., and Kakutani, T. (2012) Centromere-targeted *de novo* integrations of an LTR retrotransposon of *Arabidopsis lyrata*. *Gene Dev. 26*: 705–713.

Vanderlyn, L. (1949) The heterochromatin problem in cytogenetics as related to other branches of investigation. *Bot. Rev. 15*: 507–582.

Verdel, A., Jia, S., Gerber, S., Sugiyama, T., Gygi, S., Grewal, S.I., and Moazed, D. (2004) RNAi-mediated targeting of heterochromatin by the RITS complex. *Science 303*: 672–676.

Vitte, C., and Bennetzen, J.L. (2006) Analysis of retrotransposon structural diversity uncovers properties and propensities in angiosperm genome evolution. *Proc. Natl. Acad. Sci. USA 103*: 17638–17643.

Wang, Q., and Dooner, H.K. (2006) Remarkable variation in maize genome structure inferred from haplotype diversity at the *bz* locus. *Proc. Natl. Acad. Sci. USA 103*: 17644–17649.

Wassenegger, M., Heimes, S., Riedel, L., and Sanger, H.L. (1994) RNA-directed *de novo* methylation of genomic sequences in plants. *Cell 76*: 567–576.

Wei, F., Stein, J.C., Liang, C., Zhang, J., Fulton, R.S., Baucom, R.S., De Paoli, E., Zhou, S., Yang, L., Han, Y., et al. (2009) Detailed analysis of a contiguous 22-Mb region of the maize genome. *PLoS Genet. 5*: e1000728.

Wei, W., Ba, Z., Gao, M., Wu, Y., Ma, Y., Amiard, S., White, C.I., Rendtlew Danielsen, J.M., Yang, Y.-G., and Qi, Y. (2012) A role for small RNAs in DNA double-strand break repair. *Cell 149*: 101–112.

Wicker, T., Buchmann, J.P., and Keller, B. (2010) Patching gaps in plant genomes results in gene movement and erosion of colinearity. *Genome Res. 20*: 1229–1237.

Wicker, T., Mayer, K.F., Gundlach, H., Martis, M., Steuernagel, B., Scholz, U., Simkova, H., Kubalakova, M., Choulet, F., Taudien, S., et al. (2011) Frequent gene movement and pseudogene evolution is common to the large and complex genomes of wheat, barley, and their relatives. *Plant Cell 23*: 1706–1718.

Wicker, T., Stein, N., Albar, L., Feuillet, C., Schlagenhauf, E., and Keller, B. (2001) Analysis of a contiguous 211 kb sequence in diploid wheat (*Triticum monococcum L.*) reveals multiple mechanisms of genome evolution. *Plant J. 26*: 307–316.

Wicker, T., Zimmermann, W., Perovic, D., Paterson, A.H., Ganal, M., Graner, A., and Stein, N. (2005) A detailed look at 7 million years of genome evolution in a 439 kb contiguous sequence at the barley *Hv-eIF4E* locus: recombination, rearrangements and repeats. *Plant J. 41*: 184–194.

Williams, T.L., Jackson, E.L., Carritte, A., and Baker, T.A. (1999) Organization and dynamics of the *Mu* transpososome: Recombination by communication between two active sites. *Genes Dev. 13*: 2725–2737.

Xu, J.H., Bennetzen, J.L., and Messing, J. (2012) Dynamic gene copy number variation in collinear regions of grass genomes. *Mol. Biol. Evol. 29*: 861–871.

Yaish, M.W., Colasanti, J., and Rothstein, S.J. (2011) The role of epigenetic processes in controlling flowering time in plants exposed to stress. *J. Exp. Bot. 62*: 3727–3735.

Yan, X., Martinez-Ferez, I.M., Kavchok, S., and Dooner, H.K. (1999) Origination of *Ds* elements from *Ac* elements in maize: evidence for rare repair synthesis at the site of *Ac* excision. *Genetics 152*: 1733–1740.

Yang, L., and Bennetzen, J.L. (2009) Distribution, diversity, evolution, and survival of *Helitrons* in the maize genome. *Proc. Natl. Acad. Sci. USA 106*: 19922–19927.

Yin, H., Zhang, X., Liu, J., Wang, Y., He, J., Yang, T., Hong, X., Yang, Q., and Gong, Z. (2009) Epigenetic regulation, somatic homologous recombination, and abscisic acid signaling are influenced by DNA polymerase epsilon mutation in *Arabidopsis*. *Plant Cell 21*: 386–402.

Yoder, J.A., Walsh, C.P., and Bestor, T.H. (1997) Cytosine methylation and the ecology of intragenomic parasites. *Trends Genet. 13*: 335–340.

Yuan, Y.W., and Wessler, S.R. (2011) The catalytic domain of all eukaryotic cut-and-paste transposase superfamilies. *Proc. Natl. Acad. Sci. USA 108*: 7884–7889.

Zaratiegui, M., Castel, S.E., Irvine, D.V., Kloc, A., Ren, J., Li, F., de Castro, E., Marin, L., Chang, A.Y., Goto, D., et al. (2011) RNAi promotes heterochromatic silencing through replication-coupled release of RNA Pol II. *Nature 479*: 135–138.

Zaratiegui, M., Irvine, D.V., and Martienssen, R.A. (2007) Noncoding RNAs and gene silencing. *Cell 128*: 763–776.

Zeh, D.W., Zeh, J.A., and Ishida, Y. (2009) Transposable elements and an epigenetic basis for punctuated equilibria. *BioEssays 31*: 715–726.

Zuckerkandl, E. (1992) Revisiting junk DNA. *J. Mol. Evol. 34*: 259–271.

Zuckerkandl, E., and Cavalli, G. (2007) Combinatorial epigenetics, "junk DNA," and the evolution of complex organisms. *Gene 390*: 232–242.

Index

Page numbers followed by "f" and "t" indicate figures and tables.

Plant Transposons and Genome Dynamics in Evolution, First Edition. Edited by Nina V. Fedoroff.